CONTROLLED ATMOSPHERE STORAGE OF FRUITS AND VEGETABLES

A.K. THOMPSON

*Formerly of the Windward Islands Banana Development
and Exporting Company, St Lucia, West Indies,
and former Professor, Postharvest Technology Department,
Cranfield University at Silsoe, Bedford, UK*

CAB INTERNATIONAL

CAB INTERNATIONAL
Wallingford
Oxon OX10 8DE
UK

Tel: +44 (0) 1491 832111
Fax: +44 (0) 1491 833508
E-mail: cabi@cabi.org

CAB INTERNATIONAL
198 Madison Avenue
New York, NY 10016-4314
USA

Tel: +1 212 726 6490
Fax: +1 212 686 7993
E-mail: cabi-nao@cabi.org

A catalogue record for this book is available from the British Library, London, UK.

Library of Congress Cataloging-in-Publication Data
Thompson, A.K. (A. Keith)
    Controlled atmosphere storage of fruits and vegetables / A.K. Thompson.
        p.   cm.
    Includes bibliographical references and index.
    ISBN 0-85199-267-6 (alk. paper)
    1. Fruit—Storage.   2. Vegetables—Storage.   3. Protective atmospheres.
I. Title.
SB360.5.T48   1998
635'.0468—dc21   98-6344
                                                                      CIP

ISBN 0 85199 267 6

Typeset in Garamond by AMA Graphics Ltd, UK
Printed and bound in the UK by Biddles Ltd, Guildford and King's Lynn

Contents

Preface

Over the last 80 years or so an enormous volume of literature has been published on the subject of controlled atmosphere storage of fruits and vegetables. It would be the work of a lifetime to begin to do those results justice in presenting a comprehensive and focused view, interpretation and digest for its application in commercial practice. Such a review would be useful to those engaged in the commerce of fruits and vegetables who could utilize this technology and reap its benefits in terms of the reduction of postharvest losses, maintenance of their nutritive value and organoleptic characteristics. The potential use of controlled atmosphere storage as an alternative to the application of preservation and pesticide chemicals is of continuing interest.

In order to facilitate the task of reviewing the literature I have had to rely on a combination of reviewing original publications as well as consulting reviews and learned books. The latter are not always entirely satisfactory since they may not give their source of information and I may have inadvertently quoted the same work more than once. Much reliance has been made on conference proceedings especially the International Controlled Atmosphere Research Conference held every few years in the USA, the European Co-operation in the Field of Scientific and Technical Research (COST 94) which held postharvest meetings throughout Europe between 1992 and 1995, the International Society for Horticultural Science's regular international conferences and especially on the Commonwealth Agricultural Bureau Abstracts.

Different views exist on the usefulness of controlled atmosphere storage. Blythman (1996) described controlled atmosphere storage as a system which 'amounts to deception' from the consumer's point of view. The reason behind this assertion seems to be that the consumer thinks that the fruits and vegetables that they purchase are fresh and that controlled atmosphere storage technology 'bestows a counterfeit freshness'. Also she claims that storage may change produce in a detrimental way and cites changes in texture of apples, 'potatoes that seem watery and fall apart when cooked and bananas that have no flavour'. Some of these contentions are true and need addressing, but others are over simplifications of the facts. Another view was expressed by David Sainsbury in 1995 and reported in the press as 'These techniques [controlled atmosphere storage] could halve the cost of fruit to the customer. It also extends the season of availability, making good eating-quality fruit available for extended periods at reasonable costs'.

The purpose of this book is primarily to help the fresh produce industry in storage and transport of fruit and vegetables, and to provide an easily accessible reference source for those studying agriculture, horticulture, food science and technology and food marketing. It will also be useful to researchers in this area, giving an overview of our present knowledge of controlled atmosphere storage which will indicate areas where there is a need for further research.

Acknowledgements

I would like to thank Dr Graeme Hobson for an excellent review and correction of the manuscript; David Johnson and Dr John Stow for useful references and comments on part of the manuscript; David Bishop, Allan Hilton, Dr R.O. Sharples, Devon Zagory for photographs and tables and especially Tim Bach of Cronos for permission to include photographs and text from their controlled atmosphere container manual; Pam Cooke and Lou Ellis for help in typing.

Chapter 1

Introduction

The maintenance or improvement of the postharvest quality and the post-harvest life of fresh fruits and vegetables is becoming increasingly important. This has been partly as a response to a free market situation where the supply of good quality fruits and vegetables constantly exceed demand. Therefore to maintain or increase market share there is increasing emphasis on quality. Also consumer expectation in the supply of all types of fresh fruits and vegetables throughout the year is increasingly taken for granted. This latter expectation is partly supplied by long term storage of many crops but also long-distance transport. Controlled atmosphere storage has been shown to be a technology which can contribute to these consumer requirements in that in certain circumstances, with certain varieties of crop the marketable life can be greatly increased.

This book seeks to evaluate the current technology used in controlled atmosphere storage and its applicability and restrictions for the use in a variety of crops in different situations. While it is not exhaustive in reviewing the enormous quantity of science and technology which has been developed and published on the subject, it will provide an access into controlled atmosphere for those applying the technology in commercial situations. The book can also be used as a basis for determination of researchable issues in the whole area of controlled atmosphere storage.

DEFINITION OF TERMS

While there is no formal definition of controlled atmosphere storage it can be assumed to be the control of the levels of certain gases around and therefore within fresh fruits and vegetables. In this book controlled atmosphere storage refers to the constant monitoring and adjustment of the CO_2 and O_2 levels within gas tight stores or containers. The gas mixture will constantly change due to metabolic activity of the respiring fruits and vegetables in the store and leakage of gases through doors and walls. The gases

1

are therefore measured periodically and adjusted to the predetermined level by the introduction of fresh air or nitrogen or passing the store atmosphere through a chemical to remove CO_2.

Bishop (1996) defined controlled atmosphere storage as 'A low O_2 and/or high CO_2 atmosphere created by natural respiration or artificial means . . . controlled by a sequence of measurements and corrections throughout the storage period'. There are different types of controlled atmosphere storage depending mainly on the method or degree of control of the gases. Some researches prefer to use the terms 'static controlled atmosphere storage' and 'flushed controlled atmosphere storage' to define the two most commonly used systems (D.S. Johnson 1997, personal communication). 'Static' is where the product generates the atmosphere and 'flushed' is where the atmosphere is supplied from a flowing gas stream which purges the store continuously. Systems may be designed which utilize flushing initially to reduce the O_2 content then either injecting CO_2 or allowing it to build up through respiration, and then the maintenance of this atmosphere by ventilation and scrubbing. Scrubbing is the selective removal of CO_2 from the atmosphere by adsorption or absorption. In some cases this is referred to as product generated controlled atmospheres or injected controlled atmospheres.

Modified atmosphere packaging or modified atmosphere storage is where the fruit or vegetable is enclosed within sealed plastic film, which is slowly permeable to the respiratory gases. The gases will change within the package thus producing lower concentrations of O_2 and higher concentrations of CO_2 than exists in fresh air. Bishop (1996) defines modified atmosphere as 'an atmosphere of the required composition is created by respiration, or mixed and flushed into the product enclosure. This mixture is expected to be maintained over the storage life and no further measurement or control takes place'.

Modified atmosphere storage can also refer to stores or containers, which are gas tight but have a panel set within them which is slowly permeable to gases. In both modified atmosphere storage and modified atmosphere packaging the level of the gases around the fruit or vegetable will depend on:

● mass of fruit or vegetable within the pack or container
● the temperature of the fruit or vegetable and the surrounding air
● the type and thickness of plastic film or membrane used
● whether moisture condenses on the film or membrane surface
● external airflow around the film or membrane.

Since there are so many variable and interacting factors it has been necessary to use mathematical modelling techniques in order to predict the levels of gases around the fruit or vegetable (see Chapter 6).

The levels of O_2 and CO_2 within the store or package atmosphere are commonly given in terms of percentage levels. In odd cases (e.g. Beaudry and Gran 1993) it is expressed as kPa which is a pressure measurement where 1 pascal = 1 newton m^{-2}. This relates to the partial pressure of the gases around the fruit and is approximately related to the percentage of the pure gas since 1 atmosphere = 10^5 newton m^{-2} = 100 kPa.

Store humidity is referred to as percentage rh (relative humidity to that which is saturated) but it is also referred to as VPD (vapour pressure deficit) which relates the gaseous water in the atmosphere in relation to its maximum capacity at a given temperature. As an example, at 12°C with 70% rh the vapour pressure deficit is 3.8 mm of mercury and at 12°C and 80% rh the vapour pressure deficit is 2.7 mmHg and at 12°C and 95% rh the vapour pressure deficit is 0.6 mmHg. The vapour pressure is determined from the dry bulb and wet bulb readings by substitution in an equation of the form:

$$e = e'_w - Ap(T - T')$$

where e is the vapour pressure; e'_w is the saturation vapour pressure at the temperature (T') of wet bulb; p is the atmospheric pressure; T is the temperature of the dry bulb; T' is the temperature of the wet bulb; A is a 'constant' which depends on the rate of ventilation of the psychrometer, the latent heat of evaporation of water and the temperature scale in which the thermometers are graduated (Regnault, August and Apjohn quoted by Anon, 1964a).

Examples of the relationship between relative humidity and vapour pressure at different temperatures are given in Table 1.1.

Physiological Disorders

There is a whole range of disorders, that are not primarily associated with infection with microorganisms, which can occur to fresh produce during storage. These are collectively referred to as physiological disorders or physiological injury. It is not the function of this book to provide detailed information on the range of physiological injuries which fresh produce may suffer during controlled atmosphere storage. However, to indicate the symptoms and some of the causes of disorders mentioned in this book a few are described. A more comprehensive discussion of the subject can be found in Fidler et al. (1973) and Snowdon (1990, 1992). Li et al. (1973) showed that partial pressures of O_2 below 1% caused physiological injury during storage, owing presumably to anaerobic fermentation. They indicated that O_2 level for prolonged storage of tomatoes at 12–13°C was about 2–4% O_2.

Superficial scald
Scald is a physiological disorder that can develop on apples and pears during storage and has been associated with ethylene levels in the store. Superficial scald in apples and pears is where the skin turns brown during

Table 1.1. The relationship between relative humidity and vapour pressure.

Depression of the wet bulb in °C

Dry bulb	0.0	0.5	1.0	1.5	2.0	2.5	3.0	7.0	9.0
Vapour pressure (millibars)									
20°C	23.4	22.3	21.3	20.3	19.3	18.3	17.4	10.3	7.1
16°C	18.2	17.3	16.4	15.5	14.6	13.8	13.0	6.8	4.0
14°C	16.0	15.1	14.3	13.5	12.7	11.9	11.1	5.4	2.7
12°C	14.0	13.2	12.5	11.7	10.9	10.2	9.5	4.1	1.6
10°C	12.3	11.5	10.8	10.1	9.4	8.7	8.0	2.9	0.6
Relative humidity (%)									
20°C	100	96	91	87	83	78	74	44	30
16°C	100	95	90	85	81	76	71	37	30
14°C	100	95	90	84	79	74	70	33	25
12°C	100	94	89	83	78	73	68	29	11
10°C	100	94	88	82	77	71	65	24	5

storage (see Fig. 17). Scald susceptibility in Delicious apples was found to be strain-dependent. While storage in 0.7% O_2 effectively reduced scald in Starking and Harrold Red fruits picked over a wide range of maturity stages, it did not adequately reduce scald in Starkrimson fruits after 8 months of storage (Lau and Yastremski, 1993).

Van der Merwe (1996) suggested –0.5°C and 0–1% CO_2 with 1.5% O_2 for 7 months. Granny Smith apples were stored at –0.5° for 6 months in normal atmosphere, for 9 months in 1.5% O_2 and 0% CO_2. After storage the fruits were ripened at 20° for 7 days before evaluation for superficial scald. In a normal atmosphere, all control fruits developed scald. In controlled atmosphere storage only a few apples developed scald (van Eeden *et al.*, 1992).

Scald in apples can be controlled by a pre-storage treatment with an antioxidant. Ethoxyquin (1,2-dihydro-2,2,4-trimethylquinoline-6-yl ether) marketed as Stop-Scald or DPA (diphenylamine) marketed as No Scald or Coraza should be applied directly to the fruit within a week of harvesting. In the USA the Government approved the postharvest application of these two chemicals to apples with maximum residues of 3 ppm for ethoxyquin and 10 ppm for DPA (Hardenburg and Anderson, 1962). Residue levels of DPA in apples were found to vary depending on the application method and the position of the fruit in the pallet box.

Core flush or brown core
This disorder has been described on several cultivars of apple where it develops during storage as a brown or pink discoloration of the core. The flesh remains firm. It has been associated with CO_2 injury but may also be

related to chilling injury and senescent breakdown. The effects of O_2 shock treatment on physiological disorders has also been described. Johnson and Ertan (1983) found that the quality of apples stored at 4°C in 1% oxygen was markedly better than in 2%; the fruits were also free of core flush (brown core) and other physiological disorders. Wang (1990) reviewed the effects of CO_2 on brown core of apples and concluded that it is due to exposure to high levels of CO_2 at low storage temperatures.

This involved storing apples in 0% O_2 for the first 10 days of storage and was shown to prevent core flush in early-picked Jonathan apples from highly affected orchards. No damage due to anaerobic respiration was observed in any of the treatments (Resnizky and Sive, 1991).

Bitter pit
The incidence and severity of bitter pit is influenced also by the dynamic balance of minerals in different parts of the fruit as well as the storage temperature and levels of oxygen and carbon dioxide in the store (Sharples and Johnson, 1987). The addition of lecithin (phosphatidyl choline) to the post-harvest application of calcium can enhance its effect in controlling bitter pit in apples (Sharples *et al.*, 1979). At 18°C apple fruits treated with calcium (4% calcium chloride) had a slightly lower respiration rate, but this effect was greatly enhance when lecithin (1%) was added. At 3°C lecithin treated apples had reduced ethylene production but no effect on carbon dioxide production (Watkins *et al.*, 1982).

Chilling injury
The freezing point of fruits and vegetables is just below the freezing point of water. For example apples will freeze at about −1.5°C, bananas at about −0.8°C, mangoes at about −1°C, grapes at about −2.2°C and dates at about −16°C. The actual freezing point of the crop will vary between cultivars or even the conditions in which the crop is grown. The reason for this is because of soluble solids dissolved in the water in the cell sap and it varies not only with type of fruit or vegetable but can also vary between individuals and even different parts of the same fruit or vegetable. Where crops are exposed to temperature below the freezing point of cell sap they are damaged, and this damage is called 'freezing injury'. Generally, over the range of ambient temperature down to the temperature at which the crop will freeze, the lower the temperature, the longer the storage life. However, certain crops are subject to what is commonly called 'chilling injury' and for such crops the above statement must be modified. Chilling injury is where crops develop temperature associated physiological disorders or abnormalities when exposed to temperatures above those which would cause them to freeze. Chilling injury may be apparent as failure to ripen in climacteric fruit, different forms of external or internal discoloration or predisposition to microorganism infection (Thompson, 1971). Crop susceptibility to chilling injury is influenced by such factors as exposure time, crop cultivar and

the conditions in which the crop was grown. Intermittent warming and temperature preconditioning have been shown to reduce the effects of chilling injury in certain crops (Wang, 1982). The exact mechanism by which chilling injury affects the crop has still not been fully determined. It has been shown to be concerned with loss of membrane integrity and ion leakage from cells and changes in enzyme activity (Wang, 1982) but exactly why some crops are susceptible and some resistant is still a major research topic.

Many crops suffer from chilling injury when stored at low temperatures. These temperatures are above the freezing point of the crop and the injuries they cause may not produce symptoms until the crop is exposed to higher temperatures. If the crop is stored at a low temperature in a refrigerated room it may be in good condition at the time of removal but show chilling injury symptoms when it is being exposed to higher temperatures which often occur during marketing. On the other hand if the crop is to be processed directly after cold storage then chilling injury symptoms may not have time to develop. This could result in different, but correct, storage recommendations for the same crop. It could also be true that an experimenter working on the storage conditions of that crop may have ended his experiment after the storage period and concluded that the crop stored well at that temperature.

Application of calcium to avocados can reduce their susceptibility to chilling injury during subsequent storage (Hofman and Smith, 1993).

Vascular streaking
This is a disorder of cassava where the vascular bundles in the root turn a dark blue to black colour during storage. The symptoms can develop within a day or so of harvesting and the disorder has been associated with O_2 level in the atmosphere and other possible causes (Thompson and Arango, 1977).

HISTORY

Controlled Atmosphere Storage in the UK

Sharples (1989) in his review in *Classical Papers in Horticultural Science* stated that '[Franklin] Kidd and [Cyril] West can be described as the founders of modern controlled atmosphere storage.' Sharples describes the background to their work and how it came about. The excellent paper by Dalrymple (1967) on the development of controlled atmosphere storage, which has been much quoted in the compilation of the current review, stated 'The real start of controlled atmosphere storage had to await the later work of two British scientists [Kidd and West] who started from quite a different vantage point.' It therefore seems logical to start with the work of

Kidd and West and work backwards and forwards to achieve an under-standing of the history of controlled atmosphere storage.

In Britain during World War I, concern was expressed, by the British Government, about food shortages. It was decided that one of the methods of addressing the problem should be through research, and the Food Investigation Organisation was formed at Cambridge in 1917 under the direction of W.B. Hardy (Sharples, 1989), later to be knighted and awarded the fellowship of the Royal Society. In 1918 (Anon, 1919) the work being carried out at Cambridge was described 'as a study of the normal physiology, at low temperatures, of those parts of plants which are used as food. The influence of the surrounding atmosphere, of its content of O_2, CO_2 and water vapour was the obvious point to begin at, and such work has been taken up by Dr F. Kidd. The composition of the air in fruit stores has been suspected of being of importance and this calls for thorough elucidation. Interesting results in stopping sprouting of potatoes have been obtained, and a number of data with various fruits proving the importance of the composition of the air.'

One problem which was identified by the Food Investigation Organisation was the high levels of wastage which occurred during the storage of apples. Franklin Kidd and Cyril West were working at that time at the Botany School in the University of Cambridge on the effects of CO_2 levels in the atmosphere on seeds (Kidd and West, 1917a, 1917b) and Kidd was also working on the effects of CO_2 and O_2 on sprouting of potatoes (Kidd, 1919). Kidd and West transferred to the Low Temperature Laboratory for Research in Biochemistry and Biophysics (later called the Low Temperature Research Station) at Cambridge in 1918 and conducted experiments on what they termed 'gas storage' of apples (Sharples, 1986). By 1920 they were able to set up semi-commercial trials at a farm at Histon in Cambridgeshire to test their laboratory findings in small scale commercial practice. From this work they published a series of papers on various aspects of storage of apples in mixtures of CO_2, O_2 and nitrogen. The publications included Kidd and West (1925, 1934, 1935a, 1935b, 1938, 1939, 1949). They also worked on pears, plums and soft fruit (Kidd and West, 1930).

The Food Investigation Organisation was subsequently renamed. 'The first step towards the formation of the [Food Investigation] Board was taken by the Council of the Cold Storage and Ice association' (Anon, 1919). The committee given the task of setting up the Food Investigation Board consisted of Mr W.B. Hardy, Professor F.G. Hopkins, Professor J.B. Farmer FRS and Professor W.M. Bayliss to prepare a memorandum 'surveying the field of research in connection with cold storage'. The establishment of a 'Cold Storage Research Board' was approved. Since the title did not describe fully the many agents used in food preservation it was renamed the Food Investigation Board with the following term of reference 'To organise and control research into the preparation and preservation of food'.

Work carried out by the Food Investigation Board at Cambridge under Dr F.F. Blackman FRS (Anon, 1920) describes experiments on controlled atmosphere storage of strawberries at various temperatures by Kidd and West at the Botany School. Results were summarized as follows 'Straw berries picked ripe may be held in cold store (temperature 1°–2°C) in a good marketable condition for 6–7 days. Unripe strawberries do not ripen normally in cold storage, neither do they ripen when transferred to normal temperatures after a period of cold storage. The employment of certain artificial atmospheres in the storage chambers has been found to extend the storage life of strawberries. For example, strawberries when picked ripe can be kept in excellent condition for the market for 3–4 weeks at 1°C–2°C if maintained:

1. in atmospheres of O_2, soda lime being used to absorb the CO_2 given off in respiration;
2. in atmospheres containing reduced amounts of O_2 and moderate amounts of CO_2 obtained by keeping the berries in a closed vessel fitted with an adjustable diffusion leak. Under both these conditions of storage the growth of parasitic and saprophytic fungi is markedly inhibited, but in each case the calyces of the berries lose their green after two weeks.'

Controlled atmosphere storage at low temperature of plums, apples and pears was described as 'has been continuing' by Anon (1920), with large scale gas storage tests on apples and pears. It was reported that ripening of plums stored in total nitrogen almost completely inhibited ripening. Plums can tolerate, for a considerable period, an almost complete absence of O_2 without being killed or developing an alcoholic or unpleasant flavour.

Anon (1920) describes work by Kidd and West at the John Street store of the Port of London Authority on Worcester Pearmain and Bramley's Seedling apples at 1°C and 85% rh; 3°C and 85% rh; and 5°C and approximately 60% rh. Sterling Castle apples were stored in about 14% CO_2 and 8% O_2 from 17 September 1919 to 12 May 1920. Ten per cent of the fruit were considered unmarketable at the end of November for the controls whereas the gas-stored fruit had the same level of wastage 3 months later (by the end of February).

Besides defining the appropriate gas mixture required to extend the storage life of selected apple cultivars, Kidd and West were able to demonstrate an interaction. They showed that the effects of the gases in extending storage life varied with temperature in that at 10°C gas storage increased the storage life of fruit by 1.5 to 1.9 times longer than those stored in air, while at 15°C the storage life was the same in both gas storage or in air. They also showed that apples were more susceptible to low temperature breakdown when stored in controlled atmospheres than in air (Kidd and West, 1927a).

In 1929 the Ditton Laboratory was established close to the East Malling Research Station in Kent with J.K. Hardy as Superintendent by the Empire Marketing Board (Fig. 1). At that time it was an out station of the Low

Fig. 1. The Ditton Laboratory at East Malling in Kent. The photograph was taken in June 1996 after the controlled atmosphere storage work had been transferred to the adjacent Horticulture Research International.

Temperature Research Station at Cambridge. The research facilities were comprehensive and novel with part of the station designed to simulate the refrigerated holds of ships in order to carry out experiments on sea freight transport of fruit. Cyril West was appointed Superintendent of the Ditton Laboratory in 1931. West retired in 1948 and R.G. Tomkins was appointed Superintendent (later the title was changed to Director) until his retirement in 1969. At that time the Ditton Laboratory was incorporated into the East Malling Research Station as the Fruit Storage Section (later the Storage Department) with J.C. Fidler as head. R.O. Sharples and D.S. Johnson were subsequent successors to the post. The laboratory continued to function as a centre of excellence in controlled atmosphere storage research until 1992 when new facilities were constructed in the adjacent East Malling Research Station and the research activities were transferred to the new Jim Mount Building. In 1990 East Malling Research Station had become part of Horticulture Research International.

The Covent Garden Laboratory was set up as part of the Ditton Laboratory in 1925. It was situated in London close to the wholesale fruit market with R.G. Tomkins as Superintendent. Anon (1958) describes some of their work on pineapples and bananas as well as pre-packaging work on tomatoes, grape, carrots and rhubarb.

Controlled atmosphere stores have been constructed in many parts of the world especially in the USA and Canada. The first three controlled atmosphere stores were commissioned in South Africa in 1978, and by 1989, controlled atmosphere storage volume had increased to a total of 230 000 bulk bins, catering for over 40% of the annual apple and pear crop (Eksteen *et al.*, 1989).

Controlled Atmosphere Storage in the USA

In the USA in the mid 1930s Robert M. Smock conducted controlled atmosphere storage research on apples, pears and stone fruit (Smock, 1938). Sharples (1989) credits Smock with the 'birth of controlled atmosphere storage technology to North America'. In fact it was apparently Smock who coined the term 'controlled atmosphere storage' as he felt it better described the technology than the term 'gas storage' which was used previously by Kidd and West. The term controlled atmosphere storage was not adopted in Britain until 1960 (Fidler *et al.*, 1973).

G.R. Hill Jr reported work carried out at Cornell University in 1913 in which the firmness of peaches had been retained by storage in inert gases or CO_2. He also observed that the respiration rate of the fruit was reduced and did not return to normal for a few days after storage in a CO_2 atmosphere.

C. Brooks and J.S. Cooley working for the US Department of Agriculture stored apples in sealed containers in which the air was replaced three time each week with air plus 5% CO_2. After 5 weeks storage they noted that the fruit were green, firm and crisp, but were also slightly alcoholic and had 'a rigor or an inactive condition from which they do not entirely recover'.

J.R. Magness and H.C. Diehl in 1924 described a relationship between apple softening and CO_2 concentration in that an atmosphere containing 5% CO_2 slowed the rate of softening with a greater effect at higher concentrations, but at 20% CO_2 the flavour was impaired.

Work on controlled atmosphere storage which had been carried out at the University of California at Davis was reported by Overholser (1928). This work included a general review and some preliminary results on Fuerte avocados. In 1930 Overholser left the University and was replaced by F.W. Allen who had been working on storage and transportation of fresh fruits in artificial atmospheres. Allen began work on controlled atmosphere storage of Yellow Newtown apples. Yellow Newtown, like the Cox's Orange Pippin and Bramley's grown in England, was subject to low temperature injury at temperatures higher than 0°C. These experiments (Allen and McKinnon, 1935) led to a successful commercial trial on Yellow Newtown apples in 1933 at the National Ice and Cold Storage Company in Watsonville. Thornton (1930) carried out trials where the concentration of CO_2 tolerated by selected fruit, vegetables and flowers was examined at six temperatures over the range from 32 to 77°F (0–25°C). To illustrate the commercial importance of this type of experiment the project was financed by the Dry Ice Corporation of America.

In 1935 R.M. Smock worked at the University of California at Davis on apples, pears, plums and peaches (Allen and Smock, 1938). In 1936 and 1937 Allen spent some time with Kidd and West at Ditton and then continued his work at Davis, while in 1937 Smock moved to Cornell University. At Cornell, Smock continued to work on controlled atmosphere storage of apples (Smock and Van Doren, 1938; Smock, 1938). Smock also spent time that year with Kidd and West at Ditton. The controlled atmosphere storage work at Cornell included strawberries and cherries (Van Doren *et al.*, 1941). Smock's work in New York State University was facilitated in 1953 with the completion of large new storage facilities designed specifically to accommodate studies on controlled atmosphere storage.

A detailed report of the findings of the Cornell group was presented in a comprehensive bulletin (Smock and Van Doren, 1941) which gave the results of research on atmospheres, temperatures and varietal response of fruit as well as store construction and operation. In addition to the research efforts in Davis and Cornell, several other groups in the USA were also studying controlled atmosphere storage. Controlled atmosphere storage research of a variety of fruit and vegetables was described by Miller and Brooks (1932) and Miller and Dowd (1936), work on apples was described by Fisher (1939), on citrus fruit by Stahl and Cain (1937) and Samisch (1937).

The real expansion in controlled atmosphere storage in the USA began in the early 1950s. In addition to a pronounced growth in commercial operation in New York State University, controlled atmosphere stores were constructed in New England in 1951; in Michigan and New Jersey in 1956; in Washington, California and Oregon in 1958; and in Virginia in 1959 (Dalrymple, 1967).

By the 1955/56 season, total controlled atmosphere storage holdings had grown to about 814,000 bushels (a volume equivalent to 8 gal or 2200 cubic inches), some 684,000 bushels or 84% being in New York and the rest in New England. In the spring of that year Dalrymple did a study of the industry in New York and found that typically controlled atmosphere stores were owned by large and successful fruit farmers. The average total controlled atmosphere storage holdings per farm were large – averaging 31,200 bushels and ranging from 7500 to 65,000 bushels. Average storage room held some 10,800 bushels. A little over three-quarters of the capacity represented new construction while the other quarter was remodelled from regular storage. A large proportion of capacity, about 68%, was rented out to other farmers or speculators.

Early Knowledge of the Effects of Gases on Crops

The effects of gases on harvested crops has probably been known for centuries. In eastern countries fruits were taken to temples, where incense was burned, to improve ripening. Bishop (1996) indicated that there was evidence that Egyptians and Samarians used sealed limestone crypts for crop storage in the second century BC. He also quotes from the Bible questioning whether the technology might have been used in Old Testament Egypt during the seven plagues wrought by God in order to facilitate the release of the children of Israel. Dilley (1990) mentions the storage of fresh fruit and vegetables in tombs and crypts. This was combined with the gas tight construction of the inner vault so that the fruit and vegetables would consume the O_2 and thus help to preserve the corps. An interpretation of this practice would indicate that a knowledge of the respiration of fruit pre-dates the work described in the 19th century (Dalrymple, 1967). Wang (1990) quotes a Tang dynasty eighth century poem which described how lychees were shown to keep better during long distance transport when they were sealed in the hollow centres of bamboo stems with some fresh leaves. Burying fruit and vegetables is a centuries old practice (Dilley, 1990). In Britain crops were stored in pits which would have restricted ventilation and thus, in some cases, improved their storage life.

The earliest documented scientific study of controlled atmosphere storage was in France by J.E. Bernard in 1819 (Dalrymple, 1967), who showed that harvested fruit absorbed O_2 and gave out CO_2. He also showed that fruit stored in atmospheres containing no O_2 did not ripen, but if they were held for only a short period and then placed in air they continued to ripen. These experiments showed that storage in zero O_2 gave a life of 28 days to a month for peaches, prunes and apricots and about 3 months for apples and pears. Zero O_2 was achieved by placing a paste composed of water, lime and iron sulphate in a sealed jar which, as Dalrymple pointed out, would also have absorbed CO_2.

In 1856 B. Nyce built a commercial cold store in Cleveland, USA, using ice to keep it below 34°F. In the 1860s he experimented with modifying the CO_2 and O_2 in the store by making it air tight. This was achieved by lining the store with air tight casings made from common iron sheets, thickly painting the edges of the metal and having tightly fitted doors. It was claimed that 4000 bushels of apples were kept in good condition in the store for 11 months, but he mentioned that some fruit were injured in a way that Dalrymple (1967) interprets as being possibly CO_2 injury. The carbonic acid level was so high (or the O_2 level in the store was so low) in the store that a flame would not burn. He also used calcium chloride to control the moisture level in the mistaken idea that low humidity was necessary (Dalrymple, 1967).

Dalrymple (1967) states that R.W. Thatcher and N.O. Booth working in Washington State University around 1903 studied storage in several different gases. They found that 'the apples which had been in CO_2 were firm of flesh, possessed the characteristic apple color, although the gas in the jar had a slight odor of fermented apple juice, and were not noticeably injured in flavor.' The apples stored in hydrogen, nitrogen, O_2 and sulphur dioxide did not fare so well. They subsequently studied the effects of CO_2 on raspberries, blackberries and loganberries and 'found that berries which softened in three days in air would remain firm for from 7 to 10 days in carbon dioxid (*sic*)'.

S.H. Fulton in 1907 observed that fruit could be damaged where large amounts of CO_2 were present in the store, but strawberries were 'damaged little, if any . . . by the presence of a small amount of CO_2 in the air of the storage room'. Thatcher (1915) published a paper in which he described work in which he experimented with apples sealed in boxes containing different levels of gases and concluded that CO_2 greatly inhibited ripening.

CONCLUSION

The scientific basis for the application of controlled atmosphere technology to the storage of fresh fruit and vegetables has been the subject of considerable research throughout most of the 20th century. It has been shown to be effective in extending the postharvest life and quality of a wide range of fresh fruit and vegetables. Controlled atmosphere storage is most effective when combined with temperature control and may not be effective at all at room temperatures. Therefore a thorough knowledge of the conditions of precise gas content and temperature is essential for the success of this technology. Where controlled atmosphere storage is not applied correctly it can have a detrimental effect on the crops being stored.

Increasing commercial use of this technology over recent years has been in response to market demand for supplies of fresh fruit and vegetables of all kinds at all times. Another driving force for the application of this technology is the consumer demands for less chemical residues within the food they eat.

Chapter 2

Current Use of Controlled Atmosphere Storage for Fruit and Vegetables

Controlled atmosphere storage is now used worldwide on a variety of fresh fruits and vegetables. Even in countries like Mexico where crops are not widely stored, partly because of long harvesting seasons and limited facilities, controlled atmosphere storage is used in the northern states of Chihuahua and Coahuila for apples. There were estimated to be about 50 controlled atmosphere rooms there, storing about 33,000 tonnes of apples per year (Yahia, 1995). Controlled atmosphere storage has been the subject of an enormous number of biochemical, physiological and technological studies, in spite of which it is still not known precisely why it works. The actual effects that varying the levels of O_2 and CO_2 in the atmosphere have on crops varies with such factors as:

- the species of crop
- the cultivar of crop
- the concentration of the gases in the store
- the crop temperature
- the stage of maturity of the crop at harvest
- the degree of ripeness of the climacteric fruit
- the growing conditions before harvest
- the presence of ethylene in the store.

There are also interactive effects of the two gases, so that the effect of the CO_2 and O_2 in extending the storage life of a crop may be increased when they are combined. The effects of O_2 on postharvest responses of fruit, vegetables and flowers were reviewed and summarized by Thompson (1996) as follows:

- reduced respiration rate
- reduced substrate oxidation

- delayed ripening of climacteric fruit
- prolonged storage life
- delayed breakdown of chlorophyll
- reduced rate of production of ethylene
- changed fatty acid synthesis
- reduced degradation rate of soluble pectins
- formation of undesirable flavour and odours
- altered texture
- development of physiological disorders.

Thompson (1996) also reviewed some of the effects of increased CO_2 levels on stored fruits and vegetables as follows:

- decreased synthetic reactions in climacteric fruit
- delaying the initiation of ripening
- inhibition of some enzymatic reactions
- decreased production of some organic volatiles
- modified metabolism of some organic acids
- reducing the rate of breakdown of pectic substances
- inhibition of chlorophyll breakdown
- production of 'off-flavour'
- induction of physiological disorders
- retarded fungal growth on the crop
- inhibition of the effect of ethylene
- changes in sugar content (potatoes)
- effects on sprouting (potatoes)
- inhibition of postharvest development
- retention of tenderness
- decreased discoloration levels.

The recommendations for the optimum storage conditions have varied over time due mainly to improvements in the control technology over the levels of gases within the stores. Bishop (1994) showed the evolution of storage recommendations by illustration of the recommendation for the storage of the cultivar Cox's Orange Pippin since 1920 (Table 2.1).

Table 2.1. Recommended storage conditions for Cox's Orange Pippin apples all at 3.5°C (Bishop, 1994).

% O_2	% CO_2	Storage time in weeks	Approximate date of implementation
21	0	13	–
16	5	16	1920
3	5	21	1935
2	< 1	27	1965
1.25	< 1	31	1980
1	< 1	33	1986

Controlled atmosphere storage is still mainly applied to stored apples but studies of other fruit and vegetables have shown it has wide application, and an increasing number of crops are being stored and transported under controlled atmosphere conditions. The technical benefits of controlled atmosphere storage have been amply demonstrated for a wide range of flowers, fruit and vegetables but the economic implications of using this comparatively expensive technology have often limited its commercial application. However, with technological developments, more precise control equipment and the reducing cost, controlled atmosphere storage is being used commercially for an increasing range of crops.

CONTROLLED ATMOSPHERE STORE CONSTRUCTION

An excellent review of the design, construction, operation and safety considerations is given by Bishop (1996). In fresh fruits and vegetable stores the CO_2 and O_2 levels will change naturally through the respiration of the crop because of the reduced or zero gas exchange afforded by the store walls. Levels of the respiratory gases in non-controlled atmosphere stores have shown these changes. For example Kidd and West (1923) showed that the levels of CO_2 and O_2 in the holds of ships carrying stone fruit and citrus fruit arriving in the UK from South Africa at 3.9 to 5°C were 6 and 14%, respectively. At 2–3.3°C the levels were 10.0% CO_2 and 10.5% O_2. A.K. Thompson (Batu, 1995, unpublished data) has shown that the level of CO_2 in a banana ripening room can rise from near zero to 7% in 24 h.

Current commercial practice can be basically the same as that described in the early part of the century. The crop is loaded into an insulated store-room whose walls and door have been made gas tight. The first controlled atmosphere stores were constructed in the same way as refrigerated stores from bricks or concrete blocks with a vapour proof barrier and an insulation layer which was coated on the inside with plaster. In order to ensure that the walls were gas tight for controlled atmosphere storage they were lined with sheets of galvanized steel (Fig. 2). The bottoms and tops of the steel sheets were embedded in mastic (a kind of mortar composed of finely ground oolitic limestone, sand, litherage and linseed oil) and, where sheets abutted on the walls and ceiling, a coating of mastic was also applied. Modern controlled atmosphere stores are made from metal-faced insulated panels (usually polyurethane foam) which are fitted together with gas tight patented locking devices. The joints between panels are usually taped with gas tight tape or painted with flexible plastic paint to ensure that they are gas tight.

Major areas of the store where leaks can occur are the doors. These are usually sealed by having rubber gaskets around the perimeter which correspond to another rubber gasket around the doorjamb or frame (Fig. 3) so that when the door is closed the two meet to seal the door. In addition a

Fig. 2.   The inside of a traditional controlled atmosphere store in the UK showing the internal walls covered with galvanized sheets to render them gas tight. (Courtesy of Dr R.O. Sharples.)

flexible soft rubber hose may be hammered into the inside between the door and its frame to give a double rubber seal to ensure gas tightness. In other designs screw jacks are spaced around the periphery of the door. Some modern controlled atmosphere stores and fruit ripening rooms have inflatable rubber door seals (Fig. 4).

   With the constant changes and adjustments in the store temperature and concentrations of the various gases inside the sealed store, variation in pressure can occur. Where there is a pressure difference between the store air and the outside air there can be a difficulty in retaining the store in completely gas tight condition. Stores are therefore fitted with pressure release valves (Fig. 5), but these can make the maintenance of the precise gas level difficult, especially the O_2 level in the ultra low O_2 store. An expansion bag may be fitted to the store to overcome this problem of pressure differences. The bags are gas tight and partially inflated and are placed outside the store with the inlet to the bag inside the store. If the store air volume increases then this will automatically further inflate the bag and when the pressure in the store is reduced then air will flow from the bag to the store. The inlet of the expansion bag should be situated before the cooling coils of the refrigeration unit in order to ensure the air from expansion bag is cooled before being returned to the store.

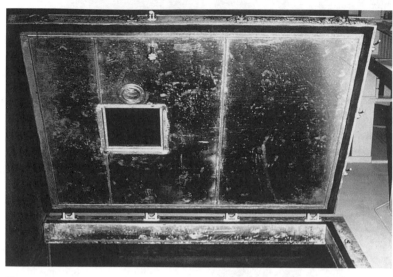

(b)

(a)

Fig. 3. Door of a traditional controlled atmosphere storage in the UK. (a) Sliding type showing the rubber gasket. (Courtesy of Dr R.O. Sharples.) (b) Hinged type. (Courtesy of Dr R.O. Sharples.)

Fig. 4. A pressurized banana ripening room in 1997 at Mack Multiples in Paddock Wood in Kent with inflatable rubber seals to ensure the doors are gas tight.

Fig. 5. Pressure release valve used in apple controlled atmosphere stores in the UK. (Courtesy of Dr R.O. Sharples.)

Testing how gas tight a store is remains crucial to the accurate maintenance of the required CO_2 and O_2 levels. One way of measuring this was
described by Anon (1974). A blower or vacuum cleaner is connected to a
store ventilator pipe and a manometer connected to another ventilator pipe.
It is essential that the manometer is held tightly in the hole so that there are
no leaks. The vacuum cleaner or blower starts to blow air into the store and
is maintained until a pressure of 200 Pa has been achieved and then the
vacuum cleaner or blower is switched off. The time for the pressure inside
the store to fall is then taken. Anon (1974) recommended that the time
taken to fall from 187 to 125 Pa should not be less than 7 min in order to
consider the store sufficiently gas tight. If the store does not reach this
standard efforts should be made to locate and seal the leaks.

Temperature Control

The main way of preserving fruits and vegetables in storage or during long
distance transport is by refrigeration, and controlled atmospheres are considered a supplement to increase or enhance the effect of refrigeration.
There is evidence that the inhabitants of the island of Crete were aware of
the importance of temperature in the preservation of food, even as early as
2000 BC (Koelet, 1992). Mechanical refrigeration was developed in 1755 by
the Scotsman William Cullen who showed that by evaporating ether under
reduced pressure, caused by evacuation, the temperature of the water in the
same vessel was reduced, thus forming ice. Cullen then patented a machine
for refrigerating air by the evaporation of water in a vacuum. Professor Sir
John Leslie, also a Scot, developed the principle in 1809 and invented a
differential thermometer and a hygrometer which led him to discover a
process of artificial congelation. Leslie's discovery improved upon Cullen's
equipment by adding sulphuric acid to absorb the water vapour and he
developed the first ice-making machine. In 1834 an American, Jacob
Perkins, was granted a British patent (number 6662) for a vapour compression refrigeration machine. Carle Linde, in Germany, developed an
ammonia compression machine. This type of mechanical refrigeration
equipment was employed in many of the early experiments on fruits and
vegetable stores and was used in the first sea freight shipment of chilled
beef from Argentina in 1879.

Controlled atmosphere storage is a method of supplementing refrigeration to increase the postharvest life of crops. In fact there is evidence that
controlled atmosphere storage is only successful when applied at low
temperatures (Kidd and West, 1927a, 1927b and others). Standard refrigeration units are therefore integral components of controlled atmosphere
stores. Temperature control is achieved by having pipes containing a refrigerant inside the store. Ammonia or chlorofluorocarbons are common

refrigerants but new refrigerants are being developed which have potentially less detrimental environmental effects than the latter. These pipes pass
out of the store, the liquid is cooled and passed into the store to reduce the
air temperature of the store as it passes over the cooled pipes. A simple
refrigeration unit consists of an evaporator, a compressor, a condenser and
an expansion valve. The evaporator is the pipe which contains the refrigerant mostly as a liquid at low temperature and low pressure, and is the part
of the system which is inside the store. Heat is absorbed by the evaporator
causing the refrigerant to vaporize. The vapour is drawn along the pipe
through the compressor, which is a pump that compresses the gas into a
hot high-pressure vapour. This is pumped to the condenser where the gas is
cooled by passing it through a radiator. The radiator is usually a network of
pipes open to the atmosphere. The high-pressure liquid is passed through a
series of small-bore pipes which slows down the flow of liquid so that a
high pressure builds up. The liquid then passes through an expansion
valve, which controls the flow of refrigerant and reduces its pressure. This
reduction in pressure results in a reduction in temperature causing some of
the refrigerant to vaporize. This cooled mixture of vapour and liquid refrigerant passes into the evaporator so completing the refrigeration cycle. In
most stores a fan passes the store air over the coiled pipes containing the
refrigerant which helps to cool the air quickly and distribute it evenly
throughout the store.

In commercial practice for controlled atmosphere stores the store
temperature is initially reduced to 0°C for a week or so whatever the subsequent storage temperature will be. Also, controlled atmosphere stores are
normally designed to a capacity which can be filled in 1 day, so fruit are
loaded directly into store and cooled the same day. In the UK, the average
controlled atmosphere store size was given as about 100 tonnes with varia-
tions between 50 and 200 tonnes, in continental Europe about 200 tonnes
and in North America about 600 tonnes (Bishop, 1996). In the UK, the
smaller sized rooms are preferred because they facilitate the speed of
loading and unloading.

Humidity Control

Most fruits and vegetables which are kept in controlled atmosphere storage
require a high relative humidity (rh), generally the closer to saturation the
better, so long as moisture does not condense on the crop. The amount of
heat absorbed by the cooling coils of the refrigeration unit is related to the
temperature of the refrigerant they contain and the surface area of the coils.
If the refrigerant temperature is low compared to the store air temperature
then water will condense on the evaporator. This removal of moisture
from the store air reduces its relative humidity which results in the stored

crop losing moisture by evapotranspiration. In order to reduce crop desiccation the refrigerant temperature should be kept close to the store air temperature. However, this must be balanced with the removal of the respiratory heat from the crop, temperature leakage through the store insulation and doors and heat generated by fans otherwise the crop temperature cannot be maintained. There remains the possibility of having a low temperature differential between the refrigerant and the store air by increasing the area of the cooling surface and this may be helped by adding such devices as fins to the cooling pipes or coiling them into spirals. In a study on evaporator coil refrigerant temperature, room humidity, cool-down and mass loss rates were compared in 1200 commercial bin apple storage rooms. Rooms in which evaporator coil ammonia temperatures were dictated by cooling demand required a significantly longer time to achieve desired humidity levels than rooms in which evaporator coil temperatures were controlled by a computer (Hellickson *et al.*, 1995). The overall mass loss rate of apples stored in a room in which the cooling load was dictated by refrigerant temperature was higher than in a room in which refrigerant temperature was maintained at approximately 1°C during the fruit cool-down period (Hellickson *et al.*, 1995).

A whole range of humidifying devices can also be used to replace the moisture in the air which has been condensed out on the cooling coils of refrigeration units. These include spinning disc humidifiers where water is forced at high velocity onto a rapidly spinning disc. The water is broken down into tiny droplets which are fed into the air circulation system of the store. Sonic humidifiers utilize energy to detach tiny water droplets from a water surface which are fed into the store's air circulation system.

A technique which retains high humidity within the store is via secondary cooling so that the cooling coils do not come into direct contact with the store air. One such system is the 'jacketed store'. These stores have a metal inner wall inside the store's insulation with the cooling pipes cooling the air in that space. This means a low temperature can be maintained in the cooling pipes without causing crop desiccation and the whole wall of the store becomes the cooling surface.

Ice bank cooling is also a method of secondary cooling where the refrigerant pipes are immersed in a tank of water so that the water is frozen. The ice is then used to cool water and the water is converted to a fine mist which is used to cool and humidify the store air (Lindsey and Neale, 1977; Neale *et al.*, 1981).

Where crops such as onions are stored under controlled atmosphere conditions they require a low humidity (about 70% rh). This can be achieved by having a large differential between the refrigerant and air temperature of about 11–12°C with natural air circulation or between 9 and 10°C where air is circulated by a fan.

Gas Control Equipment

Burton (1982) indicated that in the very early gas stores (controlled atmosphere stores) the levels of CO_2 and O_2 were controlled by making the store room gas tight with a controlled air leak. This meant that the levels of the two gases were maintained at approximately equal levels of about 10%. Even in more recent times the CO_2 level in the store was controlled while the level of O_2 was not and was calculated as: $O_2 + CO_2 = 21\%$ (Fidler and Mann, 1972). Kidd and West (1927a, 1927b) recognized that it was desirable to have more precise control over the level of gases in the store and to be able to control them independently. In subsequent work the levels of CO_2 and O_2 were monitored and adjusted by hand and considerable variation in the levels could occur. Kidd and West (1923) showed that the levels monitored in a store varied between 1 and 23% for CO_2 and between 4 and 21% for O_2 over a 6-month storage period. In many early stores Orsat apparatus was used to measure the levels of store gases. The apparatus contained O_2 and CO_2 absorbing chemicals. As the two gases were absorbed there was a change in air volume which was proportional to the volumes of the gases. This method is very time consuming in operation and cannot be used to measure volumes in a controlled atmosphere flow through system. The atmosphere in a modern controlled atmosphere store is constantly analysed for CO_2 and O_2 levels using an infrared gas analyser to measure CO_2 and a paramagnetic analyser for O_2 (Fig. 6). The prices of the analytical equipment have fallen over the years and they can be used to measure the gas content in the store air constantly. They need to be calibrated with mixtures of known volumes of gases. Another system, which is used mainly for research into controlled atmosphere storage, is to mix the gases accurately in the desired proportions and then ventilate the store at an appropriate rate with the desired mixture. This method tends to be too expensive for commercial application.

Oxygen Control

The way that a modern controlled atmosphere storage system is operated is that when the O_2 has reached the level required for the particular crop being stored it is maintained at that level by frequently introducing fresh air from outside the store. Usually tolerance limits are set at plus or minus 0.15% for O_2 levels below 2%, and plus or minus 0.3% for O_2 levels of 2% and above (Sharples and Stow, 1986). This means that if an O_2 concentration of 1% is required, when the O_2 level is 0.85%, air is vented until it reaches 1.15%. With continuing equipment development the precision with which the set levels of CO_2 and O_2 can be maintained is increasing. With recent developments in the control systems used in controlled atmosphere stores it is possible to control O_2 levels close to the theoretical minimum.

(a)

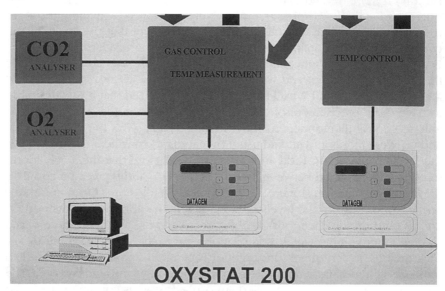

(b)

(c)

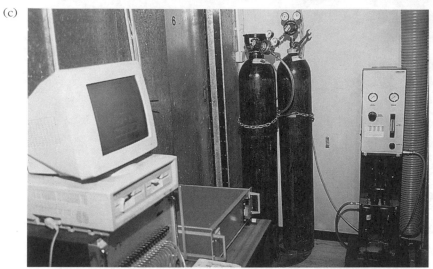

Fig. 6. (*and opposite*)   Analysers used in modern controlled atmosphere stores. (a) Diagram showing a typical system. (Courtesy of David Bishop.) (b) David Bishop system fitted in experimental stores at Silsoe College, Cranfield University in the UK. (Courtesy of Allan Hilton.) (c) An experimental system in use at Silsoe College, Cranfield University in the UK. (Courtesy of Allan Hilton.)

This is because modern systems can achieve a much lower fluctuation in gas levels and ultra low O_2 storage (levels around 1%) is now common. Schouten (1997) described a system which he called dynamic control of ULO based on headspace analysis of ethanol levels which are maintained at less than 1 ppm where O_2 levels were maintained at 0.3–0.7%.

Carbon Dioxide Control

The same precise control described above for O_2 also applies to CO_2. Sharples and Stow (1986) state that the CO_2 level in the store should be maintained at plus or minus 0.5% of the recommended level. When a predetermined level is reached the atmosphere is passed through a chemical which removes CO_2 and then back into the store. This is called 'active scrubbing'. The chemicals used in controlling CO_2 levels in a store include the following:

Calcium hydroxide scrubbers
Calcium hydroxide reacts irreversibly with CO_2 to produce calcium carbonate, water and heat. When the CO_2 level in the store is above that which is required a fan draws the store atmosphere through the room containing the bags of lime (Fig. 7) until the required level is reached. The amount of lime required depends on the type and variety of the crop and the storage

Fig. 7. A simple lime scrubber for CO_2 used in many traditional controlled atmosphere stores in the UK. (Courtesy of Dr R.O. Sharples.)

temperature. For example Koelet (1992) states that for 1 tonne of apple fruit the following amount of high calcium lime is needed every 6–10 weeks:

Cox's Orange Pippin	75 kg
Boskoop	7.5 kg
Lombards	7.5 kg
Golden Delicious	7.5 kg
Melrose	7.5 kg
Winston	7.5 kg

Bishop (1996) also calculated the amount of lime required to absorb CO_2 from controlled atmosphere stores on a theoretical basis. He calculated that 1 kg of lime will adsorb 0.59 kg of CO_2, but a practical capacity he estimated at 0.4 kg of CO_2 kg^{-1} of lime. On that basis the requirement of lime for Cox's Orange Pippin apples stored in less than 1% CO_2 with 2% O_2 was 5% of the fruit weight or 50 kg of lime. For other apple varieties, storage regimes and periods, Bishop indicates that probably less lime would be required. Thompson (1996) quotes a general requirement for apples of only 25 kg for 6 months. After scrubbing the air should re-enter the store just before it passes over the cooling coils of the refrigeration unit. There may also be some dehydration of the air as it is passed through the lime and the air may need humidification before being reintroduced to the store.

Renewable scrubbers

Many stores use renewable scrubbers to remove excess CO_2 from stores. These have the advantage of being compact and are particularly suitable for use in controlled atmosphere transport systems. They consist of two containers of material which can absorb CO_2. Store air is passed through one of these containers when it is required to reduce the CO_2 in the store atmosphere. Activated charcoal (active carbon) or a molecular sieve (aluminium calcium silicate) are commonly used for this purpose. When they have been saturated with CO_2 a device switches over to the second container. The first container then has fresh air blown through it to detach the CO_2 so that it is available to be used as a scrubber when the second container has been saturated. Where a molecular sieve is used it is necessary to heat it during the purging cycle. Activated charcoal can be purged simply by fresh air. Koelet (1992) described a system where a ventilator sucks air from the store through a plastic tube to an active carbon filter scrubber. When the active carbon becomes saturated with CO_2 the tube is disconnected from the store and outside air is passed over it to desorb the CO_2. When this process is complete the active carbon scrubber is connected to a 'lung' which is a balloon of at least 5 m^3 capacity, depending on the size of the equipment, which contains an atmosphere with low O_2. This process is to ensure that when the active carbon scrubber is reconnected to the store it does not contain high O_2 levels.

Temperature swing scrubbers

A 40% solution of ethanolamine can also be used in such a system and can be regenerated by heating the chemical to 110°C. However, ethanolamine is corrosive to metals and is not commonly used for controlled atmosphere stores.

Other types of scrubber

Various other types of CO_2 scrubbers have been designed, such as water scrubbers and sodium hydroxide scrubbers, but they are not commonly used. Water scrubbers work by passing the store air through a water tower which is pumped outside the store where it is ventilated with fresh air to remove the CO_2 dissolved in it inside the store. However, O_2 is dissolved in the water while it is being ventilated and will therefore provide a store atmosphere of about 5% O_2 and 3% CO_2. Sodium hydroxide scrubbers work by bubbling the store air through a saturated solution, at about 50 l min^{-1}, when there is excess CO_2 in the store. This method requires about 14 kg sodium hydroxide for each 9 m^3 of apples per week and, since the chemical reaction which absorbs CO_2 is not reversible, the method can be expensive. Selective diffusion membrane scrubbers have been used commercially

(Marcellin and LeTeinturier, 1966) but they require a large membrane surface area in order to maintain the appropriate gaseous content around the fruit.

Alternatively, the CO_2 absorbing chemical may be placed inside the store where it can keep the level low (usually about 1%) and is called 'passive scrubbing'. This method of controlled atmosphere store is referred to as 'product generated', since the gas levels are produced by the crops' respiration.

The time taken for the levels of these two gases to reach the optimum (especially for the O_2 to fall from the 21% in normal air) can reduce the maximum storage life of the crop. It is common therefore to fill the store with the crop, seal the store and inject nitrogen gas until the O_2 has reached the required level and then maintain it in the way described above. The nitrogen may be obtained from large liquid nitrogen cylinders. Malcolm and Gerdts (1995) in a review described three main methods used to produce nitrogen gas used for controlled atmosphere applications. These were cryogenic distillation, pressure swing adsorption (PSA) system, and membrane systems. Hollow fibre technology is also used to generate nitrogen to inject into the store (Fig. 8). The walls of these fibres are differentially permeable to O_2 and nitrogen. Compressed air is introduced into these fibres and by varying the pressure it is possible to regulate the purity of the nitrogen coming out of the equipment and produce an output of almost pure nitrogen. Another type of nitrogen generator is called 'pressure swing adsorption'. This has an air compressor which passes the compressed air through a molecular sieve which traps the O_2 in the air and allows the nitrogen to pass through. It is a duel circuit system so that when one circuit is providing nitrogen for the store, the other circuit is being renewed. The O_2 content at the output of the equipment will vary with the throughput. For a small machine with a throughput of 5 N m^3 h^{-1} the O_2 content will be 0.1%, at 10 N m^3 h^{-1} it will be 1% and at 13 N m^3 h^{-1} it will be 2% (Thompson, 1996). In China a carbon molecular sieve nitrogen generator has been developed (Shan-Tao and Liang, 1989). This was made from fine coal powder which was refined and formed to provide apertures similar to nitrogen (3.17 Å) but smaller than O_2 (3.7 Å). When air is passed through the carbon molecular sieve under high pressure, the O_2 molecules are absorbed onto the carbon molecular sieve and the air passing through has an enriched nitrogen level. A pressurized water CO_2 scrubber was described by Vigneault and Raghavan (1991) which could reduce the time required to reduce the O_2 content in a controlled atmosphere store from 417 h to 140 h.

For crops which have only a very short marketable life, such as strawberries, the CO_2 level may also be increased to the required level by direct injection of CO_2 from a pressurized gas cylinder.

Membrane System

Marcellin and LeTeinturier (1966) reported a system of controlling the gases in store by diffusion through a silicone rubber diffuser. In subsequent trials they reported that they were able to maintain an atmosphere of 3% CO_2, 3% O_2 and 94% nitrogen with the simple operation of only gas analysis as the regular necessary maintenance. In later work a silicone membrane system

(a)

(b)

Fig. 8.   Nitrogen generators in use at a commercial controlled atmosphere store in Kent, UK (a) in 1988, (b) in 1995.

for controlled atmosphere storage was shown to maintain 3.5 to 5% CO_2 and 1.5–3% O_2, where 5% and 3%, respectively, were expected. After 198 days of storage of cabbage, total mass loss was 14% under controlled atmosphere compared to 40% in air (Gariepy *et al.*, 1984). In a commercial scale (472 tonnes) experiment, an Atmolysair system controlled atmosphere room was compared with a conventional cold room for storing cabbage, cultivar Winter Green, for 32 weeks. The Atmolysair system controlled atmosphere room atmosphere was 5–6% CO_2, 2–3% O_2, 92% nitrogen and traces of other gases, with a temperature of 1.3°C; the cold room was maintained at 0.3°. The average trimming losses were less than 10% for the Atmolysair system controlled atmosphere room and exceeded 30% for the cold room (Raghavan *et al.*, 1984). The use of a carbon molecular sieve nitrogen generator in controlled atmosphere storage was described by Zhou *et al.* (1992a).

Rukavishnikov *et al.* (1984) described a 100 tonne controlled atmosphere store under a 150–300 µ polyethylene film, with windows of membranes selective for O_2 and CO_2 permeability made of polyvinyl trimethyl silane or silicon-organic polymers. Apples and pears stored at 1–4°C under the covers had 93 and 94% sound fruit, respectively after 6–7 months, whereas the fruit in air storage at similar temperatures had over 50% losses after 5 months.

A new design procedure for a silicon membrane controlled atmosphere store, based on parametric relationships, was suggested by Gariepy *et al.* (1988) for selecting the silicone membrane area for long-term controlled atmosphere storage of leeks and celery.

A 'modified gaseous components system' for controlled atmosphere storage using a gas separation membrane was designed, constructed and tested by Kawagoe *et al.* (1991). The membrane is permeable to CO_2, O_2 and nitrogen but to different degrees, thus the levels of the three gases vary. A distinctive feature of the system is the application of gas circuit selection to obtain the desired modified gaseous components. It was shown to be effective in decreasing CO_2 content and increasing O_2 content, but nitrogen introduction to the chamber was not included in the study. Future work on this aspect was proposed.

PRE-CONTROLLED ATMOSPHERE STORAGE TREATMENTS

In most cases it is advisable to place fruits and vegetables in store as quickly as possible after harvest. In work on kiwifruit, Tonini and Tura (1997) showed that storage in 4.8% CO_2 combined with 1.8% O_2 reduced rots caused by infections with *Botrytis cinerea*. If the fruit was cooled to –0.5°C immediately after harvest then the effect was greatest, the quicker the controlled atmosphere storage conditions were established. With a delay of 30 days the controlled atmosphere storage conditions were ineffective in controlling the rot. However, in certain cases it has been shown that pre-storage treatments can be beneficial, and some of these have been proved to be successful in fruit and vegetable storage, as described below.

High Temperature

Prior to storage of crops such as potatoes, yam, sweetpotato, garlic and onion, and to some extent citrus fruits, their storage period may be enhanced by such processes as curing and drying (Thompson, 1996). These practices are standard treatments on some crops and applied as routine whatever type of subsequent storage is carried out. For other crops pre-storage exposure to high temperature is less well established and still in an experimental stage. For example, Granny Smith apples were kept at 46°C for 12 h, 42°C for 24 h, or 38°C for 72 or 96 h before storage at 0°C for 8 months in 2–3% O_2 with 5% CO_2. These heat treated fruit were firmer at the end of storage and had a higher soluble solids to acid ratio and a lower inci-dence of superficial scald than fruits not heat treated. Pre-storage regimes with longer exposures to high temperatures of 46°C for 24 h or 42°C for 48 h resulted in fruit damage after storage (Klein and Lurie, 1992).

Chemicals

Pre-treatment of fruit with chemicals prior to storage can affect their storage characteristics. There is currently a growing consumer preference for less application of chemical to the food we eat. Growth regulating chemicals have been applied to trees to increase fruit quality and yield. One such chemical, which has been the subject of considerable debate in the news media, is called daminozide, Alar, B9 or B995 (N-dimethylaminosuccinamic acid). In a comparison between preharvest and postharvest application of daminozide to Cox's Orange Pippin apples, immersion of fruits in a solution containing 4.25 g l^{-1} for 5 min delayed the rise in ethylene production at 15°C by about 2 days, whereas orchard application of 0.85 g l^{-1} caused delays of about 3 days. Both modes of application depressed the maximal rate of ethylene production attained by ripe apples by about 30%. Dami-nozide treated fruit were shown to be less sensitive to the application of ethylene than untreated fruit, but this response varied between cultivars (Knee and Looney, 1990). Graell and Recasens (1992) sprayed Starking Delicious apple trees with daminozide (1000 mg l^{-1}) in mid-summer in 1987 and 1988. Controls were left unsprayed. Fruits were harvested on 12 and 24 September 1987, and 7 and 17 September 1988, and stored in controlled atmosphere storage (3% O_2 and 4% CO_2 at 0–1°C). Two controlled atmos-phere storage rooms were used: a low ethylene controlled atmosphere room, fitted with a continuous ethylene removal system, with a store ethylene concentration ranging from 4 to 15 μl l^{-1}; and a high ethylene controlled atmosphere room, where ethylene was not scrubbed and store ethylene concentration was more than 100 μl l^{-1}. In both seasons, after 8 months of storage in high ethylene controlled atmosphere, fruits harvested earlier were less ripe and had developed more scald than fruits harvested later. Scald is a physiological disorder which can develop on apples and pears during storage and has been associated with ethylene levels in the store. Daminozide sprayed and earlier harvested fruits which had been stored in low ethylene controlled atmosphere were firmer and more acid after storage than unsprayed, later-harvested fruits stored in high ethylene controlled atmosphere. Generally, after a 7-day post-storage holding period at 20°C, the differences between treatments were maintained. The soluble solids content of fruit samples was similar for all treatments. Low ethylene controlled atmosphere stored fruits developed less scald than high ethylene controlled atmosphere stored fruits in the 1987/88 season; however, scald development was similar in low ethylene controlled atmosphere and high ethylene controlled atmosphere stored fruits in the 1988/89 season. Daminozide has been withdrawn from the market in several countries (J. Love, 1995, personal communication) because of suggestions that it might be carcinogenic.

For crops which are to be stored for long periods, either in controlled atmosphere or air storage, it is commonly necessary to treat them with a chemical which will prevent microorganisms attacking them. The use of these chemicals is controlled by legislation with maximum residue levels stipulated for the edible portions of the fruits and vegetables. Other chemicals are applied which prevent the development of physiological disorders and sprouting of root crops during storage. Controlled atmosphere storage is one way that can reduce or replace the chemicals used for these purposes (see later sections).

INTERRUPTED CONTROLLED ATMOSPHERE STORAGE

Where controlled atmosphere storage has been shown to have detrimental side-effects on fruits and vegetables the possibility of alternating controlled atmosphere storage with air storage has been studied. Results have been mixed with positive, negative effects and in some cases no effect.

Neuwirth (1988) described storage of Golden Delicious apples during January to May in controlled atmospheres of 1% CO_2 with 3% O_2, 3% CO_2 with 3% O_2 or 5% CO_2 with 5% O_2 at 2°C. These regimes were interrupted by a 3 week period of ventilation with natural air, beginning on 15 January, 26 January, 17 February or 11 March, after which the controlled atmosphere treatment was reinstated. Ventilation at the time of the climacteric (late February to early March) produced large increases in the production of both CO_2 and volatile flavour substances, but ventilation at other times had little effect. After controlled atmosphere conditions were restored, respiration and the production of flavour substances declined again, sometimes to below the level of fruit stored in continuous controlled atmospheres.

Storage of bananas at high temperatures (as may happen in producing countries) causes physiological disorders and unsatisfactory ripening. In trials with the cultivar Poyo (from Cameroon), storage at 30–40°C was interrupted by one to three cooling periods (20°C) of 12 h either in air or in atmospheres with 50% O_2 or 5% O_2. Cooling periods reduced high temperature damage, especially when fruits stored at 30°C received three cooling periods in 50% O_2 (Dick and Marcellin, 1985).

Parsons *et al.* (1974) interrupted controlled atmosphere storage (3% O_2 combined with either 0, 3 or 5% CO_2) of tomatoes each week by exposing them to air for 16 h. The storage temperature was 55°F (13°C). This interrupted storage had no measurable effect on the storage life of the fruit but increased the level of decay which developed on fruit when removed from storage to higher temperatures to simulate shelf life.

Intermittent exposure of the avocado cultivar Haas to 20% CO_2 increased their storage life at 12°C and reduced chilling injury during storage at 4°C compared to those stored in air at the same temperatures (Marcellin and Chevez, 1983).

Anderson (1982) described experiments where peaches and nectarine were stored at 0°C in 5% CO_2 with 1% O_2 which was interrupted every 2 days by removing the fruit to 18–20°C in air. When subsequently ripened, fruits in this treatment had little of the internal breakdown found in fruits stored in air at 0°C.

CARBON DIOXIDE SHOCK TREATMENT

Treating fruits and vegetables with high levels of CO_2 prior to storage can have beneficial effects on their subsequent storage life. Hribar *et al.* (1994) described experiments where Golden Delicious apple fruits were held in 15% CO_2 for 20 days prior to storage under ultra low O_2 conditions (1.5% O_2 and 1.5% CO_2) for 5 months. The CO_2 treated fruits retained a better green skin colour during storage which was shown to be due to inhibition of carotenoid production. Other effects of the CO_2 shock treatment were that it increased fruit firmness after storage, ethanol production was greater due to anaerobic respiration, but titratable acidity was not affected. Pesis *et al.* (1993) described experiments where apple fruits were treated with 95% CO_2, 4% nitrogen and 1% O_2 at 20°C for 24–48 h for Golden Delicious and 24–96 h for Braeburn, then transferred to either 20°C in air or 5°C for 6 weeks in air followed by a shelf life of 10 days at 20°C. High CO_2 pretreatment induced CO_2, ethylene, ethanol and ethyl acetate production in Golden Delicious fruits during storage at 20°C and fruits became softer and yellower but more tasty than untreated fruits after 2 weeks at 20°C. For

Braeburn fruits CO_2 and ethylene production were reduced and volatiles were induced by the high CO_2 pre-treatment during shelf life at 20°C following 6 weeks of cold storage, and treated fruits remained firmer but yellower than control fruits.

In earlier experiments, two cultivars of table grape were stored in controlled atmospheres, Waltham Cross in 1–21% O_2 and 0–5% CO_2 and Barlinka in 1–21% O_2 and 5% CO_2 both at –0.5°C for 4 weeks by Laszlo (1985). Some of the fruit were pre-treated with CO_2 shock treatment (10% CO_2 for 3 days) before storage in air at 10°C for 1 week. With both cultivars the incidence of berry decay (mainly due to infections by *Botrytis cinerea*) was highest with fruits that had been subjected to the CO_2 shock treatment. The incidence of berry cracking was less than 1% in Waltham Cross but in Barlinka it was higher with fruits that had been subjected to the CO_2 shock treatment, high O_2 concentrations and controls without SO_2.

In a trial with Passe Crassane pears stored in air at 0°C for 187 days internal browning (a physiological disorder which can develop during storage) was prevented by shock treatment with 30% CO_2 for 3 days at intervals of 14–18 days. Also the fruits that had been subjected to CO_2 shock treatment had excellent quality after being ripened at 20°C (Marcellin *et al.*, 1979).

In experiments in which 10–30% CO_2 was applied periodically during storage for 201 days, followed by after-ripening, the incidence of internal browning was greatly reduced. With Comice pears stored at 0°C for 169 days followed by after-ripening for 7 days, 15% CO_2 treatment every 14 days greatly reduced the incidence of scald and internal browning (Marcellin *et al.*, 1979). Park *et al.* (1970) showed that polyethylene film packaging and/or CO_2 shock markedly delayed ripening, preserved freshness and reduced spoilage and core browning of pears during storage.

Fuerte avocado fruits were stored for 28 days at 5.5°C, under 2% O_2 and 10% CO_2. They were then treated with 25% CO_2 for 3 days commencing one day after harvest followed by normal atmosphere (CO_2 shock) and normal atmosphere (control). Fruits from the controlled atmosphere and CO_2 shock treatments showed a lower incidence of physiological disorders (mesocarp discoloration, pulp spot and vascular browning) than fruits from the control. Total phenols tended to be lower in CO_2 shock fruits than in fruits from other treatments (Bower *et al.*, 1990). Wade (1979) showed that intermittent exposure of unripe avocados to 20% CO_2 reduced chilling injury when they were stored at 4°C.

Pesis and Sass (1994) showed that exposure of feijoa fruits (*Acca sellowiana*) of the cultivar Slor to total nitrogen or CO_2 for 24 h prior to storage induced the production of aroma volatiles including acetaldehyde, ethanol, ethyl acetate and ethyl butyrate. The enhancement of flavour was mainly due to the increase in volatiles and not to changes in total soluble solids or the total soluble solids to acid ratio.

Eaks (1956) showed that high CO_2 in the storage atmosphere could have detrimental effects on cucumbers, in that it appeared to increase their susceptibility when stored at low temperatures.

TOTAL NITROGEN OR HIGH NITROGEN STORAGE

Since fresh fruit and vegetables are living organisms, they require O_2 for aerobic respiration. Where this is not available to individual cells or available below a threshold level the fruit or vegetable or any part of the organism can go into anaerobic respiration, the end products of which are organic compounds which can affect the flavour (see section on effects of controlled atmosphere storage on flavour). It was reported by Anon (1920), that storage of plums in total nitrogen almost completely inhibited ripening. Plums were said to be able to tolerate, for a considerable period, an almost complete absence of O_2 without being killed or developing an alcoholic or unpleasant flavour. Parsons et al. (1964) successfully stored several fruit and vegetables at 33°F (1.1°C) in either total nitrogen or 1% O_2 combined with 99% nitrogen. During 10 days of storage of lettuce both treatments reduced the physiological disorder russet spotting and butt discoloration compared to those stored in air without affecting their flavour. At 60°F (15.5°C) ripening of both tomatoes and bananas was retarded when they were stored in total nitrogen, but their flavour was poorer only if they were held in these conditions for longer than 4 days or for over 10 days in 99% nitrogen combined with 1% O_2. In strawberries, 100 and 99% nitrogen was shown to reduce mould growth during 10 days of storage at 33°F (1.1°C), with little or no effect on flavour. Decay reduction was also observed on peaches stored in either 100 or 99% nitrogen at 60°F (1.1°C), but off-flavours were detected after 4 days in 100% nitrogen, but none in those stored in 99% nitrogen. Storing potato tubers in anaerobic conditions of total nitrogen prevented accumulation of sugars at low temperature (Harkett, 1971).

Treatment of fruits in atmospheres of total nitrogen prior to storage were shown to retard the ripening of tomatoes (Kelly and Saltveit, 1988) and avocados (Pesis et al., 1993). Avocado cultivar Fuerte fruits were treated with 97% nitrogen gas for 40 h and then stored at 17°C for 7 days by Dori et al. (1995). They showed that fruit softening was retarded and the onset of the increase in ethylene production with ripening was delayed in the nitrogen treated fruits. In earlier work Pesis et al. (1993) exposed Fuerte avocado fruits to 97% nitrogen for 24 h at 17°C then stored them at 2°C for 3 weeks followed by shelf life at 17°C. Pre-treatment with nitrogen reduced chilling injury symptoms significantly, fruit softening was also delayed and they had lower respiration rates and ethylene production during cold storage and shelf life.

RESIDUAL EFFECTS OF CONTROLLED ATMOSPHERE STORAGE

There is considerable evidence in the literature that storing fruits and vegetables in controlled atmosphere storage can affect their subsequent shelf or
marketable life. Hill (1913) described experiments on peaches stored in
increased levels of CO_2 and showed that their respiration rate was reduced,
not only during exposure, but he also showed that respiration rate only
returned to the normal level after a few days in air. Bell pepper fruits
exposed to 1.5% O_2 for 1 day exhibited suppressed CO_2 production and O_2
consumption for at least 24 h after transfer to air (Rahman *et al.*, 1993b).
Berard (1985) showed that cabbage stored at 1°C and 92% rh in 2.5% O_2
with 5% CO_2 had reduced losses during long-term storage compared to
those stored in air, but also the beneficial effects persisted after removal
from controlled atmospheres.

Goulart *et al.* (1990) showed that when the black raspberry cultivar
Bristol was stored at 5°C in 2.6, 5.4 or 8.3% O_2 with 10.5 or 19.6% CO_2 the
weight loss was greatest after 3 days for fruit stored in air. When fruits were
removed from controlled atmospheres after 3 days and held for 4 days in air
at 1°C, those which had been stored in 15% CO_2 had less deterioration than
any other treatment except for controls. Deterioration was greatest in the
fruit which had previously been stored in the three O_2 and the 10% CO_2
treatments. Fruits removed after 7 days and held for up to 12 days at 1°C
showed least deterioration with the 15% CO_2 treatment.

The climacteric rise in respiration of cherimoya fruit was delayed by
storage in 15 or 10% O_2 and fruits kept in 5% O_2 did not show a detectable
climacteric rise and did not produce ethylene. All fruits ripened normally
after being transferred to air storage at 20°C; however, the time needed to
reach an edible condition differed with O_2 level and was inversely proportional to O_2 concentration during storage. The actual data showed that fruit
took 11, 6 and 3 days to ripen following 30 days of storage at 5, 10 and 20%
O_2, respectively (Palma *et al.*, 1993).

Fruit firmness can be measured by inserting a metal probe into a fruit
and measuring its resistance to the insertion. This is called a pressure test
and the greater the resistance the firmer and the more immature the fruit.
The plum cultivars Santa Rosa and Songold were partially ripened to a firmness of approximately 4.5 kg pressure then kept at 0.5°C in 4% O_2 with 5%
CO_2 for 7 or 14 days. This treatment kept the fruits in an excellent condition
for an additional 4 weeks when they were removed to normal atmospheres
at 7.5°C (Truter and Combrink, 1992).

Mencarelli *et al.* (1983) showed that storage of courgettes in low O_2
protected them against chilling injury, but on removal to air storage at the
same chilling temperature the protection disappeared within 2 days.

Day (1996) indicated that minimally processed fruit and vegetables
stored in high levels of O_2 (70% plus) deteriorated more slowly on removal
than those freshly prepared.

Khanbari and Thompson (1997) stored potato cultivars Record, Saturna and Hermes in different conditions and found that there was almost complete sprout inhibition, low weight loss and maintenance of a healthy skin for all cultivars stored in 9.4% CO_2 with 3.6% O_2 at 5°C for 25 weeks. When tubers from this treatment were stored for a further 20 weeks in air at 5°C the skin remained healthy and they did not sprout while the tubers which had been previously stored in air or other controlled atmosphere combinations sprouted quickly. The fry colour of the crisps made from these potatoes was darker than the industry standard, but when they were reconditioned, tubers of Saturna produced crisps of an acceptable fry colour while crisps from the other two cultivars remained too dark. This residual effect of controlled atmosphere storage could have major implications in that it presents an opportunity to replace chemical treatments in controlling sprouting in stored potatoes.

The reverse was the case with bananas which had been initiated to ripen by exposure to exogenous ethylene and then immediately stored in 1% O_2 at 14°C. They remained firm and green for a 28-day storage period but then ripened almost immediately when transferred to air at 21°C (Liu, 1976b). However, Wills et al. (1982) showed that pre-climacteric bananas exposed to low O_2 took longer to ripen when subsequently exposed to air than fruits kept in air for the whole period.

Hardenburg et al. (1977) showed that apples stored in controlled atmosphere storage for 6 months and then displayed for 2 weeks at 21°C were firmer and more acid, and had a lower respiration rate than those that had previously been stored in air. Storage of Capsicum annum cultivar Jupiter fruits for 5 days at 20°C in 1.5% O_2 resulted in post-storage respiratory suppression of CO_2 production for about 55 h after transfer to air and a marked reduction in the oxidative capacity of isolated mitochondria. Mitochondrial activity was suppressed for 10 h after transfer to air but within 24 h had recovered to values comparable to those of mitochondria from fruits stored continuously in air (Rahman et al., 1995).

FRUIT RIPENING AND CONTROLLED ATMOSPHERE STORAGE

The level of CO_2 and O_2 in the surrounding environment of climacteric fruit can affect their ripening rate. Controlled atmosphere storage has been shown to suppress the production of ethylene by fruit. The biosynthesis of ethylene in ripening fruit was shown in early work by Gane (1934) to cease in the absence of O_2. Wang (1990) reviewed the literature on the effects of CO_2 and O_2 on the activity of enzymes associated with fruit ripening, and cited many examples of the activity of these enzymes being reduced in controlled atmosphere storage. This is presumably, at least partly, due to many of these enzymes requiring O_2 for their activity.

Quazi and Freebairn (1970) showed that high CO_2 and low O_2 delayed the high production of ethylene associated with the initiation of ripening in bananas, but the application of exogenous ethylene was shown to reverse this effect. Wade (1974) showed that bananas could be ripened in atmospheres of reduced O_2, even as low as 1%, but the peel failed to degreen, which resulted in ripe fruit which were still green. Similar effects were shown at O_2 levels as high as 15%. Since the degreening process in Cavendish bananas is entirely due to chlorophyll degradation (Seymour et al., 1987), the controlled atmosphere storage treatment was presumably due to suppression of this process. Hesselman and Freebairn (1969) showed that ripening of bananas, which had already been initiated to ripen by ethylene, was slowed in a low O_2 atmosphere.

Goodenough and Thomas (1980) and Goodenough and Thomas (1981) also showed suppression of degreening of fruits during ripening, in this case it was with tomatoes ripened in 5% CO_2 combined with 5% O_2. Their work, however, showed that this was due to a combination of suppression of chlorophyll degradation and the suppression of the synthesis of carotenoids, lycopene and xanthophyll. Jeffery et al. (1984) also showed that lycopene synthesis was suppressed in tomatoes stored in 6% CO_2 combined with 6% O_2.

Ethylene biosynthesis was studied in Jonathan apple fruits stored at 0°C under controlled atmosphere storage conditions of raised CO_2 concentrations (0–20%) and low (3%) and high (15%) O_2 concentrations. Fruits were removed from storage after 3, 5 and 7 months. Internal ethylene concentration, 1-aminocyclopropane 1-carboxylic acid levels and ethylene forming enzyme activity were determined in fruits immediately after removal from storage and after holding at 20°C for 1 week. Ethylene production by fruits was inhibited by increasing CO_2 concentration from 0 to 20% at both low and high O_2 concentrations. 1-aminocyclopropane 1-carboxylic acid levels were similarly reduced by increasing CO_2 concentrations even in low O_2; low O_2 enhanced 1-aminocyclopropane 1-carboxylic acid accumulation but only in the absence of CO_2. Ethylene forming enzyme activity was stimulated by CO_2 up to 10% but was inhibited by 20% CO_2 at both O_2 concentrations. The inhibition of ethylene production by CO_2 may therefore be attributed to its inhibitory effect on 1-aminocyclopropane 1-carboxylic acid synthase activity (Levin et al., 1992). In other work, storing pre-climacteric fruits of apple cultivars Barnack Beauty and Wagner at 20°C for 5 days in 0% O_2 with 1% CO_2 (99% nitrogen) or in air containing 15% CO_2 inhibited ethylene production and reduced 1-aminocyclopropane 1-carboxylic acid concentration and ethylene forming enzyme activity compared with storage in normal atmospheres (Lange et al., 1993).

Low O_2 and increased CO_2 concentrations were shown to have an effect on the decrease in ethylene sensitivity of Elstar apple fruits and Scania carnation flowers during controlled atmosphere storage (Woltering et al., 1994). Storage of kiwifruit in 2–5% O_2 with 0–4% CO_2 reduced ethylene production and 1-aminocyclopropane 1-carboxylic acid oxidase activity (Wang *et al.*, 1994). Ethylene production was lower in Mission figs stored at 15–20% CO_2 concentrations compared to those kept in air (Colelli *et al.*, 1991).

During the 1990s there has been an increasing demand for all the fruit being offered for sale in a supermarket to be of exactly the same stage of ripeness so that it has an acceptable and predictable shelf life. This has led to the development of a system called 'pressure ripening' which is used mainly for bananas but is applicable to any climacteric fruit. The system involves direction of the circulating air in the ripening room being channelled through boxes of fruit so that ethylene gas, which initiates ripening, is in contact equally with all the fruit in the room. At the same time the CO_2, which can impede ripening initiation is not allowed to concentrate around the fruit (Fig. 9).

CONTROLLED ATMOSPHERE TRANSPORT

A large and increasing amount of fresh fruit and vegetables is transported by seafreight refrigerated (reefer) containers (Thompson and Stenning, 1994). These reefers are increasingly of standard size and capacity (Table 2.2).

Controlling the levels of some of the gases in reefer containers has been used for many years to increase the marketable life of fresh produce, especially fruits and vegetables, during transport. Many commercial systems have been manufactured and an example is given in Fig. 10. Champion (1986) reviewed the state of the art of controlled atmosphere transport as it existed at that time. He listed the companies who cooperated in the production of the paper, which gives some indication of the large number of companies who were involved in the commercial application of this technology, as follows:

- CA (Container) Systems, UK
- Fresh Box Container, West Germany
- Finsam International, Norway
- Industrial Research, The Netherlands
- AgriTech Corporation, USA
- Synergen, UK
- Transfresh Corporation, USA
- Transfresh Pacific, New Zealand
- TEM, USA
- Franz Welz, Austria.

Many of those companies have since gone out of business and others have been started. Champion (1986) also defined the difference between controlled atmosphere containers and modified atmosphere containers. The

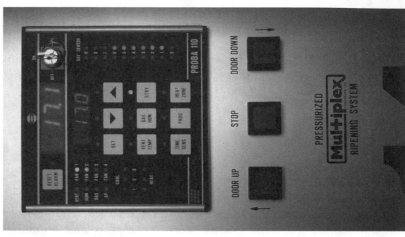

Fig. 9. Pressure ripening rooms at Mack Multiples in 1997. (a) Two-tier room stacked with bananas for ripening. (b) Control panel on the outside of each room.

Table 2.2. Sizes and capacities of reefer containers (source: Seaco Reefers).

Type	External dimensions (mm)	Internal dimensions (mm)		
20 foot (RM2)				
Length	6096	5501	Capacity	28.06 m²
Width	2438	2264	Tare	3068 kg
Height	2591	2253	Maximum payload	21,932 kg
			ISO payload	17,252 kg
40 foot (RM4)				
Length	12,192	11,638	Capacity	59.81 m²
Width	2438	2264	Tare	4510 kg
Height	2591	2253	Maximum payload	27,990 kg
			ISO payload	25,970 kg
40 foot (RM5) 'high cube'				
Length	12,192	11,638	Capacity	68.03 m²
Width	2438	2264	Tare	4620 kg
Height	2896	2557	Maximum payload	27,880 kg
			ISO payload	25,860 kg

Fig. 10. Diagram of controlled atmosphere reefer container used in 1993.

latter has the appropriate mixture injected into the sealed container at the beginning of the journey with no subsequent control, which means that in containers being used to transport fresh fruits and vegetables the atmosphere will constantly change during transport. Controlled atmosphere containers have some mechanism for measuring the changes in gases and adjusting them to a pre-set level.

Dohring (1997) stated that the world fleet has increased fourfold since 1993 and in 1997 consisted of 38,000 reefer containers with only some 1000 providing control of humidity, CO_2 and O_2. The degree of control over the gases in a container is affected by how gas tight the container is; some early systems had a leakage rate of 5 m^3 h^{-1} or more, but current systems can be below 1 m^3 (Garrett, 1992). Much of the air leakage is through the door and fitting plastic curtains inside the door could reduce the leakage, but they were difficult to fit and maintain in practice. A system introduced in 1993 had a single door instead of the double doors of reefers containers, which are easier to make gas tight. Other controlled atmosphere containers are fitted with a rail from which a plastic curtain is fitted to make the container more gas tight. The systems used to generate the atmosphere in the containers falls into three categories (Garrett, 1992):

●   The gases that are required to control the atmosphere are carried with the container in either a liquid or solid form.
●   Membrane technology is used to generate the gases by separation.
●   The gases are generated in the container and recycled with pressure absorption technology and swing absorption technology.

The first method involves injecting nitrogen into the container to reduce the level of O_2 with often some enhancement of CO_2 (Anon, 1997). It was claimed that such a system could carry cooled produce for 21 days compared with an earlier model, using nitrogen injection only, which could be used only on journeys not exceeding 1 week. The gases were carried in the compressed liquid form in steel cylinders at the front of the container, with access from the outside. Oxygen levels were maintained by injection of nitrogen if the leakage into the container was greater than the utilization of O_2 through respiration by the stored crop. If the respiration of the crop was high the O_2 could be replenished by ventilation.

In containers which use membrane technology, the CO_2 is generated by the respiration of the crop and nitrogen is injected to reduce the O_2 level. The nitrogen is produced by passing the air through fine porous tubes, made from polysulphones or polyamides, at a pressure of about 5–6 bar. These will divert most of the oxygen through the tube walls leaving mainly nitrogen which is injected into the store (Sharples, 1989). A controlled atmosphere reefer container, which has controls that can give a more precise control over the gaseous atmosphere, was introduced in 1993. The specifications are given in Table 2.3.

Table 2.3 Specifications for a refrigerated controlled atmosphere container (Freshtainer Intac 401)

	External dimensions (mm)	Internal dimensions (mm)	Door		Internal dimensions (mm)
Length	12,192	11,400		Capacity	66.6 m^2
Width	2438	2280	2262	Tare	5446 kg
Height	2895	2562	2519	Maximum payload	24,554 kg

Temperature range at 38°C ambient is −25 to +29°C (± 0.25°C)
O_2 down to 1% (+1% or −0.5%) up to 20 l h^{-1} removal
CO_2 0–80% (+0.5% or −1%) up to 180 l h^{-1} removal
Humidity 60–98% (± 5%)
Ethylene removal rate 120 l h^{-1} (11.25 mg h^{-1})
Water recycled to maintain high humidity

The containers use ventilation to control O_2 levels and a patented molecular sieve to control CO_2. The molecular sieve will also absorb ethylene and has two distinct circuits which are switched at predetermined intervals so that while one circuit is absorbing, the other is being regenerated. The regeneration of the molecular sieve beds can be achieved when they are warmed to 100°C to drive off the CO_2 and ethylene. This system of regeneration is referred to as temperature swing where the gases are absorbed at low temperature and released at high temperature. Regeneration can also be achieved by reducing the pressure around the molecular sieve which is called pressure swing. During the regeneration cycle the trapped gases are usually ventilated to the outside, but they can be directed back into the container if this is required. The levels of gas, temperature and humidity within the container are all controlled by a computer which is an integral part of the container. It monitors the levels of oxygen from a paramagnetic analyser and the CO_2 from an infrared gas analyser and adjusts the levels to those which have been pre-set in the computer (Fig. 11).

Dohring (1997) claimed that 'avocados, stone fruit, pears, mangoes, asparagus and tangerine made up over 70% of container volumes in recent years. Lower value commodities, e.g. lettuce, broccoli, bananas and apples, make up a greater percentage of the overall global produce trade volumes but cannot absorb the added CA costs in most markets.' Harman (1988) suggested the use of controlled atmosphere containers for transport and storage for the New Zealand fruit industry. Lallu et al. (1997) described experiments on the transport of kiwifruit in containers where the atmosphere was controlled by either nitrogen flushing and Purafil to absorb ethylene or lime and Purafil. The former maintained CO_2 levels of approximately 1% and O_2 at 2–2.5% while in the latter the CO_2 levels were 3–4% and O_2 levels increased steadily to 10%. A control container was included in the shipments and on arrival the ethylene levels in the three containers were

Fig. 11. A computer controlled atmosphere storage system in use at East Kent Packers' controlled atmosphere stores in 1993.

less than 0.02 µl l⁻¹. It was concluded that controlled atmosphere containers can result in benefits to fruit quality.

Champion (1986) mentions the Tectrol gas sealing specifications. Simulated commercial export of mangoes using the Transfresh system of controlled atmosphere container technology was described in Ferrar (1988). Avocados and bananas were stored in two Freshtainer controlled atmosphere containers (Fig. 12), 40 feet long and controlled by microprocessors. The set conditions for avocados were: $7.4 \pm 0.5°C$ for 8 days followed by $5.5 \pm 0.5°C$ for 7 days, with $2 \pm 0.5\%$ O_2 and a maximum of $10 \pm 0.5\%$ CO_2. Those for bananas were: $12.7 \pm 0.5°C$ for 8 days followed by $13.5 \pm 0.5°C$ for 11 days, with $2.0 \pm 0.5\%$ O_2 and a maximum of $7 \pm 0.5\%$ CO_2. Temperatures and container atmospheres were continually monitored and fruit quality was assessed after the predetermined storage period. The results confirmed that the containers were capable of very accurate control within the specified conditions. Also that fruit quality was better than that for controls which were avocados held in a cold store for 2 weeks at $5.5°C$ in normal atmosphere and bananas held at $13.5 \pm 0.5°C$ in an insulated container (Eksteen and Truter, 1989). Peacock (1988) described a controlled atmosphere transport experiment. Mango fruits were harvested from three commercial sites in Queensland when the content of total soluble solids was judged to be 13–15%. Fruit were de-stalked in two of the three sources and washed, dipped for 5 min in 500 ppm benomyl at 52°C, cooled and sprayed with prochloraz (200 ppm), dried and sorted, and finally size graded and packed in waxed fibreboard cartons. Some cartons contained

Fig. 12.   A commercial controlled atmosphere container in use in 1993.

polypropylene inserts, which cupped the fruits, other fruits were packed on an absorbent pad, but the majority were packed on expanded polystyrene netting. After packing, fruits were transported by road to a precooling room overnight (11°C). Pulp temperatures were 18–19°C the following morning; after 36 h of transport to Brisbane and overnight holding in a conventional cool room the fruits were placed in a controlled atmosphere shipping container. At loading, the pulp temperature was 12°C. Fruits were stored in the container at 13°C with 5% O_2 and 1% CO_2 for 18 days. Ripening was significantly delayed in controlled atmosphere storage (although there were problems with CO_2 control, related to the efficiency of the scrubbing system). There was virtually no anthracnose infection, confirming the effectiveness of prochloraz in its control, but very high levels of stem end rot were observed. On removal from the controlled atmosphere container, fruits immediately began to ooze sap and the loss continued for at least 24 h, the sap tended to cause the absorbent pads and polystyrene netting to stick to the fruits.

A UK company, Cronos Containers, supply inserts, called the Cronos Controlled Atmosphere System (manufactured under licence from BOC) which can convert a standard reefer container to a controlled atmosphere container. The installation may be permanent or temporary, and is self-contained taking some 3 h the first time. If the equipment has already been installed in a container and subsequently dismantled then reinstallation can be achieved in about 1 h. The complete units measure 2 m × 2 m × 0.2 m which means that 50 of them can be fitted into a 40 foot dry container for transport. This facilitates management of the system. It also means that they take up little of the cargo space when fitted into the container, only 0.8 m³. The unit operates alongside the container's refrigeration system and is capable of controlling, maintaining and recording the levels of oxygen, CO_2 and relative humidity to the levels and tolerances pre-set into a programmable controller. Ethylene can also be removed from the container by scrubbing. This level of control is greater than any comparable controlled atmosphere storage system, increasing shipping range and enhancing the quality of fresh fruit, vegetables, flowers, fish, meat, poultry and similar products. The system is easily attached to the container floor and bulkhead, and takes power from the existing reefer equipment with minimal alteration to the reefer container. The design and manufacture is robust to allow operation in the harsh (marine) environments that will be encountered in typical use. The system fits most modern reefers and is easy to install, set up, use and maintain. A menu-driven programmable controller provides the interface to the operator who simply has to pre-set the required percentages of each gas to levels appropriate for the product in transit. The controls are located on the front external wall of the container, and once set up a display will indicate the measured levels of O_2, CO_2 and relative humidity.

The system consists of a rectangular aluminium mainframe onto which the various sub-components are mounted (Fig. 13). The compressor is located at the top left of the mainframe and is driven by an integral electric motor supplied from the control box. Air is extracted from the container and pressurized (up to 4 bar) before passing through the remainder of the system. A pressure relief valve is incorporated in the compressor along with inlet filters. Note also that a bleed supply of external air is ducted to the compressor and is taken via the manifold with a filter mounted external to the container.

Fig. 13. Cronos controlled atmosphere reefer container system. (Courtesy of Tim Bach.)

The compressed air is cooled prior to passing through the remainder of the system. A series of coils wound around the outside of the air cooler radiate heat back into the container. The compressed air then passes into this component which removes the pressure pulses produced by the compressor and provides a stable air supply. The water trap passing into the controlled atmosphere storage system then removes moisture. Water from this component is ducted into the water reservoir and used to increase the humidity when required. Two activated alumina drier beds are used in this equipment, each located beneath one of the nitrogen and CO_2 beds. Control valves are used to route the air through parts of the system as required by the conditioning process. Mesh filters are fitted to the outlet which vents nitrogen and CO_2 back into the container, and to the outlet which vents oxygen to the exterior of the container. Nitrogen and CO_2 beds are located above the drier sieve beds and contain zeolite for the absorption of nitrogen and CO_2.

Ethylene Bed

If required ethylene can be removed from the container air. A single ethylene absorption bed is used which contains a mixture of activated alumina and Hisea material (a clay mineral-based system patented by BOC). Air from the container is routed to the bed that remains pressurized for several hours and then depressurized via the O_2 venting lines. The O_2 flow is then routed through the bed for 20 min in order to scrub it. After this the process is repeated.

Humidity Injection System

To increase the relative humidity within the container, an atomized spray of water is injected as required into the main airflow through the reefer. The water supply is taken from a reservoir located at the base of the mainframe which is fed from the reefer defrost system and the water trap. Air from the instrument air buffer is used to form the spray by drawing up the water as required. A further stage of moisture removal is carried out using drier beds that are charged with activated alumina. From here the air is routed either directly to the nitrogen and CO_2 beds, or via the ethylene bed depending on the type of conditioning needed. The reefer process tends to decrease the humidity within the container, with water being discharged from the defrost equipment. When the controlled atmosphere storage system is used, this water is drained into a 5-l reservoir located within the mainframe and used to increase the humidity if required. Air from the instrument air buffer is used to draw the water into an atomizing injection system located in the main reefer airflow. The control valve is operated for a short period, and once the additional water spray has been mixed in with the air in the container the humidity level is measured and the valve operated again if required.

Carbon Dioxide Injection System

Four 10-l bottles of CO_2 are located in the mainframe. To increase the level of CO_2 within the container the gas is vented via a regulator. CO_2 is retained in the molecular sieve along with the nitrogen. Normally this would be returned to the container, but if it is required to remove CO_2 then the flow of gas (which is mainly CO_2 at this point) is diverted for the last few seconds to be vented outside the container. This supply of CO_2 is intended to last for the duration of the longest trip. A check to ensure the bottles are full is included as one of the pre-trip checks.

Oxygen removal is accomplished using a molecular sieve which is pressurised at up to 3 bar. Nitrogen and CO_2 are retained in the bed. The separated oxygen is taken from the sieve bed to the manifold at the bottom of the mainframe and then vented outside the container. It is possible to reduce the level of oxygen to around 4% although this does depend on a maximum air leakage into the container and in certain cases O_2 levels of 2% or even 1.5% can be maintained (Tim Bach, personal communication, 1997).

Oxygen Addition

To add O_2, air from outside the container is allowed to enter by opening a control valve for a short period. Once the additional air has been mixed in with the air in the container the oxygen level is measured and the valve operated again if required.

Gas Return Flow

The nitrogen (and CO_2) retained in the main sieve beds is returned to the alumina drier beds in order to recharge them and remove moisture from them. Finally, the gas is returned to the main container via a valve located at the bottom of the mainframe. If the CO_2 is to be removed, a modified process occurs.

The levels of oxygen, CO_2 and relative humidity are continually monitored and the values passed back to the display. A supply of gas is taken from the general atmosphere within the container via a small pump located within the control box. The gas is passed onto an oxygen transducer and then a CO_2 transducer before being returned to the container. The location of these components is shown in Fig. 14. A transducer, located in the airflow through the mainframe and connected into the control box, measures relative humidity.

The above sequence of operation is carried out according to instructions provided by the display. In particular, the measured values of oxygen, CO_2 and relative humidity are compared in the display with those pre-set by the operator and the gas conditioning cycle adjusted accordingly. Once set up the process is automatic and no further intervention is required by the

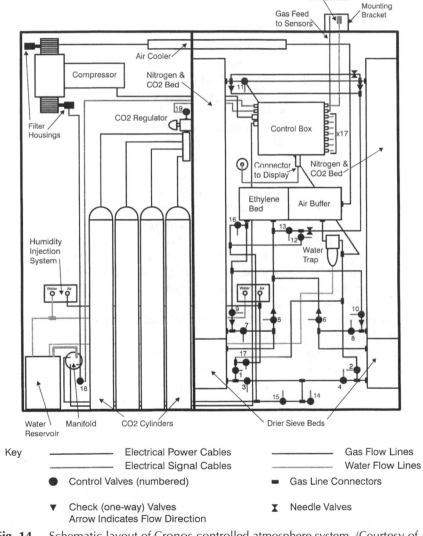

Fig. 14. Schematic layout of Cronos controlled atmosphere system. (Courtesy of Tim Bach.)

operator unless a fault occurs. Further detail on setting up the system is provided in the next section of the manual.

In addition to providing the operator interface the display has a communications socket (shown in Fig. 14) which allows data to be transferred (e.g. to/from an external computer). The most frequent use of this facility will be the downloading of the data logged during normal operation of the system – the levels of O_2, CO_2 and relative humidity as periodically recorded. Data is output as an ASCII file or one of the common spreadsheet packages (e.g. Excel). It is also possible to use the external computer to modify system settings and carry out other diagnostic operations.

Various other methods have been used for modifying the atmosphere during fruit and vegetable transport including hypobaric containers (see below). Alvarez (1980) described experiments where papaya fruits were subjected to sub-atmospheric pressure (20 mmHg, 10°C and 90–98% rh) for 18–21 days during shipment in hypobaric containers from Hilo in Hawaii, to Los Angeles and New York. Both ripening and disease development was inhibited. Fruits ripened normally after removal from the hypobaric containers, but abnormal softening unrelated to disease occurred in 4–45% of fruits of one packer. Hypobaric stored fruits had 63% less peduncle infection, 55% less stem end rot and 45% fewer fruit surface lesions than those stored in a refrigerated container at normal atmospheric pressure. Postharvest fungicide wax applications further decreased disease incidence.

Modified atmosphere packaging has been developed for particular transport situations. For long distance transport of bananas a system was developed and patented by the United Fruit Company called Banavac (Badran, 1969). The system uses polyethylene film bags, 0.04 mm thick, in which the fruit are packed (typically 18.14 kg) and a vacuum is applied and the bags are sealed. Typical gas contents that develop during transport at 13–14°C in the bags through fruit respiration are about 5% CO_2 and 2% O_2. A UK company developed a system in 1997 which involves enclosing a large trolley of food in a six-sided insulating blanket (a material with the trade name of Tempro) which is closed with heavy duty Velcro. This was shown to maintain the product temperature within 1°C, at a starting temperature of 0°C, during a 22-h distribution period. The inner walls of the insulating blanket have pockets in which dry ice can be placed to further help to maintain product temperature.

HYPOBARIC STORAGE

Exposing fruits to pressures higher than normal or lower than normal, have been used in the postharvest technology of fresh fruits and vegetables. Exposing food products, including fresh fruits and vegetables to high-pressure storage (up to 9000 atm) was recently reviewed (Anon, 1997). This is a technique mainly used for the control of microorganisms and involves exposure for only brief periods, but was reported as still in the developmental stage. Robitaille and Badenhop (1981) designed vessels suitable for storing mushrooms (*Agaricus bisporus*) at pressures up to 35 atm and found that neither pressurization nor gradual depressurization over 6 h injured mushrooms. Pressure did not affect respiration rate but significantly reduced weight loss during storage.

Hypobaric storage, on the other hand, is a method where fruits and vegetables are kept under a partial vacuum. As the pressure is reduced within the store the partial pressure of O_2 and thus its availability to the crop in the store is reduced in proportion to the pressure reduction. A major engineering problem with hypobaric storage is that the humidity must be kept high otherwise the fruits and vegetables will lose excessive moisture. The reason for this is that the lower the atmospheric pressure the lower the boiling point of water, which means that the water in the fruits and vegetables will be increasingly likely to be vaporized. The air being introduced into the store must therefore be as close as possible to saturation (100% rh). If it is less than this, serious dehydration of the crop can occur. The water vapour in the store atmosphere has to be taken into account when calculating the partial pressure of O_2 in the store. To do this the relative humidity must be measured and, from a psychometric chart, the vapour pressure deficit can be calculated. This is then included in the following equation:

$$\frac{P_1 - VPD \times 21}{P_0} = \text{partial pressure of } O_2 \text{ in the store}$$

where P_0 = outside pressure at normal temperature, P_1 = pressure inside the store, and VPD = vapour pressure deficit inside the store.

This reduced pressure in the store is achieved by a vacuum pump evacuating the air from the store. The vacuum pump constantly changes the store atmosphere because fresh air is constantly introduced from the outside atmosphere thus removing gases given out by the crop. The air inlet and the air evacuation from the store are balanced to achieve the required low pressure within the store. The store needs to be designed to withstand low pressures without imploding. To overcome this, stores have to be strongly constructed of thick steel plate normally with a curved interior. The control of the O_2 level in the store can be very accurately and easily achieved and simply measured by measuring the pressure inside the store with a vacuum gauge. Hypobaric storage also has the advantage of constantly removing ethylene gas from the store which prevents it building up to levels which could be detrimental to the crop. The effects of hypobaric storage on fruits and vegetables has been reviewed by Salunkhe and Wu (1975) and by Burg (1975) where they show considerable extension in the storage life of a wide range of crops when it is combined with refrigeration compared to refrigeration alone. In other work (Hughes *et al.*, 1981) these extensions in storage life under hypobaric conditions have not been confirmed. Capsicums stored at 8.8°C in either 152, 76 or 38 mmHg did not have an increased subsequent storage life compared to those stored in air under the same conditions (Table 2.4). The hypobaric stored capsicums also had a significantly higher weight loss during storage than the air stored ones.

Hypobaric storage of bananas was shown to increase their storage life. When bananas were stored at 14°C their storage life was 30 days at 760 mmHg, but when the pressure was reduced to 80 or 150 mmHg the fruit remained unripe for 120 days (Apelbaum *et al.*, 1977). When these fruits were subsequently ripened they were said to be of very good texture, aroma and taste. Awad *et al.* (1975) showed that banana fruits kept in hypobaric atmosphere for 3 h daily showed no climacteric and remained green during the 15 days of the experiment. Storage of pineapple under hypobaric condition can extend the storage life up to 30–40 days (Staby, 1976). Hypobaric storage at 0.1 atm (76 mmHg) extended the storage life of tomatoes to 7 weeks, compared with 3–4 weeks in cold storage. Similar results were obtained with cucumbers and sweet peppers. With parsley hypobaric storage extended the storage life from 5 to 8 weeks without appreciable losses in protein, ascorbic acid or chlorophyll contents (Bangerth, 1974).

Hypobaric storage of grapefruit at 380 mmHg (the lower limit of the experimental equipment) and 4.5°C had no effect on the incidence of chilling injury (Grierson, 1971). Wu *et al.* (1972) showed that subatmospheric pressure storage inhibited the ripening of tomatoes and thus

Table 2.4. Effects of controlled atmosphere storage, plastic film wraps and hypobaric storage on the mean percentage sound capsicum fruit after 20 days at 8.8°C followed by 7 days at 20°C (Hughes *et al.*, 1981).

Storage treatment	% Sound fruit[1]
Air control	43[ab]
Controlled atmosphere storage	
0.03% CO_2 with 2% O_2	42[ab]
3.00% CO_2 with 2% O_2	39[b]
6.00% CO_2 with 2% O_2	21[c]
9.00% CO_2 with 2% O_2	29[bc]
Storage in plastic film wrap	
Sarenwrap	67[a]
Vitafilm PWSS	57[a]
Polyethylene (Pe301)	69[a]
VF 71	63[a]
Hypobaric storage	
152 mm Hg	46[ab]
76 mm Hg	36[b]
38 mm Hg	42 [ab]

[1]Figures followed by the same letter were not significantly different ($P = 0.05$).

extended storage life. Inhibition was proportional to the reduction in pressure. Physiological changes associated with ripening were delayed. Tomatoes could be stored at 102 mmHg for 100 days and then ripened at 646 mmHg in 7 days.

A more recent system of hypobaric storage controls water loss from produce without humidifying the inlet air with heated water (Burg, 1993). This is achieved by slowing the evacuation rate of air from the storage chamber to a level where water evaporated from the produce by respiration exceeds the amount of water required to saturate the incoming air. Burg used this technique with roses stored at 2°C and a reduced pressure of 3.33 $\times 10^3$ N m^2 stored for 21 days with or without humidification at a flow rate of 80–160 cm^3 min^{-1}. The vase life of these roses was not significantly different to that of freshly harvested roses.

SAFETY

The use of controlled atmosphere storage has health and safety implications. One factor that should be taken into account is that the gases in the atmosphere could possibly have a stimulating effect on microorganisms. Also the levels of these gases could have detrimental effects on the workers operating the stores. The Health and Safety Executive (1991) showed that work in confined spaces could be potentially dangerous and entry must be strictly controlled preferably through some permit system. Also it is recommended that anyone entering such an area should have proper training and instruction in the precautions and emergency breathing apparatus. Stringent procedures need to be in place with a person on watch outside and the formulation of a rescue plan. Anon (1974) also indicated that when a store is sealed anyone entering it must wear breathing apparatus. Warning notices should be placed at all entrances to controlled atmosphere stores and an alarm switch located near the door inside the chamber in the event that someone may be shut in. There should be a release mechanism so that the door can be opened from the inside. When the produce is to be unloaded from a store the main doors should be opened and the circulating fan run at full speed for at least 1 h before unloading is commenced. In the UK the areas around the store chamber must be kept free of impedimenta in compliance with the appropriate Agricultural Safety Regulations.

Ethylene

Ethylene should more correctly be called ethene but modern commercial use favours the term ethylene. It is a colourless gas with a sweetish odour and taste, has asphyxiate and anaesthetic properties and is flammable. Its flammable limits in air are 3.1–32% volume for volume and its autoignition temperature is 543°C. Care must be taken when the gas is used for fruit ripening to ensure that levels in the atmosphere do not reach 3.1%. As added precautions all electrical fittings must be of a 'spark-free' type and warning notices relating to smoking and fire hazards must be displayed around the rooms.

Oxygen

Oxygen levels in the atmosphere of 12–16% can affect human muscular coordination, increase respiration rate and affect one's ability to think clearly. At lower levels vomiting and further impediment to coordination and thinking can occur. At levels below 6% human beings rapidly lose consciousness, and breathing and heart beats stop (Bishop, 1996). Oxygen levels of over 21% can create an explosive hazard in the store and great care needs to be taken when high O_2 modified atmosphere packaging is used.

Carbon Dioxide

Being in a room with higher than ambient CO_2 levels can be hazardous to health. The limits for CO_2 levels in rooms for human occupation was quoted by Bishop (1996) from the Health and Safety Executive, publication EH40-95 *Occupational Exposure Limits*, as 0.5% CO_2 for continuous exposure and 1.5% CO_2 for 10 min exposure.

CONCLUSION

The application of controlled atmosphere technology to the fresh fruit and vegetable industry requires inputs of other technologies including special treatments of the crop before it is put into store, temperature and humidity control. Also the equipment used in the control of the atmosphere inside stores is constantly being developed to provide more accurate control. The control methods include monitoring and changing the atmosphere constantly, but pre-mixing the gases before introduction into the store or even reduced pressure store are other methods which have been used.

Besides the common application of the technology in reducing the rate of metabolism of the crop, special techniques are used where short-term exposure to very high levels of CO_2 or atmospheres devoid of O_2 can have beneficial effects where they would be toxic with more prolonged exposure.

The literature is ambivalent of the quality and rate of deterioration of fruit and vegetables after they have been removed from controlled atmosphere storage. Most of the available information indicates that their storage life is adequate for marketing, but, in certain cases, the marketable life is even better than freshly harvested produce. The physiological mechanism that could explain this effect has so far not been demonstrated.

Chapter 3

The Effect of Controlled Atmosphere Storage on Flavour, Quality and Physiology

FLAVOUR

Controlled atmosphere storage can affect the eating quality of fruits and vegetables. Both positive and negative effects on their eating quality have been cited in the literature. Controlled atmosphere stored fruits were shown to retain a good flavour longer than fruits stored in air (Reichel, 1974) and in most commodities controlled atmosphere storage generally maintains better flavour than storage in air (Wang, 1990). However, the stage of ripeness of fruit when storage begins has the major effect on its flavour, sweetness, acidity and texture. It is therefore often difficult to specify exactly what effect the controlled atmosphere storage has on fruit since the effect on flavour may well be confounded with the effect on ripening. An example of this was the storage of tomatoes in low concentrations of O_2 that had less effect on fruit, which were subsequently ripened, than the stage of maturity at which they were harvested (Kader *et al.*, 1978). In another experiment the muskmelon cultivar Earl's Favourite was stored at 10°C in one of the following conditions:

● under conventional conditions (control)
● at low pressure of 80–100 mmHg
● in sealed polyethylene film packs
● in a controlled atmosphere with 2.46% O_2, 0.98% CO_2 and 96.56% nitrogen.

The best flavour was found in fruits stored in a controlled atmosphere (Zhao and Murata, 1988).

Stoll (1976) showed that Louise Bonne pears were still of good quality after 3–4 months in cold storage but were of similar good quality after 5–6 months in controlled atmosphere storage. The best flavoured fruit were from controlled atmosphere storage after a maximum of 4–4.5 months. Blednykh *et al.* (1989) showed that a high CO_2 percentage in the atmosphere (CO_2 at 8–16% and O_2 at 5–8%) allowed 3–3.5 months of storage of Russian cultivars of cherry with maintenance of high market quality and good flavour.

In a 6-year experiment using the apple cultivars Golden Delicious, Auralia, Spartan and Starkrimson, the effects of four harvesting dates and four durations of cold (3°C, 85–90% relative humidity, rh) or controlled atmosphere storage (13% O_2 with 8% CO_2 or 3% O_2 with 3% CO_2) on flavour were studied. Flavour was best 1–2 months after harvest under all storage conditions but particularly with the later harvesting dates. For prolonged storage early harvesting and controlled atmosphere storage proved best in terms of flavour maintenance. In a second experiment fruit of 11 cultivars (including Golden Delicious, Jonathan, Gold Spur, Idared and Boskoop) were stored and after 180 days in cold storage fruit of all cultivars except Boskoop had poor flavour. After 180 days in controlled atmosphere storage only Gloster 69 and Jonathan fruit had good flavour (Kluge and Meier, 1979). Low O_2 (0.5%) storage of Abbe Fetel pears at −0.5°C resulted in the lowest incidences of scald and brown core, but also resulted in losses in aroma and flavour. It was suggested that this may be overcome by raising the O_2 level to 1% (Bertolini *et al.*, 1991). Golden Delicious, Idared and Gloster cultivars of apple, were stored for 100 or 200 days at 2°C, 95% rh in 15–16% O_2 with 5–6% CO_2, then moved to 5, 10 or 15°C (all at approximately 60% rh) for 16 days to determine shelf life. Flavour improvement occurred only in fruit that were removed from controlled atmosphere storage after 100 days. Apples removed from controlled atmosphere storage after 200 days showed a decline in flavour during subsequent storage at 5, 10 or 15°C (Urban, 1995).

Spalding and Reeder (1974a) showed that limes (*Citrus latifolia*) stored for 6 weeks at 10°C at 0.3, 0.2 or 0.1 atm, in a controlled atmosphere or 5% O_2 with 7% CO_2 or 21% O_2 with 7% CO_2 had acceptable juice content, and limes from all treatments had acceptable flavour. Fruit held in controlled atmosphere storage lost more acid and sugar than those held in conventional cold storage. It was reported that the flavour of fruit stored at low O_2 with high CO_2 was inferior to that from conventional air storage (from an English summary from Ito *et al.*, 1974).

Truter *et al.* (1994) showed that at −0.5°C, storage in 1.5% CO_2 with 1.5% O_2 or 5% CO_2 with 2% O_2 apricots had an inferior flavour compared to those stored in air after 6 weeks (Table 3.1).

Table 3.1. Effects of controlled atmosphere storage and time on the flavour of apricots which were assessed by a taste panel 3 months after being canned, where a score of 5 or more was acceptable.

Storage period (weeks)	CO_2 concentration	O_2 concentration	Flavour score	
			Cultivar Bulida	Cultivar Peeka
None	–	–	7.5	4.6
4	1.5	1.5	6.3	6.3
5	1.5	1.5	5.0	5.0
6	1.5	1.5	1.3	1.3
4	5.0	2.0	6.3	5.0
5	5.0	2.0	5.5	5.4
6	5.0	2.0	2.9	5.4
4	0	21	6.7	5.8
5	0	21	4.2	5.4
6	0	21	2.1	4.6

OFF-FLAVOUR

Some of the effects on flavour of fruit and vegetables, which have been reported, are the result of anaerobic respiration. For example off-flavour developed in intact lettuce heads exposed to 20% CO_2, which was associated with increased concentrations of ethanol and acetaldehyde (Mateos *et al.*, 1993). Ke *et al.* (1991a) described experiments where fruits of Granny Smith and Yellow Newtown apples, 20th Century pear and Angeleno plums were kept in air and in 0.25 or 0.02% O_2 with the consequent development of an alcoholic off-flavour. Mattheis *et al.* (1991) stored Delicious apple fruit in 0.05% O_2 plus 0.2% CO_2 at 1°C for 30 days and found that they developed high concentrations of ethanol and acetaldehyde these included ethyl propanoate, ethyl butyrate, ethyl 2-methylbutyrate, ethyl hexanoate, ethyl heptanoate and ethyl octanoate. The increase in the emission of these compounds was accompanied by a decrease in the amounts of other esters requiring the same carboxylic acid group for synthesis. Karaoulanis (1968) showed that oranges stored in atmospheres containing 10–15% CO_2 had increased alcohol content while those stored in 5% CO_2 did not. He also showed that grape stored in 12% CO_2, 21% O_2 and 67% nitrogen for 30 days had only 17 mg 100 g^{-1} fresh weight while those stored in 25% CO_2, 21% O_2 and 54% nitrogen had 170 mg 100 g^{-1} at the same time. It was found by Magness and Diehl (1924) that the flavour of apples was impaired when CO_2 exceeded 20%. In contrast plums were shown to tolerate an almost complete absence of O_2 for a considerable period without being killed or developing an alcoholic or unpleasant flavour (Anon, 1920).

The alcohol content of strawberry fruit increased with the length of storage and with higher concentrations of CO_2; 20% CO_2 caused severe injury after 30 days (Woodward and Topping, 1972). Ke *et al.* (1991b) stored fruits of the peach cultivar Fairtime in air or in 0.25 or 0.02% O_2 at 0 or 5°C for up to 40 days. Low O_2 atmospheres did not significantly influence changes in skin colour, flesh firmness and soluble solids content, but retarded titratable acidity decline and pH rise. Flavour was affected by ethanol and acetaldehyde accumulated in peaches kept in 0.02% O_2 at 0 or 5°C or in 0.25% O_2 at 5°C. The fruits kept in air or 0.25% O_2 at 0°C for up to 40 days and those stored in 0.02% O_2 at 0°C or in air, 0.25 or 0.02% O_2 at 5°C for up to 14 days had good to excellent taste, but the flavour of the fruits stored at 5°C for 29 days was unacceptable. Delate and Brecht (1989) showed that exposure of sweetpotatoes to 2% O_2 plus 60% CO_2 resulted in less sweetpotato flavour and more off-flavour. Strawberries could be stored in 1.0, 0.5 or 0.25% O_2 or air plus 20% CO_2 at 0 or 5°C for 10 days without detrimental effects on quality. Fruit could also be stored in 0% O_2 or 50% or 80% CO_2 for up to 6 days without visual injury. The trained taste panel found slight off-flavour in G3 strawberries kept in 0.25% or 0% O_2 which correlated with ethanol, ethyl acetate and acetaldehyde in juice. Transfer of fruit to air at 0°C for several days after treatment would reduce ethanol and acetaldehyde levels leading to an improvement in final sensory quality (Ke et al., 1991b). Pajaro strawberries were held in air, or air enriched with 10, 20 or 30% CO_2 for 5 days at 5°C, followed by an additional 4 days in air at the same temperature. Ethanol and acetaldehyde accumulation was very slight, although sensory evaluation of the fruits showed that off-flavours were present at transfer from controlled atmosphere, but not after the following storage in air (Colelli and Martelli, 1995). With storage at 5°C, CO_2 concentrations in sealed polyethylene and perforated polyethylene packs during storage for 28 days were 1–4%; no off-flavours were found (Zhao and Murata, 1988). Corrales-Garcia (1997) found that avocados stored at 2 or 5°C for 30 days in air, 5% CO_2 with 5% O_2 or 15% CO_2 with 2% O_2 had a higher ethanol and acetaldehyde content for fruits stored in air than the fruits in controlled atmosphere storage.

VOLATILE COMPOUNDS

Controlled atmosphere storage has also been shown to affect volatile compounds which are produced by fruit and give them their characteristic flavour and aroma. Hatfield (1975) showed that Cox's Orange Pippin apples after storage in 2% O_2 at 3°C for 3.5 months and subsequent ripening at 20°C produced smaller amounts of volatile esters than they did when ripened directly after harvest or after storage in air. This was correlated with a marked loss of flavour. The inhibition of volatile production could be relieved considerably if the apples were first kept in air at 5–15°C after storage, before being transferred to 20°C. With this lower post-storage temperature, fruit also softened less, lost less acid and remained greener. Terpenoid compounds such as linalool and its epoxide and farnesene have been shown to develop in some apple cultivars (Dimick and Hoskins, 1983). Controlled atmosphere storage of apples in either 2% oxygen in 98% nitrogen or 2% oxygen, 5% CO_2 and 93% nitrogen showed that few organic volatile compounds were produced during the storage period (Hatfield and Patterson, 1974). Even when the fruit were removed from storage they did not synthesize normal amounts of esters during ripening. Willaert *et al.* (1983) isolated 24 aroma compounds from Golden Delicious apples and, using headspace analysis, showed that the relative amounts of 18 of these components declined considerably during controlled atmosphere storage. Hansen *et al.* (1992) showed that volatile ester production by Jonagold apple fruits was reduced after prolonged, controlled atmosphere storage in low O_2. After removal from storage, large differences were seen in the emanation of esters of different biogenetic origin. The alcohol moiety in esters appeared to influence the shape of the production curves. The propyl esters developed in a different way from other straight-chained esters, indicating that the biochemical origin of propanol may differ from other straight-chained alcohols. However, all straight-chained esters were produced in positive correlation to the O_2 concentration during storage. Esters with a methyl-branched alcohol moiety or carboxylic acid moiety were produced in the largest amount at intermediate O_2 concentrations (3–5%). A series of esters with the alcohol 2-methylbut-2-enol was produced in negative correlation to O_2 concentration. Yahia (1989) showed that McIntosh and Cortland apple fruits stored in 100% O_2 at 3.3°C for 4 weeks, did not enhance the production of aroma volatiles compared to storage in air at the same temperature.

Table 3.2. Effect of controlled atmosphere storage on the acidity and firmness of Spartlett pears stored at 0°C and their taste after 7 days of subsequent ripening at 20°C (Meheriuk, 1989b).

CO_2	O_2	Storage time (days)	Acidity (mg 100 ml^{-1})	Taste (score:1 = like, 9 = dislike)	Firmness (newtons)
0	21	60	450	6.1	62
5	3	60	513	7.6	61
2	2	60	478	7.6	60
0	21	120	382	4.3	55
5	3	120	488	6.7	58
2	2	120	489	6.9	58
0	21	150	342	2.9	51
5	3	150	451	7.7	58
2	2	150	455	7.1	58
0	21	180	318	–	40
5	3	180	475	–	62
2	2	180	468	–	61

Aroma volatiles in Golden Delicious apples were suppressed during storage at 1°C for up to 10 months in atmospheres containing high levels of CO_2 (3%) or low levels of O_2 (1 or 3%) compared to storage in air (Brackmann, 1989). Jonagold apples were stored at 0°C in air or in a controlled atmosphere (1.5% O_2 with 1.5% CO_2) for 6 months. Controlled atmosphere storage significantly decreased production of volatile compounds (esters, alcohols and hydrocarbons) by half (Girard and Lau, 1995). The measurement of the organic volatile production of the apples was over a 10 day period at 20°C after removal from the cold store. In McIntosh and Cortland apples most organic volatile compounds were produced at lower rates during ripening after controlled atmosphere storage than those produced from fruits ripened immediately after harvest (Yahia, 1989).

Grapes of the cultivar Agiorgitiko were stored at 23–27°C for 10 days either in 100% CO_2 or in air, and 114 compounds were identified in the CO_2 stored fruits, compared with only 60 in the fruits stored in air (Dourtoglou et al., 1994).

ACIDITY

The acid levels in fruits and vegetables can affect their flavour and acceptability. Storage of pears in controlled atmosphere conditions did not have a deleterious effect on fruit flavour compared to those stored in air and were generally considered better by a taste panel (Meheriuk, 1989b). This was in spite of controlled atmosphere storage generally resulting in significantly higher acid levels in fruit (Table 3.2).

Controlled atmosphere storage significantly reduced the loss of acidity in stored apples (Girard and Lau, 1995). In a detailed study, Batu (1995) also confirmed changes in acidity in relation to controlled atmosphere storage and modified atmosphere packaging. During a 60-day storage period at temperatures of 13, 15 and 20°C tomatoes harvested at the mature green or pink stages of maturity all became less acid. Those stored in sealed plastic film bags generally had a lower rate of loss in acidity than those stored without film wraps (Fig. 15).

Generally the fruit stored in the less permeable films had a lower rate of acid loss. The acidity levels of fruits sealed in polypropylene film (where the O_2 levels in the polypropylene film fluctuated between 0.5 and 8% at equilibrium and 11–13% for CO_2) were similar to control fruit and both were lower than the fruits sealed in 30 or 50 μ polyethylene film.

Titratable acidity of tomatoes in controlled atmosphere storage generally increased during the first 20 days at both 13 and 15°C (Fig. 16). After 20 days acidity levels tended to decrease until 70 days of storage. Although there was no correlation between the O_2 or CO_2 concentrations and acidity levels of fruits during controlled atmosphere storage, the acidity values of tomatoes stored in 6.4% CO_2 with 5.5% O_2 were the highest among the

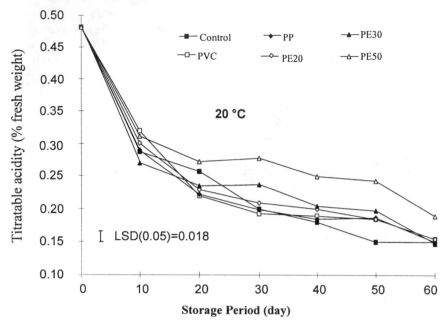

Fig. 15. Changes of titratable acidity values of tomatoes harvested at pink stage of maturity and sealed in various thickness of various packaging films and stored at 20°C (Batu, 1995).

treatments and the lowest was in 9.1% CO_2 with 5.5% O_2. There was a significant ($P = 0.05$) difference between the control fruits and fruits exposed to all levels of CO_2 during storage while no significant difference occurred among the tomatoes exposed to different levels of CO_2. The acidity values of tomatoes kept in controlled atmosphere storage at 13°C were very similar to fruits stored at 15°C. Titratable acidity of tomatoes in controlled atmosphere storage increased during the first 20 days at both 13 and 15°C (see Figs 15 and 16). After 20 days, acidity levels tended to decrease until 70 days of storage.

In other work, Kays (1991) described an experiment where oranges of the cultivar Valencia lost less acid during storage at 3.5°C in 5% CO_2 with 3% O_2 than those in 0°C in air. Kollas (1964) found that total acid concentration in apples of the cultivar McIntosh was much greater for fruits in controlled atmosphere storage (3.3°C combined with 5% CO_2 with 3% O_2) than those stored in air at 0°C. Kollas speculated on why there were higher acid levels in fruit after storage in controlled atmospheres than in air, and concluded that it was likely to be due to lower oxidation but that significant rates of CO_2 fixation may occur.

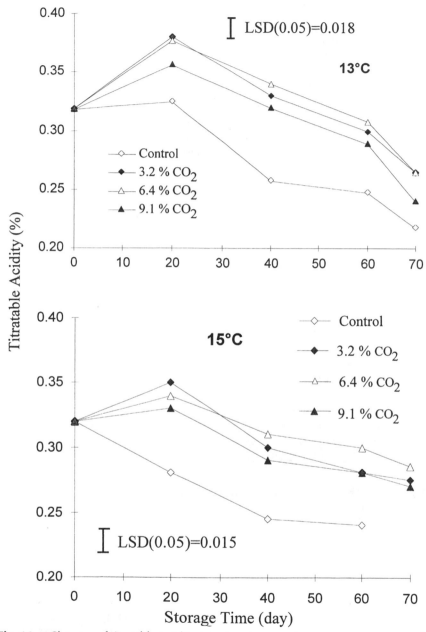

Fig. 16. Changes of titratable acidity tomatoes harvested at mature green stage of maturity and kept in controlled atmosphere storage (all with 5.5% O_2) for 60 days storage time at 13 and 15°C plus 10 days in air at 20°C (Batu, 1995).

NUTRITION

Controlled atmosphere storage has been shown to hasten the loss of ascorbic acid compared to air storage. The ascorbic acid content of tomato cultivars Punjab Chuhara and Punjab Kesri decreased as the CO_2 concentration in the storage atmosphere increased and increased as the storage period was lengthened (Singh *et al.*, 1993). Strawberry (cultivar Elvira) and blackberry (cultivar Thornfree) fruits were stored at 0–1°C in high CO_2 atmospheres of up to 20% for strawberries and 30% for blackberries combined with either 1–3% or > 14% O_2. Loss of ascorbic acid was highest in the higher CO_2 treatment, and degradation after 20 days storage was more rapid with the low O_2 treatments. This degradation was even more pronounced during simulated shelf life following storage (Agar *et al.*, 1994c). Wang (1983) showed that ascorbic acid levels of Chinese cabbage were not affected in storage in 10 or 20% CO_2, but in 30 or 40% CO_2 the rate of loss of ascorbic acid was much higher than for cabbage stored in air. Ogata *et al.* (1975) stored okra at 1°C in air or 3% O_2 combined with 3, 10 or 20% CO_2 or at 12°C in air or 3% O_2 combined with 3% CO_2. At 1°C there was no effects of any of the controlled atmosphere storage treatments on ascorbic acid retention, but at 12°C this treatment resulted in lower ascorbic acid retention. In low O_2 high CO_2 atmospheres storage of Satsuma mandarin at 1–4°C the ascorbic acid contents of the flesh and peel gradually declined but the dehydroascorbic acid content of the flesh rose. Such changes were smaller at high O_2 levels. Fruit held in controlled atmosphere storage lost more acid and sugar than that held in conventional cold storage (from an English summary from Ito *et al.*, 1974).

Bangerth (1974) showed that hypobaric storage at 2–3°C improved retention of ascorbic acid, chlorophyll and protein in stored parsley. Haruenkit and Thompson (1996) showed that storage of pineapples in O_2 levels below 5.4% helped to retain ascorbic acid levels but generally has little effect on total soluble solids (°Brix) (Table 3.3).

Table 3.3. Effects of controlled atmosphere storage on ascorbic acid, °Brix and acidity (w/v) of Smooth Cayenne pineapples stored at 8°C for 3 weeks and then 5 days at 20°C (Haruenkit and Thompson, 1996).

Gas composition (%)			Ascorbic acid mg 100 ml^{-1}	°Brix	Titratable acidity
O_2	CO_2	N_2			
1.3	0	98.7	7.14	10.4	0.70
2.2	0	97.8	8.44	10.2	0.66
5.4	0	94.6	0.76	11.3	0.76
1.4	11.2	87.4	9.16	8.6	0.53
2.3	11.2	86.5	7.94	10.0	0.58
20.8	0	79.2	0.63	10.3	0.73
LSD ($P = 0.05$)			5.28	1.5	0.08

Table 3.4. Effects of controlled atmosphere storage and storage in air on the retention of vitamins in leeks at 0°C and 95% rh (Kurki, 1979).

	Initial level before storage	After 4 months storage	
		Air	10% CO_2 with 1% O_2
Ascorbic acid (mg 100 g^{-1})	37.2	24.1	20.2
Vitamin A (IU 100 g^{-1})	2525	62	1350

Kurki (1979) showed some greater loss in ascorbic acid in controlled atmosphere storage compared to air storage, but controlled atmosphere storage gave much better retention of vitamin A than air storage (Table 3.4).

Vidigal *et al.* (1979) in controlled atmosphere storage studies with tomatoes found that the ascorbic acid levels increased during storage at 10°C.

DISCOLORATION

Henderson and Buescher (1977) showed that *Phaseolus vulgaris* beans broken during harvesting and handling for processing, developed a brown discoloration on the exposed surfaces. O_2 levels of 5% or less controlled this browning but caused off-flavour in the canned products. Elevated CO_2 levels were not injurious to bean quality as long as O_2 was maintained at 10% or higher. High CO_2 concentrations have also been reported to inhibit browning of snap beans (*Phaseolus vulgaris*) at sites of mechanical injury by Costa *et al.* (1994). In studies of litchis, fruit were packed in punnets overwrapped with plastic film with 10% CO_2 or vacuum packed and stored for 28 days at 1°C followed by 3 days shelf life at 20°C. It was found that skin browning was reduced in both of these plastic film packages compared to unwrapped fruit but their taste and flavour was unacceptable, while they remained acceptable for unwrapped fruits (Ahrens and Milne, 1993). Lipton and Mackey (1987) showed that Brussels sprouts of the cultivars Lunette, Rampart and Valiant were stored for 2, 3 or 4 weeks at 2.5, 5 or 7.5°C at different controlled atmosphere combinations. Storage in 0.5% O_2 occasionally induced a reddish-tan discoloration of the heart leaves and frequently an extremely bitter flavour in the non-green portion of the sprouts.

RESPIRATION RATE

Kidd and West (1927a) showed that the respiration rate of apples in controlled atmosphere storage (12% CO_2 combined with 9% O_2) at 46.5°F (8°C) was between 54 and 55% of fruits stored in air at the same temperature. The effect of reduced O_2 and increased CO_2 on respiration rate can be affected by temperature. Izumi *et al.* (1996b) showed that controlled atmosphere storage (10% CO_2 with 0.5% O_2) of freshly cut carrots reduced their respiration rate by about 55% at 0°C, about 65% at 5°C and 75% at 10°C. Kubo *et al.* (1989b) showed that the respiratory responses to high CO_2 varied according to the crop and its stage of development. The relationship between O_2 and CO_2 levels and respiration is not a simple one (Table 3.5). It is also cultivar dependent as in the comparison between Golden Delicious and Cox's Orange Pippin apples where the respiration rate is suppressed more in the former than the latter in controlled atmosphere storage (Fidler *et al.*, 1973). Respiration rate was also decreased by low O_2 (3%) compared to levels in air in cauliflowers (Romo Parada *et al.*, 1989).

Pal and Buescher (1993) described experiments in which bananas, guavas, onions, Navel oranges, Russet potatoes, pink tomatoes, carrots and pickling cucumbers were held in air or in controlled atmospheres containing 10, 20 or 30% CO_2. Short-term exposure to CO_2 reduced respiration in ripening bananas, pink tomatoes and pickling cucumbers, increased respiration in potatoes (20 and 30% CO_2) and carrots (30% CO_2 only), and had no effect on respiration in guavas, oranges and onions. Changes in respiration was found seldom to coincide with changes in ethylene evolution. Evolution of ethylene from guavas and tomatoes was substantially reduced by all levels of CO_2. However, 30% CO_2 accelerated ethylene evolution in bananas, carrots, cucumbers, onions and potatoes, possibly due to an early injury response.

Robinson *et al.* (1975) showed for a range of fresh fruits and vegetables that the respiration rate was generally, but not always, lower during storage at 3% O_2 than in air at various temperatures (Table 3.6). Weichmann (1973) showed that the respiration rate of five cultivars of carrots in store at 1.5°C in atmospheres containing 0.03, 2.5, 5.0 or 7.5% CO_2, but with no regulation of the O_2 content, increased with increasing CO_2 content in the atmosphere. Weichmann (1981) found that horseradish stored in 7.5% CO_2 had a higher respiration rate than those stored in air.

Table 3.5. Rates of respiration of some cultivars of apples stored in different controlled atmosphere conditions (Source: Fidler *et al.*, 1973).

Cultivar	Storage conditions			Respiration rate* (CO_2)	Respiration rate* (O_2)
	°C	CO_2%	O_2%		
Bramley's Seedling	3.5	8–10	11–13	40–45	40
Cox's Orange Pippin	3.5	5	16	62	57
Cox's Orange Pippin	3.5	5	3	42	40
Cox's Orange Pippin	3.5	< 1	2.5	80	55
Golden Delicious	3.5	5	3	20	20
Delicious	0	5	3	18	–
Jonathan	3.5	7	13	33	38
McIntosh	3.5	5	3	35	–

*Litres per 1000 kg day^{-1}.

Green bananas were held in humidified gas streams comprising air (control); high CO_2 (5% CO_2, 20% O_2, 75% nitrogen); low O_2 (0% CO_2, 3%, O_2, 97% nitrogen); and high CO_2 low O_2 (5% CO_2, 3% O_2, 92% nitrogen). Ripening in all three controlled atmosphere storage combinations was delayed at least two, eight and 12 times, respectively, compared with the control. These three gas streams also reduced the rates of O_2 uptake by the fruit but increased the total O_2 uptake over the period before the beginning of the respiratory climacteric. It is suggested that low O_2 limited the operation of the Krebs cycle between pyruvate and citrate and 2-oxoglutarate and succinate. No similar effect of high CO_2 was apparent (McGlasson and Wills, 1972). Wade (1974) showed that the respiratory climacteric and the accumulation of total soluble solids in the pulp were induced in banana fruits by Ethephon (a source of ethylene) at O_2 concentrations of 3–21%. However, the induction of peel colour change by Ethephon was retarded or prevented at O_2 concentrations of 10% or less. Thus pulp tissue ripened whilst the peel remained green. Respiration rate, total soluble solids content of pulp, and peel colour were not affected by Ethephon at O_2 concentrations of 1% or less. Awad et al. (1975) showed that green banana fruits immersed for 2 min in Ethephon at 500 ppm had their climacteric advanced by 5 days whereas fruits treated with gibberellic acid at 100 ppm had their climacteric delayed by 2 days, compared with the control.

The output of CO_2 from the strawberry cultivar Cambridge Favourite held at 4.5°C in air or 1, 2 or 5% O_2 fell to a minimum after 5 days. Thereafter the rate increased, more rapidly in air (in which rotting was more prevalent) than in 1 or 2% O_2. Fruits stored at 3°C in air or 5, 10, 15 or 20% CO_2 remained in good condition for 10 days, and all concentrations of CO_2 reduced rotting due to infections with Botrytis (Woodward and Topping, 1972).

When apples and melons were stored in a high CO_2 atmosphere (60% CO_2 with 20% O_2 with 20% nitrogen), respiration rate fell to about half the initial level. Ripening tomatoes and bananas also showed a reduction in respiration in response to high CO_2, but showed little response when tested before the climacteric. Exposure to high CO_2 produced little or no effect in Citrus natsudaidai, lemons, potatoes, sweetpotatoes or cabbage, but reduced the respiration rate in broccoli. High CO_2 stimulated respiration in lettuces, aubergines and cucumbers (Kubo et al., 1989a). Storage of strawberries in 10–30% CO_2 or 0.5–2% O_2 were shown to slow their respiration rate (Hardenburg et al., 1990). Knee (1973) showed that CO_2 could inhibit an enzyme (succinate dehydrogenase) in the tricarboxylic acid cycle which is part of the crop's respiratory pathway.

Andrich et al. (1994) showed some evidence that the respiration rate of Golden Delicious apples was affected differentially by CO_2 concentration in storage at 21°C and 85% rh, depending on the O_2 concentration in the store. The effect was that in anaerobic conditions, or near anaerobic conditions,

Table 3.6. Effects of temperature and reduced O_2 level on the respiration rate and storage life of selected fruit and vegetables (Robinson *et al.*, 1975).

| | CO$_2$ production[a] (mg kg^{-1} h^{-1})[b] | | | | | | | | |
| | In air | | | | | In 3% O$_2$ | | | Water |
Temperature (°C)	0	5	10	15	20	0	10	20	loss[c]
Asparagus	28	44	63	105	127	25	45	75	3.6[d]
Beans, broad	35	52	87	120	145	40	55	80	(2.1)
Beans, runner		28	36	54	90	15	25	46	(1.8)
Beetroot, storing	4	7	11	17	19	6	7	10	1.6
Beetroot, bunching with leaves	11	14	22	25	40	7	14	32	(1.6)
Blackberries, Bedford Giant	22	33	62	75	155	15	50	125	0.5
Blackcurrants, Baldwin	16	27	39	90	130	12	30	74	–
Brussels sprouts	17	30	50	75	90	14	35	70	(2.8)
Cabbage, Primo	11	26	30	37	40	8	15	30	1.0
Cabbage, January King	6	13	26	33	57	6	18	28	–
Cabbage, Decema	3	7	8	13	20	2	6	12	0.1
Carrots, storing	13	17	19	24	33	7	11	25	1.9
Carrots, bunching with leaves	35	51	74	106	121	28	54	85	(2.8)
Calabrese	42	58	105	200	240	–	70	120	(2.4)
Cauliflower, April Glory	20	34	45	67	126	14	45	60	(1.9)
Celery, white	7	9	12	23	33	5	9	22	(1.8)
Cucumber	6	8	13	14	15	5	8	10	(0.4)
Gooseberries, Leveller	10	13	23	40	58	7	16	26	–
Leeks, Musselburgh	20	28	50	75	110	10	30	7	(0.9)
Lettuce, Unrivalled	18	22	26	50	85	15	20	55	(7.5)
Lettuce, Kordaat	9	11	17	26	37	7	12	25	–
Lettuce, Kloek	16	24	31	50	80	15	25	45	–
Onion, Bedfordshire Champion	3	5	7	7	8	2	4	4	0.02
Parsnip, Hollow Crown	7	11	26	33	49	6	12	30	(2.4)
Potato, maincrop (King Edward)	6[e]	3	4	5	6	5[g]	3	4	(0.05)
Potato, 'new' (immature)[f]	10	15	20	30	40	10	18	30	(0.5)
Pea, (in pod) early (Kelvedon Wonder)	40	61	130	180	255	29	84	160	(1.3) (cv. Onward)
Peas, main crop (Dark Green Perfection)	47	55	120	170	250	45	60	160	–

Table 3.6 *Continued*

Temperature (°C)	CO_2 production[a] (mg kg^{-1} h^{-1})[b]								Water loss[c]
	In air					In 3% O_2			
	0	5	10	15	20	0	10	20	
Peppers, green	8	11	20	22	35	9	14	17	0.6
Raspberry (Malling Jewel)	24	55	92	135	200	22	56	130	2.5
Rhubarb (forced)	14	21	35	44	54	11	20	42	2.3
Spinach (Prickly True)	50	70	80	120	150	51	87	137	(11.0) (glasshouse grown)
Sprouting broccoli	77	120	170	275	425	65	115	215	(7.5)
Strawberries (Cambridge Favourite)	15	28	52	83	127	12	45	86	(0.7)
Sweetcorn	31	55	90	142	210	27	60	120	(1.4)
Tomato (Eurocross BB)	6	9	15	23	30	4	6	12	0.1
Turnip, bunching with leaves	15	17	30	43	52	10	19	39	1.1 (without leaves)
Watercress	18	36	80	136	207	19	72	168	(35.0)

[a]These figures, which give the average rates of respiration of the samples, are a guide only. Other samples could differ, but the rates could be expected to be of the same order of magnitude, ± *c*.20%.

[b]Heat production in British thermal units (Btu) ton^{-1} h^{-1} is given by multiplying CO_2 output in mg kg^{-1} h^{-1} by 10 (1 Btu approximates to 0.25 kcal and to 1.05 kJ).

[c]Values in parentheses were determined at 15°C and 6–9 millibar water vapour pressure deficit (wvpd). Remainder at 10°C and 3–5 millibar wvpd.

[d]Water loss as a percentage of the initial weight per day per millibar wvpd.

[e]After storage for a few weeks at 0°C, sufficient time having elapsed for low-temperature sweetening. The figures for potato are mid-season (December) values and could be 50% greater in October and March.

[f]Typical rounded-off values, for tubers with an average weight of 60 g immediately after harvest. The rates are too labile for individual values to be meaningful. After a few weeks storage the approximate to the values for maincrop potatoes.

[g]After storage for a few weeks at 0°C, sufficient time having elapsed for low-temperature sweetening. The figures for potato are mid-season (December values) and could be 50% greater in October and March.

respiration rates were more affected in increasingly high CO_2 levels than for the same fruits stored in aerobic conditions.

CONCLUSION

Controlled atmosphere storage has been demonstrated to reduce the respiration rate of fruit and vegetables in certain circumstances. For certain crops in certain conditions high CO_2 or low O_2 can have either no effect or an increasing effect on respiration rates. The reasons for this variability would be many. Interactions with temperature would mean that the metabolism of the crop could be changed so that it would be anaerobic and thus higher. The same would apply where the O_2 content was too low. High levels of CO_2 can actually injure the crop, which again could affect its rate of respiration.

These effects on respiration rate could also affect the eating quality of fruit and vegetables. Generally crops stored in controlled atmospheres have a longer storage life because the rate of the metabolic processes is slower. Particularly with climacteric fruit this would slow ripening and deterioration so that when they have been stored for protracted periods they may well be less ripe than fruits stored in air. If the fruits stored in air are over ripe or suffering with physiological disorders associated with long storage (e.g. senescent breakdown) then this would affect the flavour. In fact it is a secondary effect of controlled atmosphere storage slowing metabolism and therefore affecting the eating quality of the fruit.

This effect of controlled atmosphere storage on fruit and vegetables metabolism can have another indirect affect on flavour. Some chemicals, which are produced during anaerobic respiration such as alcohols and aldehydes, would affect their eating quality.

Storage of crops *per se* can affect eating quality and there is evidence in the literature that fruit ripened after storage do not taste as good as fruit ripened directly after harvest. This is not always the case and would depend on the length of storage and the type of crop. Generally flavour volatile compounds are less after storage than in freshly harvested fruit, but there is no evidence that controlled atmosphere storage affects flavour volatiles in any way that is different to air storage.

The acidity of fruit and vegetables also has an effect on flavour. The data from storage trials show widely different effects on acidity including levels of ascorbic acid. Generally the acid levels of fruit and vegetables should be greater after controlled atmosphere storage compared to air storage because CO_2 in solution is acid and the lower acid metabolism is associated with controlled atmosphere storage.

Chapter 4

The Effect of Controlled Atmosphere Storage on Pests and Diseases

DISEASE

Reports on the effects of controlled atmosphere storage on the development of diseases of fruit and vegetables have shown mixed results. It was reported that the growth of the fungi *Rhizopus* spp., *Penicillium* spp., *Phomopsis* spp. and *Sclerotinia* spp. was inhibited by atmospheres of total nitrogen but not in atmospheres containing 99% nitrogen in combination with 1% O_2 (Ryall, 1963). In other work there is some evidence that controlled atmosphere storage can reduce disease development on crops. Disease incidence on cabbage stored in 3% O_2 with 5% CO_2 or 2.5% O_2 with 3% CO_2 was lower than those stored in air (Prange and Lidster, 1991). In celery stored at 8°C disease suppression was greatest in atmospheres of 7.5–30% CO_2 with 1.5% O_2, but there was only a slight reduction in 4–16% CO_2 with 1.5% O_2 or in 1.5–6% O_2 alone (Reyes, 1988). A combination of 1 or 2% O_2 and 2 or 4% CO_2 prevented black stem disease development in celery during storage (Smith and Reyes, 1988). Golden Delicious apple losses from shrinkage and rotting were lower in controlled atmosphere storage than in the same cold storage conditions in air (Reichel, 1974). In a review of the fungistatic effects of high CO_2 levels, Kader (1997) indicated that levels of 15–20% can retard decay incidence on cherry, blackberry, blueberry, raspberry, strawberry, fig and grape.

Storage of strawberries in 10–30% CO_2 or 0.5–2% O_2 are used to reduce disease levels, but 30% CO_2 or less than 2% O_2 was also reported to cause off-flavour to develop in some circumstances (Hardenburg *et al.*, 1990). When the strawberries were held in atmospheres with 0, 10, 20 or 30% CO_2 (21% O_2) at a temperature and a storage period simulating shipments to Chicago and New York (41°F or 5°C for 3–5 days) and subsequent handling

(60°F or 15.6°C for 1–2 days), fruit held in CO_2 enriched atmospheres had less softening and decay (mostly *Botrytis cinerea*) than that held in air. Concentrations of 20 or 30% CO_2 were most effective, but fruit held in 30% CO_2 developed persistent off-flavour. Differences in decay in fruit held in air and at high CO_2 were greater after subsequent holding in air at 60°F (15.6°C) than on removal from storage. The percentage of soft over-ripe fruit was about twice as great in lots stored in air as in the CO_2 enriched atmospheres (Harris and Harvey, 1973). Cox's Orange Pippin apples were stored in a normal cold store at 2.5°C or either in air or in 2.5% O_2 and 3% CO_2 at 3.5°C and 92% relative humidity (rh). Natural contamination of the fruits on the tree led to attack by *Pezicula* spp. and the disease was shown to be slightly retarded in the controlled atmosphere storage. Bitter pit was reduced in controlled atmosphere storage, but there were no differences in the occurrence of *Botrytis* and *Penicillium* rots between the two types of storage. After 4 months there was no internal breakdown in controlled atmosphere stored fruit but a low incidence of the disorder in the cold store. When undamaged and injured fruits were artificially inoculated at picking time prior to controlled atmosphere storage the total attack and number of rot spots per fruit were higher with *Pezicula malicorticis* than with *Glomerella cingulata*. There were no differences in the size of the individual rot spots on intact fruits, but in injured fruits *P. malicorticis* produced the larger rots. *P. malicorticis* was shown to be more active on injured fruits in controlled atmosphere storage than in normal cold storage (Schulz, 1974). Exposure of the kiwifruit cultivar Hayward to 60% CO_2 with 20% O_2 at 30 or 40°C for 1, 3 or 5 days was tested by Cheah *et al.* (1994) for the control of *Botrytis cinerea* storage rot. Spore germination and growth of *B. cinerea* were completely inhibited in culture by 60% CO_2 at 40°C and partially suppressed at 30°C. Kiwifruits were inoculated with *B. cinerea* spores, exposed to 60% CO_2 at 30 or 40°C, and stored in air at 0°C for up to 12 weeks after treatment. 60% CO_2 at 40°C reduced disease incidence from 85% in air at 20°C to about 50%, but exposure to CO_2 at 40°C for longer than 1 day adversely affected fruit ripening.

Parsons *et al.* (1974) showed considerable reduction in disease levels on tomatoes in controlled atmosphere storage with most of the effect coming from the low O_2 levels with little additional effect from increased CO_2 levels (Table 4.1).

Table 4.1. Effects of controlled atmosphere storage conditions on the decay levels of tomatoes harvested at the green mature stage (Parsons *et al.*, 1974).

Storage atmosphere	After removal from 6 weeks at 13°C (%)	Plus 1 week at 15–21°C (%)	Plus 2 weeks at 15–21°C (%)
Air (control)	65.6	93.3	98.6
0% CO_2 with 3% O_2	2.2	4.4	16.7
3% CO_2 with 3% O_2	3.3	5.6	12.2
5% CO_2 with 3% O_2	5.0	9.4	13.9

Table 4.2. The effects of CO_2 concentration during storage for 3 days at 5°C on the percentage levels of decay in strawberries and the development of decay when removed to ambient conditions (15°C) (Harris and Harvey, 1973).

Storage	0% CO_2	10% CO_2	20% CO_2	30% CO_2
3 days storage at 5°C	11	5	2	1
+ 1 day at 15°C in air	35	9	5	4
+ 2 days at 15°C in air	64	26	11	8

Other work (Harris and Harvey, 1973) has also shown that controlled atmosphere storage of strawberries not only reduced disease levels while the crop is being stored but can have an additional beneficial effect in disease reduction when it is removed for marketing (Table 4.2).

There is evidence of interactions between controlled atmosphere storage and temperature on disease development. After 24 weeks the average incidence of fungal rots was lowest on currants (*Ribes* spp.) at about 5% when kept at –0.5°C and 20 or 25% CO_2. Rotting increased to about 35% at 1°C and 25% CO_2 and to about 50% at 1°C and 20% CO_2. With 0% CO_2 at either temperature the incidence of rots was about 95% (Roelofs, 1994).

There remains the question of how controlled atmosphere storage actually reduces or controls diseases on fruits and vegetables. Parsons *et al.* (1970) showed that atmospheres containing 3% O_2 reduced decay on stored tomatoes caused by *Rhizopus* or *Alternaria*. However, both genera of fungi grew well in 3% O_2 or less *in vitro*. This led them to hypothesize that the reduction in decay was due to the controlled atmosphere storage conditions acting on the tomato fruit itself so that it developed resistance to the fungi, rather than acting only on the fungi.

The effects of controlled atmosphere storage on bacterial diseases have not been well documented. Potatoes are cured before long-term storage by exposing them to a higher temperature at the beginning of storage to heal any wounds (Thompson, 1996). Weber (1988) showed that the defence reaction of potatoes to infection by *Erwinia carotovora* subspecies *atroseptica* during the curing period was inhibited by temperatures of less than 10°C, reduced O_2 levels of less than 5%, and CO_2 levels of over 20%. CO_2 can retard bacterial growth by both increasing their lag phase before they begin to develop and the generation time. The degree of retardation increases with increasing concentrations of the gas but at these high levels *Clostridium botulinum* may survive (Daniels *et al.*, 1985).

PHYSIOLOGICAL DISORDERS

As would be expected controlled atmosphere storage has been shown to have positive, negative and no effect on physiological disorders of stored fresh fruits and vegetables. This is because of the wide and disparate range of disorders as well as the range of gas mixtures and temperatures applied in storage.

After 4 months storage of Cox's Orange Pippin apples at 2.5°C either in air or in 2.5% O_2 and 3% CO_2 there was no internal breakdown in controlled atmosphere stored fruit but a low incidence in air (Schulz, 1974). Tonini *et al.* (1993) showed that storage for 40 days at 0°C in 2% O_2 with either 5 or 10% CO_2 reduced internal breakdown of nectarine and plums compared to those stored in air. When the nectarine cultivar Flamekist fruits were stored by Lurie *et al.* (1992) in controlled atmosphere for 6 or 8 weeks, the 10% O_2 plus 10% CO_2 atmosphere prevented internal breakdown and reddening which occurred to fruits stored in air. Cooper *et al.* (1992) showed that controlled atmosphere storage with up to 20% CO_2 reduced the incidence of physiological disorders of nectarine such as woolliness and internal browning without any other adverse effect on fruit quality. Good results were obtained in some cultivars by decreased O_2 concentration to 4%. In a subsequent paper Streif *et al.* (1994) found that exposure of nectarine to 25% CO_2 prior to controlled atmosphere storage had little effect, but storage at high CO_2 levels, especially in combination with low O_2, significantly delayed ripening, retained fruit firmness and prevented both storage disorders. There was no deleterious effect on flavour with storage at high CO_2 levels.

Jankovic and Drobnjak (1994) described experiments where the apple cultivars Idared, Cacanska Pozna, Jonagold and Melrose were stored at 1°C and 85–90% rh, either in less than 7% CO_2 and 7% O_2 or air. Fruits stored in controlled atmosphere exhibited no physiological disorder whereas bitter pit was observed on cultivar Melrose under normal atmospheric conditions.

Johnson and Ertan (1983) found that the quality of apples stored at 4°C in 1% O_2 was markedly better than in 2%; the fruits were also free of core flush (brown core) and other physiological disorders. Wang (1990) reviewed the effects of CO_2 on brown core of apples and concluded that it is due to exposure to high levels of CO_2 at low storage temperatures.

Many physiological disorders in stored fruit and vegetables are associated with CO_2 toxicity (see section on CO_2 effects). Arpaia *et al.* (1985) stored the kiwifruit cultivar Hayward for up to 24 weeks in 2% O_2 and 0, 3, 5 or 7% CO_2 at 0°C. The occurrence and severity of the physiological disorder of white core inclusions under controlled atmosphere plus ethylene was highest in 5% CO_2. Two other physiological disorders were observed (translucency and graininess) and their severity was increased by the combination of high CO_2 and ethylene. Bohling and Hansen (1977) described the development of necrotic spots on the outer leaves of stored cabbage, which was largely prevented by low O_2 atmospheres in the store, but increased CO_2 had no effects. Onions stored for 162 or 224 days by Adamicki *et al.* (1977) in controlled atmosphere conditions containing 10% CO_2 and 3–5% O_2 at 1°C exhibited a physiological disorder where the epidermis and parenchyma of fleshy scales showed destruction of the cell walls. Alterations of the ultrastructure of the mitochondria were also observed. Scott and Wills (1974) showed that all Williams' Bon Chretien (Bartlett) pears which had been stored at $-1°C$ for 18 weeks in the presence of about 5% CO_2 were externally in excellent condition but were affected by brown heart. Stewart and Uota (1971) showed that lettuce during cold storage had increasing levels of a brown stain on the leaves with increasing levels of CO_2 in the storage atmosphere.

Superficial scald in apples and pears is where the skin turns brown during storage (Fig. 17). It can be controlled by a pre-storage treatment with an antioxidant such as ethoxyquin (1,2-dihydro-2,2,4-trimethylquinoline-6-yl ether) marketed as Stop-Scald or DPA (diphenylamine) marketed as No Scald or Coraza. They should be applied directly to the fruit within a week of harvesting. Postharvest treatment with ethoxyquin gave virtually complete control of scald and stem end browning on fruit stored in 8–10% CO_2 at 3.9°C and at 5.0°C, although scald tended to appear earlier on untreated fruit at the higher temperature (Knee and Bubb, 1975). Fidler et al. (1973) showed that scald development in storage could be related to the CO_2 and O_2 levels but that the relationship varied between cultivars (Table 4.3).

Very low levels of O_2 have been used commercially to control scald in apples in North America (British Columbia: 0.7% O_2; Washington: 1.0% O_2) and in pears (Oregon: 1.0% O_2) (Lau and Yastremski, 1993). Coquinot and Richard (1991) stored apples from three orchards under commercial conditions (27 tonnes of apples) in ultra low O_2 atmospheres (1.2% O_2 with 1% CO_2) with or without removal of ethylene, or in conventional controlled atmospheres (3% O_2 with 1% CO_2) with removal of ethylene; controls were stored in 10% O_2 with 5% CO_2. The results showed that scald could be controlled by the use of ultra low O_2 and ethylene removal was not necessary.

Very low O_2 storage in relation to ethanol production and control of superficial scald in Bramley's Seedling apples continues to be investigated. For example, an investigation was carried out by Johnson et al. (1993) on the effect of storing Bramley's apples in low (0.2–1.0%) O_2 with CO_2 removed completely by hydrated lime. Storing in 0.2% O_2 was too low to support aerobic respiration as was evidenced by an immediate build up of ethanol in the fruits. In 0.4 and 0.6% O_2 fruits respired normally for 150 days but ethanol accumulated thereafter. Retardation of scald development by 0.4 and 0.6% O_2 was as effective as 5% CO_2 with 1% O_2 and ethylene removal from 9% CO_2 with 12% O_2 storage provided scald-free fruits for 216 days. However, rapid loss of firmness occurred in fruits stored in all low O_2 (combined with nominally 0% CO_2) conditions after 100 days of storage and was the major limitation to storage life. It is recommended that scrubbed low O_2 storage, e.g. 5% CO_2 with 1% O_2, and ethylene removal from scrubbed/unscrubbed controlled atmosphere stores are considered as alternatives to chemical antioxidants for the control of scald. Superficial scald in Granny Smith apple fruits was reported by Gallerani et al. (1994) to be successfully controlled by storing fruits in targeted low O_2 concentrations

(1% O_2 with 2% CO_2), and that their findings enable low O_2 application to be more precisely directed towards superficial scald control. Van der Merve *et al.* (1997) showed that for the apple cultivars Granny Smith and Topred exposure of fruit to initial low oxygen stress of 0.5% O_2 for 10 days at either –0.5 or 3°C (after 7 days in air at the same temperatures) could be used as an alternative to DPA for superficial scald control during storage at 1°C and 3% CO_2 with 1% O_2.

(a)

(b)

Fig. 17.   Superficial scald. (a) Symptoms on apples in New Zealand. (b) Symptoms on pears in Turkey.

Table 4.3. Effects of controlled atmosphere storage conditions during storage at 3.5°C for 5–7 months on the development of superficial scald in three apple cultivars (Fidler et al., 1973).

Storage conditions		Cultivar		
CO_2	O_2	Wagener	Bramley's Seedling	Edward VII
0	21	100	–	–
0	6	–	89	75
0	5	100	–	–
0	4	–	85	62
0	3	9	30	43
0	2.5	–	17	43
0	2	0	–	–
8	13		24	3

There is some evidence that controlled atmosphere storage can affect the development of chilling injury symptoms in stored fruit. Flesh browning and flesh breakdown of plums was shown to occur when they were stored in air at 0°C for 3–4 weeks (Sive and Resnizky, 1979). However, when the same fruit were stored at 0°C in 2–8% CO_2 with 3% O_2 they could be stored for 2–3 months followed by 7 days in air also at 0°C and a shelf life period of 5 days at 20°C without showing the symptoms. Wade (1981) showed that storage of the peach cultivar J.H. Hale at 1°C resulted in flesh discoloration and the development of a soft texture after 37 days, but in atmospheres containing 20% CO_2 fruit had only moderate levels of damage even after 42 days. Conversely Visai et al. (1994) showed a higher incidence of chilling injury in the form of internal browning in pears (cultivar Passe Crassane) stored at 2°C in 5% CO_2 with 2% O_2 compared to fruit stored in air at 0°C. They accounted for this effect as being due to the stimulation of the production of free radicals in the fruit stored in controlled atmosphere.

The effects of O_2 shock treatment on physiological disorders have also been described. This involved storing apples in 0% O_2 for the first 10 days of storage and was shown to prevent core flush in early-picked Jonathan apples from highly affected orchards. No damage due to anaerobic respiration was observed in any of the treatments (Resnizky and Sive, 1991).

INSECT CONTROL

Controlled atmosphere storage has been used to control insects. However, the levels which are necessary to control insects within or on the fruit may be phytotoxic to the fruit themselves. Avocado (cultivar Hass), papaya (cultivar Sunrise) and mango (cultivar Keitt) fruits tolerated low O_2 ($\pm 0.5\%$) and/or very high CO_2 ($\pm 50\%$) for 1, 2 and 5 days, respectively, at 20°C and these treatments have some potential use as insecticidal atmospheres for quarantine insect control treatment on the basis of fruit tolerance, insect mortality and costs (Yahia and Kushwaha, 1995). Ke and Kader (1992b) reviewed the insecticidal effects of very low O_2 and/or very high CO_2 atmospheres at various temperatures and compared with the responses and tolerance of fresh fruits and vegetables to similar controlled atmosphere conditions. The time required for 100% mortality was shown to vary with insect species and its developmental stage, temperature, O_2 and CO_2 levels and humidity. Cantwell *et al.* (1995) showed that various treatment combinations of O_2, CO_2 and temperature were effective in achieving complete insect kill before the development of phytotoxic symptoms on stored vegetables. Storage in 10–20% CO_2 concentrations could control thrips (*Thysanoptera*). It was shown that thrips could be killed in as little as 7 days, although for the peach potato aphid (*Myzus persicae*) more typically mortality was achieved by 10–14 days exposure. Under 80–100% CO_2 concentration, complete mortality was consistently achieved for both types of insect at 0°C within 12 h. Ke and Kader (1992b) showed that the times required to completely kill specific insect species by exposing infested fruit to O_2 levels at or below 1% had potential as postharvest quarantine treatments for fruits such as Bing cherry, Red Jim nectarine, Angeleno plum, Yellow Newton and Granny Smith apples and 20th Century pear. Ke *et al.* (1994) showed that tolerances of peach and nectarine fruits to insecticidal controlled atmospheres were determined by the time before occurrence of visual injury and/or off-flavour. The tolerances of John Henry peaches, Fantasia nectarine, Fire Red peaches, O'Henry peaches, Royal Giant nectarine and Flamekist nectarine to 0.25% O_2 (balance nitrogen) at 20°C were 2.8, 4.0, 4.0, 4.4, 5.1 and 5.2 days, respectively. Fairtime peaches tolerated 0.21% O_2 with 99% CO_2 at 20°C for 3.8 days, 0.21% O_2 with 99% CO_2 at 0°C for 5 days, 0.21% O_2 at 20°C for 6 days, and 0.21% O_2 at 0°C for 19 days. Comparison of fruit tolerance on the time to reach 100% mortality of some insect species suggests that 0.25% O_2 at 20°C is probably not suitable for postharvest insect disinfestation while 0.21% O_2 with or without 99% CO_2 at 0°C merits further investigation. It was suggested by Ke and Kader (1989) that the most promising application for O_2 levels at less than 1% is to expose fresh fruits, such as pears, stone fruits, blueberries and strawberries, to low O_2 atmospheres for short periods to replace chemical treatments for postharvest insect control to meet quarantine requirements. Kerbel *et al.* (1989) showed that Fantasia nectarine were quite tolerant to short-term exposures to low O_2 and/or high CO_2 stresses, which may be effective for insect control. Yahia *et al.* (1992) showed that insecticidal O_2 concentrations (less than 0.4% O_2 with the balance being nitrogen) can be used as a quarantine insect control treatment in papaya for periods less than 3 days at 20°C without the risk of significant fruit injury. Scale, *Quadraspidiotius perniciosus*, on stored apples was completely eliminated by storing infested fruit in 2.6–3.0% O_2 combined with 2.4–2.5% CO_2 or 1.5–1.7% O_2 combined with

1.0–1.1% CO_2 at 1 or 3°C for 31–34 weeks, plus an additional week at 20°C and 50–60% rh (Chu, 1992).

Sweetpotatoes were exposed to low O_2 and high CO_2 for 1 week during curing or subsequent storage to evaluate the use of controlled atmospheres against the weevil *Cylas formicarius elegantulus*. Sweetpotato roots tolerated 8% O_2 during curing, but when exposed to 2 or 4% O_2 or to 60% CO_2 plus 21 or 8% O_2 they were unusable within 1 week after curing, mainly due to decay. Exposure of cured sweetpotatoes to 2 or 4% O_2 plus 40% CO_2, or 4% O_2 plus 60% CO_2 for 1 week at 25°C had little effect on postharvest quality. However, exposure to 2% O_2 plus 60% CO_2 resulted in increased decay, and they had less sweetpotato flavour and more off-flavour. Exposure of sweetpotatoes to levels required for insect control is not feasible during curing, but cured sweetpotatoes can tolerate controlled atmosphere treatments that have potential as quarantine procedures (Delate and Brecht, 1989). Adults of *C. formicarius elegantulus* or sweetpotato roots infested with immature stages of the pest were exposed to controlled atmospheres containing low O_2 and increased concentrations of CO_2 with a balance of nitrogen for up to 10 days at 25 and 30°C. Adults were killed within 4–8 days when exposed to 8% O_2 combined with 40–60% CO_2 at 30°C. At 25°C, exposure to 2 or 4% plus 40 or 60% CO_2 at 25°C killed all the adult insects within 2–8 days. Exposure of sweetpotato roots infested with weevils to 8% O_2 plus 30–60% CO_2 for 1 week at 30°C failed to kill all the weevils. However, no adult weevils emerged from infested roots treated with either 4% O_2 plus 60% CO_2 or 2% O_2 plus 40 or 60% CO_2 for 1 week at 25°C (Delate *et al.*, 1990).

CONCLUSION

Postharvest control of pests and diseases by controlled atmosphere storage can be used as a method of reducing the amount of chemicals applied to the crops and thus chemical residues in the food we eat. Low O_2 but especially high CO_2 levels in storage have been generally shown to have a negative effect on the growth and development of disease-causing micro-organisms. In certain cases the levels of CO_2 necessary to give effective disease control have detrimental effects on the quality of the fruit and vege-tables. There is also some evidence that fruit develop less disease on removal from high CO_2 storage than after previously being stored in air. The mechanism for reduction of diseases appears to be a reaction of the fruit rather than the low O_2 or high CO_2 directly affecting the microorganism, although there is some evidence for the latter.

Physiological disorders of fruit can result from controlled atmosphere storage. In other cases levels of disorders can be reduced. The mechanisms for reduction of disorders vary and are not well understood.

Exposing fruit and vegetables to either high levels of CO_2 or a combination of high CO_2 and low O_2 can control insects infecting fruit and vegetables. Extended exposure to insecticidal levels of CO_2 may be phytotoxic. The controlled atmosphere treatment may be applied for just a few days at the beginning of storage, which may be sufficient to kill the insects without damaging the crop.

Chapter 5

The Influence of Environmental Factors on Controlled Atmosphere Storage

The effects of increased CO_2 and reduced O_2 in combination has been shown, in certain cases, to be more effective than exposure of fruits and vegetables to either reduced O_2 or increased CO_2 alone. There are also many other factors in the crop storage environment which can influence the effects of CO_2 and O_2 on stored fruits and vegetables and some of these are discussed in this chapter.

INTERACTION BETWEEN ETHYLENE AND CARBON DIOXIDE AND OXYGEN

The effects of controlled atmosphere storage on ripening of climacteric fruit is dealt with earlier (see Chapter 2). It was suggested that most of the beneficial effects of controlled atmosphere storage in climacteric fruit are due to suppression of ethylene action (Woltering *et al.*, 1994). However, Knee (1990) pointed out that ethylene is not removed in commercial controlled atmosphere apple stores, but the technology is successful so that there must be an effect of reduced O_2 and or increased CO_2 apart from those on ethylene synthesis or ethylene action. Knee (1990) also pointed out that laboratory experiments on controlled atmosphere storage commonly use small containers which are constantly purged with the appropriate gas mixture. This means that ethylene produced by the fruits or vegetables under such conditions is constantly being removed. In contrast the concentration of gases in commercial controlled atmosphere stores is adjusted by scrubbing and limited ventilation which can allow ethylene to accumulate to high levels in the stores. Knee quotes that ethylene concentrations up to 1000 μl l^{-1} have been reported in controlled atmosphere

81

stores containing apples and $100\,\mu l\,l^{-1}$ in controlled atmosphere stores containing cabbage.

The physiological effects of ethylene on plant tissue have been known for several decades. The concentration of ethylene causing half maximum inhibition of growth of etiolated pea seedlings was also shown to be related to the CO_2 and O_2 levels in the surrounding atmosphere (Burg and Burg, 1967). With decreasing O_2 concentration the effects of ethylene were lower, while with increasing levels of CO_2 at high O_2 the effects of ethylene were higher (Table 5.1).

Burg and Burg (1967) postulated that the interaction between ethylene and CO_2 was that ethylene exerts its effect on plant tissues by attachment to a receptor site and that CO_2 can compete for that receptor site. Wang (1990) also showed that ethylene production was suppressed in controlled atmosphere storage. He describes experiments with sweet peppers stored at 13°C where exposure to controlled atmosphere combinations of 10–30% CO_2 with either 3 or 21% O_2 all suppressed ethylene production down to less than 10 nl 100 g^{-1} h^{-1} compared to fruits stored in air at 13°C which had ethylene production in the range of 40–75 nl 100 g^{-1} h^{-1}. He also showed that the ethylene levels rapidly increased to a similar level to those stored in air when removed from exposure to controlled atmosphere storage for 3 or 6 days. The concentrations of CO_2 and O_2 in such conditions should therefore inhibit ethylene biosynthesis. In contrast controlled atmosphere storage has been shown to stimulate ethylene production. In cold storage of celeriac in air, the ethylene content did not exceed 2 ppm, but the ethylene level in controlled atmosphere stores of celeriac was nearly 25 ppm after 70 days (Golias, 1987). Bangerth (1984) in studies on apple and banana fruit suggested that they became less sensitive to ethylene during prolonged storage.

INTERACTION OF TEMPERATURE WITH CARBON DIOXIDE AND OXYGEN

Kidd and West (1927a) were the first to show that the effects of the gases in extending storage life could vary with temperature. At 10°C controlled

Table 5.1. Sensitivity at 20°C of pea seedlings to ethylene in different levels of O_2, CO_2 and nitrogen (Burg and Burg, 1967).

Nitrogen (%)	O_2 (%)	CO_2 (%)	Sensitivity (μl ethylene l^{-1})
99.3	0.7	0	0.6
97.8	2.2	0	0.3
82.0	18.0	0	0.14
80.2	18.0	1.8	0.3
74.9	18.0	7.1	0.6

atmosphere storage could increase the storage life of apples compared to those stored in air, while at 15°C the storage life of the apples was the same in both gas storage or in air. The opposite effect was shown by Ogata et al. (1975). They found that controlled atmosphere storage of okra at 1°C did not increase their storage life compared to storage in air but at 12°C controlled atmosphere storage increased their postharvest life. Izumi et al. (1996a) also showed that the best controlled atmosphere conditions varied with temperature. They found that the best storage conditions for the broccoli cultivar Marathon were 0.5% O_2 with 10% CO_2 at 0 and 5°C, and 1% O_2 and 10% CO_2 at 10°C.

The relationship between controlled atmosphere storage and temperature has been shown to be complex. Rates of CO_2 production by apple fruit were progressively reduced by lowering O_2 levels from 21 to 2 and 1%. Although lowering the temperature from 4 to 2°C also reduced the respiration rate, fruits stored in 1 or 2% O_2 were shown to be respiring faster after 100 days at 0°C than at 2 or 4°C. After 192 days the air stored fruits also showed an increase in respiration rate at 0°C. These higher respiration rates preceded the development of low temperature breakdown in fruits stored in air, 2 or 1% O_2 at 0°C and in 1% O_2 at 2°C. Progressively lower O_2 concentrations reduced ethylene production whilst increasing the retention of acid, total soluble solids, chlorophyll and firmness. In the absence of low temperature breakdown the effects of reduced temperature on fruit ripening were similar to those of lowered O_2 concentrations. The quality of apples stored at 4°C in 1% O_2 was markedly better than in 2%; the fruits were also free of core flush (brown core) and other physiological disorders (Johnson and Ertan, 1983).

INTERACTION OF HUMIDITY WITH CARBON DIOXIDE AND OXYGEN

Little information is available in the literature on the effects of humidity in the store on the effects of controlled atmosphere storage. Kajiura (1973) stored *Citrus natsudaidai* at 4°C and either 98–100% rh or 85–95% rh in air mixed with 0, 5, 10 or 20% CO_2 for 50 days. It was found that at the higher humidity, high CO_2 increased the water content of the peel and the ethanol content of the juice, and produced abnormal flavour and reduced the internal O_2 content of the fruit causing water breakdown. At the lower humidity, no injury occurred and CO_2 was beneficial, its optimum level being much higher. In another trial, *C. natsudaidai* fruit stored at 4°C in 5% CO_2 with 7% O_2 had abnormal flavour development at 98–100% rh but those in 85–95% rh did not. Bramlage et al. (1977) showed that treatment of McIntosh apples with high CO_2 in a non-humidified room reduced CO_2 injury without also reducing treatment benefits compared to those treated in a humidified room.

CARBON DIOXIDE AND OXYGEN DAMAGE

Fidler *et al.* (1973) gave detailed descriptions of injury caused to different cultivars of apples stored in atmospheres containing low O_2 or high CO_2 levels. Internal injury was described as often beginning 'in the vascular tissue and then increases to involve large areas of the cortex. At first the injury zones are firm, and have a "rubbery" texture when a finger is drawn over the surface of the cut section of the fruit. Later, the damaged tissue loses water and typical cork-like cavities appear' (Fig. 18). The authors also show that the appearance of CO_2 injury symptoms is a function of concentration, exposure time and temperature. They also describe external CO_2 injury which 'initially the damaged area is markedly sunken, deep green in colour and with sharply defined edges. Later in storage the damaged tissue turns brown and finally almost black'. Injury caused as a result of low O_2 levels is due to anaerobic respiration resulting in the accumulation of the toxic by-products alcohols and aldehydes. These can result in necrotic tissue which tends to begin at the centre of the fruit (Fig. 19). It may be that the injury which is referred to as 'spongy tissue' in Alphonso and some other mango varieties may be related to CO_2 injury since they display similar morphological symptoms as those described for apples. Figure 20 shows injury to mangoes which was diagnosed as CO_2 injury.

The lower O_2 limit for apple fruits was found to be cultivar dependent, ranging from a low of approximately 0.8 kPa for Northern Spy and Law Rome to a high of approx. 1.0 kPa for McIntosh. For blueberry fruits, the lower O_2 limit increased with temperature and CO_2 partial pressure. Raising the temperature from 0 to 25°C caused the lower O_2 limit to increase from

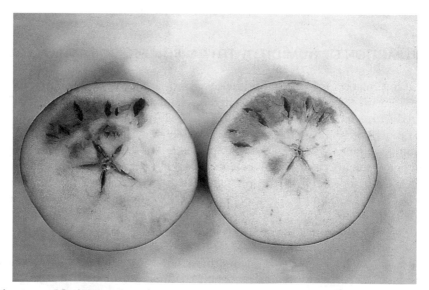

Fig. 18. CO_2 injury on apples in the UK. (Courtesy of Dr R.O. Sharples.)

about 1.8 to approximately 4 kPa. Raising CO_2 levels from 5 to 60 kPa increased the lower O_2 limit for blueberry fruits from approximately 4.5 to over 16 kPa (Beaudry and Gran, 1993). Marshall McIntosh apples held at 3°C in 2.5 to 3% O_2 with 11–12% CO_2 developed small desiccated cavities in the cortex, associated with CO_2 injury (DeEll *et al.*, 1995). Mencarelli *et al.*

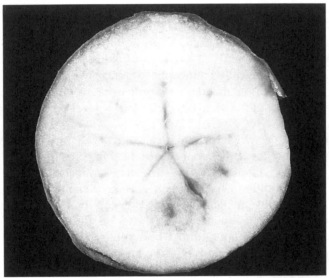

Fig. 19. O_2 damage to apples taken from a controlled atmosphere store. (Courtesy of Dr R.O. Sharples.)

Fig. 20. CO_2 injury on mangoes imported into the UK from India. (Courtesy of Allan Hilton.)

(1989) showed that high CO_2 concentrations during storage (5, 8 or 12%) resulted in CO_2 injury of aubergines, characterized by external browning without tissue softening. Wardlaw (1937) working with aubergines showed that high CO_2 can cause surface scald browning, pitting and excessive decay, and these symptoms are similar to those caused by chilling injury. Mencarelli et al. (1989) showed that the spherical shaped cultivars Black Beauty and Sfumata Rosa were more tolerant of high CO_2 concentrations than the long fruited cultivar Violetta Lunga. Low O_2 levels (0.21 and 3%) on avocados from early season harvests could lead to severe external browning ('brown cold') (Allwood and Cutting, 1994).

Anon (1978) reported that the extension in the postharvest life of bananas at 20°C in controlled atmosphere storage was less at 0.5% O_2 than at 1.5–2.5% O_2. They note that there is some evidence that ethylene production may be promoted in bananas at this low O_2 level which could account for that difference.

The level of CO_2 which can cause damage to fruit and vegetables varies between cultivar of the same crop. Burton (1974b) puts forward a hypothesis to explain these differences. The varietal or cultivar difference could be due to anatomical differences between crops rather than biochemical. He stated that 'Variability in plant material prevents precise control of intercellular atmosphere; recommendations can be designed only to avoid complete anaerobic conditions and a harmful level of CO_2 in the centre of the least permeable individual fruit or vegetable.'

High CO_2 levels (10%) have been shown to cause damage to stored onions, particularly internal browning (Gadalla, 1997). Adamicki et al. (1977) suggested that the decomposition of the cell walls was the result of the influence of hydrolytic enzymes of the pectinase group. Their preliminary comparative studies on cells of sound and physiologically disordered onions due to CO_2 injury, indicated destructive changes in the ultrastructure of the mitochondria. The mitochondria displayed fragmentation, reduction in size and change in shape from elliptical to spherical. These changes have also been observed in studies on the influence of CO_2 on the ultrastructure of pears (Frenkel and Patterson, 1974). Elevated CO_2 concentrations were shown to inhibit the activity of succinic dehydrogenase, resulting in accumulation of succinic acid, a toxicant to plant tissues (Hulme, 1956; Williams and Patterson, 1964; Frenkel and Patterson, 1974). Adamicki et al. (1977) found that the highest amino acid content was found in the physiologically disordered onion bulbs stored at 1°C with lower values found at 5°C, and they added that this may be due to greater enzyme activity, especially in physiologically disordered bulbs. Isenberg (1979) concluded that controlled atmosphere storage is an alternative to the use of sprout suppression, but stressed the need for further testing of the optimum condition of O_2 and CO_2 required for individual varieties.

Kader (1989, 1993), Saltveit (1989) and Meheriuk (1989b) reviewed some CO_2 and O_2 injury symptoms on selected tropical fruits (Table 5.2). It

Table 5.2. Threshold level of O_2 or CO_2 required to cause injury to some fruits and vegetables and typical injury symptoms (Kader, 1989, 1993; Meheriuk, 1989; Saltveit, 1989).

Crop and cultivar	CO_2 injury level	CO_2 injury symptoms	O_2 injury level	O_2 injury symptoms
Apple, Boskoop	> 2%	CO_2 cavitation	< 1.5%	
Apple, Cox's Orange Pippin	> 1%	Core browning	< 1%	Alcoholic taste
Apple, Golden Delicious	> 6% continuous > 15% for 10 days or more	CO_2 injury	< 1%	Alcoholic taint
Apple, Red Delicious	> 3%	Internal browning	< 1%	Alcoholic taste with late-picked fruit
Apple, Starking Delicious	> 3%	Internal disorders	< 1%	Alcoholic taste
Apple, Elstar	> 2%	CO_2 injury	< 2%	Coreflush
Apple Empire	> 5%	CO_2 injury	< 1.5%	Flesh browning
Apple, Fuji	> 5%	CO_2 injury	< 2%	Alcoholic taint
Apple, Gala	> 1.5%	CO_2 injury	< 1.5%	Ribbon scald
Apple, Gloster	> 1%	Core browning	< 1%	
Apple, Granny Smith	> 1% with O_2 at less than 1.5% > 3% with O_2 less than 2%	Severe core flush	< 1%	Alcoholic taint, ribbon scald and core browning
Apple, Idared	> 3%	CO_2 injury	< 1%	Alcoholic taint
Apple, Jonagold	> 5%	Unknown	< 1.5%	Unknown
Apple, Jonathan	> 5%	CO_2, flesh browning	< 1%	Alcoholic taint, core browning
Apple, Karmijn	> 3%	Core browning and low temperature breakdown	< 1%	Alcoholic taint
Apple, McIntosh	> 5% continuous > 15% for short periods	CO_2 injury, core flush	< 1.5%	Corky browning, skin discoloration, flesh browning, alcoholic taint
Apple, Melrose	> 5%	Unknown	< 2%	Alcoholic taint
Apple, Mutsu	> 5% with O_2 > 2.5%	Unknown	< 1.5%	Alcoholic taint
Apple, Rome	> 5%	Unknown	< 1.5%	Alcoholic taint
Apple, Spartan	> 3%	Coreflush, CO_2 injury	< 1.5%	Alcoholic taint
Apple, Stayman		Unknown	< 2%	Alcoholic taint
Apricot	> 5%	Loss of flavour, flesh browning	< 1%	Off-flavour development

Table 5.2. *Continued*

Crop and cultivar	CO_2 injury level	CO_2 injury symptoms	O_2 injury level	O_2 injury symptoms
Artichoke, globe	> 3%	Stimulates papus development	< 2%	Blackening of inner bracts and receptacle
Asparagus	> 10% at 3–6°C > 15% at 0–3°C	Increased elongation, weight gain and sensitivity to chilling and pitting	< 10%	Discoloration
Avocado	> 15%	Skin browning, off-flavour	< 1%	Internal flesh breakdown, off-flavour
Banana	> 7%	Green fruit softening, undesirable texture and flavour	< 1%	Dull yellow or brown skin discoloration, failure to ripen, off-flavour
Beans, green, snap	> 7% for more than 24 h	Off-flavour	< 5% for more than 24 h	Off-flavour
Blackberry	> 25%	Off-flavour	< 2%	Off-flavour
Blueberry	> 25%	Skin browning, off-flavour	< 2%	Off-flavour
Broccoli	> 15%	Persistent off-odours	< 0.5%	Off-odours, can be lost upon aeration if slight
Brussels sprouts	> 10%		< 1%	Off-odours, internal discoloration
Cabbage	> 10%	Discoloration of inner leaves	< 2%	Off-flavour, increased sensitivity to freezing
Cauliflower	> 5%	Off-flavour, aeration removes slight damage, curd must be cooked to show symptoms	< 2%	Persistent off-flavour and odour after cooking
Celery	> 10%	Off-flavour and odour, internal discoloration	< 2%	Off-flavour and odour
Cherimoya	Not determined	Not known	< 1%	Off-flavour

Table 5.2. *Continued*

Crop and cultivar	CO_2 injury level	CO_2 injury symptoms	O_2 injury level	O_2 injury symptoms
Cherry, sweet	> 30%	Brown, discoloration of skin, off-flavour	< 1%	Skin pitting, off-flavour
Cranberry	?	?	< 1%	Off-flavour
Cucumber	> 5% at 8°C > 10% at 5°C	Increased softening, increased chilling injury, surface discoloration and pitting	< 1%	Off-odours, breakdown and increased chilling injury
Custard apple	15% +	Flat taste, uneven ripening	< 1%	Failure to ripen
Durian	> 20%	Not known	< 2%	Failure to ripen, grey discoloration of pulp
Fig	> 25% (?)	Loss of flavour (?)	< 2% (?)	Off-flavour (?)
Grape	> 5%	Browning of berries and stems	< 1%	Off-flavour
Grapefruit	> 10%	Scald-like areas on the rind, off-flavour	< 3%	Off-flavour due to increased ethanol and acetaldehyde contents
Kiwifruit	> 7%	Internal breakdown of the flesh	< 1%	Off-flavour
Lemon	> 10%	Increased susceptibility to decay, decreased acidity	< 5%	Off-flavour
Lettuce, Crisphead	> 2%	Brown stain	< 1%	Breakdown at centre
Lime	> 10%	Increased susceptibility to decay	< 5%	Scald-like injury, decreased juice content
Mango	> 10%	Softening, off-flavour	< 2% (< 5%)	Skin discoloration, greyish flesh colour, off-flavour
Melon, cantaloupe	> 20%	Off-flavour and odours, impaired ripening	< 1%	Off-flavour and odours, impaired ripening
Mushroom	> 20%	Surface pitting	Near 0%	Off-flavour and odours, stimulation of cap opening and stipe elongation

Table 5.2. *Continued*

Crop and cultivar	CO_2 injury level	CO_2 injury symptoms	O_2 injury level	O_2 injury symptoms
Nectarine	> 10%	Flesh browning, loss of flavour	< 1%	Failure to ripen, skin browning, off-flavour
Olive	> 5%	Increased severity of chilling injury on olives kept at < 7°C	< 2%	Off-flavour
Onion, bulb	> 10% for short term, > 1% for long term	Accelerated softening, rots and putrid odour	< 1%	Off-odours and breakdown
Orange	> 5%	Off-flavour	< 5%	Off-flavour
Papaya	> 8%	May aggravate chilling injury at < 12°C, off-flavour	< 2%	Failure to ripen, off-flavour
Peach, clingstone	> 5%	Internal flesh browning severity increases with CO_2%	< 1%	Off-flavour in the canned product
Peach, freestone	> 10%	Flesh browning, off-flavour	< 1%	Failure to ripen, skin browning, off-flavour
Peppers, bell	> 5%	Calyx discoloration, internal browning, and increased softening	< 2%	Off-odours and breakdown
Pepper, chilli	> 20% at 5°C, > 5% at 10°C	Calyx discoloration, internal browning, and increased softening	< 2%	Off odours and breakdown
Persimmon	> 10%	Off-flavour	< 3%	Failure to ripen, off-flavour
Pineapple	> 10%	Off-flavour	< 2%	Off-flavour
Plum	> 1%	Flesh browning	< 1%	Failure to ripen, off-flavour
Rambutan	> 20%	Not known	< 1%	Increased decay incidence
Raspberry	> 25%	Off-flavour, brown discoloration	< 2%	Off-colours

Table 5.2. *Continued*

Crop and cultivar	CO_2 injury level	CO_2 injury symptoms	O_2 injury level	O_2 injury symptoms
Strawberry	> 25%	Off-flavour, brown discoloration of berries	< 2%	Off-flavour
Sweetcorn	> 10%	Off-flavour and odours	< 2%	Off-flavour and off-odours
Tomato	> 2% for mature-green, > 5% for turning, also depends on length of exposure and temperature	Discoloration, softening and uneven ripening	< 2% depending on length of exposure	Off flavour, softening and uneven ripening

should be noted that the exposure time to different gases will affect their susceptibility to injury and there may be interactions with the ripeness of the fruit, harvest maturity or the storage temperature.

HIGH OXYGEN STORAGE

An international 'High O_2 Club' has been formed to promote research into this subject area (Day, 1996). Kidd and West (1934) showed that storage of apples in pure O_2 can be detrimental. Lurie *et al.* (1991) investigated the effects of storage of Granny Smith apples in 70% O_2 to delay or prevent the darkening of colour which occurs on the sun-scalded peel of fruits. It was deduced that the colour change of sun-scalded peel during storage is due to non-oxidative, non-enzymatic processes but the high O_2 treatment showed no such effect. Yahia (1989, 1991) showed that exposure of McIntosh and Cortland apple fruits to 100% O_2 at 3.3°C for 4 weeks did not enhance the production of aroma volatiles. Pears kept in 100% O_2 showed an increase in the rate of softening, chlorophyll degradation, and ethylene evolution (Frenkel, 1975).

Various publications have shown that high O_2 levels in citrus fruit stores can affect the fruit colour. Navel and Valencia orange fruits were stored for 4 weeks at 15°C in a continuous flow of:

- air
- air with 20 ppm ethylene
- 40% O_2 with 60% nitrogen
- 80% O_2 with 20% nitrogen.

After 2 weeks of storage with high O_2 the endocarp of Navel, but not of Valencia oranges, had turned a perceptibly darker orange than fruit stored in air or in air with ethylene. The orange colour intensified with increased storage time and persisted at least 4 weeks after fruit was removed from the test atmospheres. Juice from Navel, but not from Valencia, oranges stored in 80% O_2 for 4 weeks was slightly darker than juice from oranges stored in air. Peel of both cultivars stored with ethylene was brighter orange than that of fruit stored in air or in high O_2, but, juice colour of Navel oranges was adversely affected by ethylene. Ethylene adversely affected the flavour of Navel and Valencia juice and caused the rind to become sticky (Houck et al., 1978). Hamlin, Parson Brown and Pineapple oranges were stored for 4 weeks at 15°C. Fruits stored in 80% O_2 with 20% nitrogen had the palest rind but their endocarp and juice colour were the deepest orange. Colour change was detectable after 2–3 weeks and continued developing. In 40% O_2 with 60% nitrogen, fruit response was intermediate. The respiration rate of Pineapple oranges was highest in fruits stored in 80% O_2 with 20% nitrogen, and lowest in 40% O_2 with 60% nitrogen or air. Fruits stored in different atmospheres did not differ significantly in total soluble solids content, total acidity, or pH of the juice, and juice flavour was not adversely affected (Aharoni and Houck, 1980). Ruby, Tarocco and Sanguinello blood oranges were stored for 4 weeks at 15°C in flowing air, 20 ppm ethylene in air, 40% O_2 with 60% nitrogen or 80% O_2 with 20% nitrogen, followed by storage in air. Ethylene caused the rind of Ruby and Tarocco oranges to become deeper orange while 80% O_2 caused the rind of Ruby and Sanguinello oranges to be paler orange than when stored in other atmospheres. Ethylene deepened the red of the flesh and juice of Ruby and Tarocco oranges, but high O_2 deepened the red of flesh and juice of all three cultivars very markedly. Total soluble solids content, total acidity, and pH of the juice were not affected. Fruit stored in the ethylene atmosphere developed an off-flavour (Aharoni and Houck, 1982).

Ripe green tomatoes were kept at 19°C or at 13°C followed by 19°C in air or in 100% O_2 at normal or reduced atmospheric pressure until they were orange in colour. Fruit softening was exacerbated in 100% O_2 and reduced pressure (0.25 atmospheric pressure) compared to a normal gaseous atmosphere and reduced pressure. Removal of ethylene in air made little difference to the fruit but in 100% O_2 ethylene accumulation was detrimental (Stenvers, 1977). Ripe green tomatoes were kept at 19°C or at 13°C followed by 19°C in air or in 100% O_2 at normal or reduced atmospheric pressure until they were orange in colour, when their weight loss and softness were assessed. Ripening was delayed at 0.25 atmospheric pressure, especially in air. Fruit softening was also less in air at reduced pressure, but in 100% O_2 and reduced pressure it was exacerbated. Removal of ethylene in air made little difference but in 100% O_2 it was detrimental (Stenvers, 1977). Li et al. (1973) reported that ripening of tomatoes was accelerated at 12–13°C in 40–50% O_2 partial pressures compared to those ripened in air.

At 4°C tuber respiration rate of potato tubers was lower in 3 or 1% O_2 compared to those in air and in 35% O_2. Sprouting was inhibited and sugar content increased in 1% and 35% O_2 compared to those in air, but those in 3% showed no effects. Sprouting was inhibited in 3% CO_2 or higher, but at 7% CO_2 partial deterioration occurred and there was an increase in sugar content (Hartmans *et al.*, 1990). Storage atmospheres containing 40% O_2 reduced mould infection compared to low O_2 levels, but increased sprouting and rooting of stored carrots (Abdel-Rahman and Isenberg, 1974).

Dick and Marcellin (1985) showed that bananas held in 50% O_2 during cooling periods of 12 h at 20°C had reduced high temperature damage during subsequent storage at 30–40°C.

Day (1996) indicated highly positive effects of storing minimally processed fruit and vegetables in 70 and 80% O_2 in modified atmosphere film packs. He indicated that it inhibited undesirable fermentation reactions, delayed browning (caused by damage during processing) and the O_2 levels of over 65% inhibited both aerobic and anaerobic microbial growth. The technique used involved flushing the packs with the required gas mixture before they are sealed. The O_2 level within would then fall progressively as storage proceeds due to the respiration of the fruit or vegetable contained in the pack.

High O_2 levels have also been used in the marketing of other commodities. For example Champion (1986) mentioned that O_2 levels above ambient are used to help preserve 'redness and eye appeal' of red meat.

CONCLUSION

It is crucial for the success of controlled atmosphere storage technology that the precise levels of gas are achieved and maintained within the store. Where these are too high in the case of CO_2, or too low for O_2, then the fruit or vegetable may be irrevocably damaged. The range of manifestations of the symptoms of damage varies with the types of product and the intensity of the effect. The effects of controlled atmosphere storage are not simple. In many cases their effects are dependent on other environmental factors. These include the effects of temperature where controlled atmosphere storage may be less or totally ineffective at certain temperatures on some fruits. This effect could be related to fruit metabolism and gas exchange at different temperatures. From the limited information available it would appear that the store humidity and CO_2 level in the store could interact, with high CO_2 being more toxic at high than low store humidities. The physiological mechanism to explain this effect is difficult to find. Controlled atmospheres can reduce or eliminate detrimental effects of ethylene accumulation possibly by CO_2 competing for sites of ethylene action within the cells of the fruit.

Increasing the level of O_2 above that which occurs in the atmosphere has been shown to improve the colour of some stored fruit. There is, however, some indication that high levels of O_2 in the atmosphere can increase the rate of fruit softening but reduce microbial growth. More work is required before this method of storage can be generally recommended.

Chapter 6

Modified Atmosphere Packaging

Modified atmosphere packaging can be defined as an alteration in the composition of gases in and around fresh produce by respiration and transpiration when such commodities are sealed in plastic films (G.E. Hobson, 1997, personal communication). For fresh fruits and vegetables this is commonly achieved by packing them in plastic films. Plastic film bags are made from the by-products of the mineral oil refining industry. Different units are used by different researchers and commercial packers to describe film thickness (Table 6.1).

Recent developments in modified atmosphere packaging were reviewed by Church (1994). These included low permeability films, high permeability films, O_2 scavenging technology, CO_2 scavengers and emitters, ethylene absorbers, ethanol vapour generators, tray-ready and bulk modified atmosphere packaging systems, easy opening and resealing systems, leak detection, time–temperature indicators, gas indicators, combination treatments, and predictive/mathematical modelling. The surface of the bag is slippery which can make them very difficult to stack, especially large

Table 6.1. The units used to describe the thickness of plastic films.

Gauge	mm	μ
4	0.001	1
48	0.012	12
50	0.0125	12.5
100	0.025	25
150	0.0375	37.5
200	0.05	50
400	0.1	100
500	0.125	125

bags. Low density polyethylene is commonly produced in this manner for the ubiquitous 'polythene' bag so frequently used in fresh produce packaging. The permeability of plastic films to gases and water vapour will vary with the type and thickness of the plastic used. Generally, the acceptable time which fruit can remain in a modified atmosphere varies with cultivars and storage conditions, respectively, but it is usually 2–4 weeks which meets the requirement of export shipment and marketable life.

Packing crops into plastic film bags can lead to a build up of water vapour and the respiration gases CO_2, O_2 and ethylene. The effects of these gas changes may have beneficial or detrimental effects on the crop. The beneficial effects, which can result in packing crops in plastic films, have led to this method being referred to as modified atmosphere packaging (Kader *et al.*, 1989b). Modified atmosphere storage can also refer to sealing fruit and vegetables in a gas impermeable container where respiratory gases change the atmosphere over time.

If a crop is in equilibrium with its environment the rate of gas exchange will be the same in both directions. So if the concentration of the respiration gases around the outside of the crop is changed by surrounding it with plastic film, this will change the concentrations of the gases inside the crop. The concentration of these gases will then vary with such characteristics as the type and thickness of the film used as well as such innate crop characteristics as mass of produce inside the bag, type of crop, temperature, maturity and activity of microorganisms.

Given an appropriate gaseous atmosphere, modified atmosphere packaging can be used to extend the postharvest life of crops in the same way as controlled atmosphere storage. Storage of mushrooms in modified atmosphere packaging at 10°C and 85% relative humidity (rh) for 8 days delayed maturation and reduced weight loss compared to those stored without packaging (Lopez-Briones *et al.*, 1993). Pala *et al.* (1994) studied storage in sealed low density polyethylene film of various thicknesses at 8°C and 88–92% rh. They found that non-wrapped peppers had a shelf life of 10 days compared to 29 days for those sealed in 70 μm low density polyethylene film (Table 6.2). Also this method of packaging gave good retention of sensory characteristics.

Batu and Thompson (1994) stored tomatoes at both the mature green and the pink stage of maturity either unwrapped or sealed in polyethylene films of 20, 30 or 50 μ thickness at either 13 or 20°C for 60 days. All unwrapped pink tomatoes were over-ripe and soft after 30 days at 13°C and after 10–13 days at 20°C. Pink tomatoes in polymeric films were still firm after 60 days stored at either 13 or 20°C although those in 20 μ were slightly softer than those in the other two films. Green tomatoes sealed in 20 and 30 μ film reached their reddest colour after 40 days at 20°C and 30 days at 13°C. Fruit in 50 μ film still had not reached their maximum red colour even after 60 days at both. All green tomatoes sealed in polymeric film were very firm even after 60 days of storage at 13 and 20°C.

Unwrapped tomatoes remained acceptably firm for about 50 days at 13°C and 20 days at 20°C.

The effects of polyethylene film wraps on the postharvest life of fruits may also be related to moisture conservation around the fruit as well as the change in the CO_2 and O_2 content. This was shown by Thompson et al. (1972) for plantains where fruit stored in moist coir or perforated polyethylene film bags had a longer storage life than fruits which had been stored unwrapped. But there was an added effect when fruits were stored in unperforated bags which was presumably due to the effects of the changes in the CO_2 and O_2 levels (Table 6.3). So the positive effects of storage of fresh pre-climacteric fruits in sealed plastic films may be, in certain cases, a combination of its effects on the CO_2 and O_2 content within the fruit and the maintenance of a high moisture content. The effect of moisture content is more likely a reduction in stress of the fruit which may be caused by a rapid rate of water loss in unwrapped fruit. This in turn may result in increased ethylene production to internal threshold levels which can initiate ripening.

Table 6.2. Number of days during which green pepper fruit retained either good or acceptable quality during storage at 8°C and 88–92% rh (after Pala et al., 1994).

		Sensory score*				Storage time (days)	
	Days	Appearance	Colour	Texture	Taste and flavour	Good quality	Acceptable quality
Before storage	0	9.0	9.0	9.0	9.0		
Non-wrapped control	15	4.3	4.6	2.7	4.2	7	10
20 μm LDPE	22	5.7	5.5	5.7	5.3	10	22
30 μm LDPE	20	5.8	6.0	6.0	4.5	10	20
50 μm LDPE	20	6.0	5.0	6.0	5.0	15	20
70 μm LDPE	29	7.8	7.8	7.6	7.8	27	29
100 μm LDPE	27	5.5	5.6	5.7	5.5	10	27

*1–9 where 1–3.9 = non acceptable; 4–6.9 = acceptable; 7–9 = good.
LDPE, low density polyethylene.

Table 6.3. Effects of wrapping and packing material on ripening and weight loss of plantains stored at tropical ambient conditions of 26–34°C and 52–87% rh (source: Thompson et al., 1972).

Packing material	Days to ripeness	Weight loss at ripeness
Not wrapped	15.8	17.0%
Paper	18.9	17.9%
Moist coir fibre	27.2	(3.5%)*
Perforated polyethylene	26.5	7.2%
Polyethylene	36.1	2.6%
LSD (P = 0.05)	7.28	2.81

*The fruit actually gained in weight.

Hansen (1975) showed that Jonagold apples kept in cold stores at 0.5, 2.5 or 3.5°C retained satisfactory organoleptic quality only until the end of January even at the lowest temperature. The apples could be stored at 3.5°C in polyethylene bags (with a CO_2 content of 7%) until April. In controlled atmosphere stores (6% CO_2 with 3% O_2 at 2.5°C) fruit quality was satisfactory to the end of May. Under all conditions apples of this cultivar showed no rotting, CO_2 damage or physiological disorders. Many other examples are described under individual crops in Chapter 7. Modified atmosphere packaging does not always have a beneficial effect on the postharvest life of fresh fruits and vegetables. Storing yam in polyethylene bags was shown to reduce weight loss, but have little effect on surface fungal infections and internal browning of tissue (Table 6.4). Geeson (1989) also showed that film packages with lower water vapour transmission properties can encourage rotting of the tomatoes.

Modified atmosphere packaging is used to slow the deterioration of prepared fruit and vegetables and this technique is often referred to as minimal processing. An example of this technique is described by Lee and Lee (1996) who used low density polyethylene film of 27 μm thickness for the preservation of a mixture of carrot, cucumber, garlic and green peppers. The steady state atmosphere at 10°C inside the bags was 5.5–5.7% CO_2 and 2.0–2.1% O_2.

Plastic film wraps have been developed which can give both protection as well as modified atmospheres to produce. One such packaging system is being marketed as Airbox (Fig. 21), but because of the cost, the space taken up by the package and the difficulty in packaging, it currently has only limited application to the fresh produce industry.

FILM TYPES

The actual concentration of gases in the fruit will also be affected to a limited degree by the amount of space between the fruit and the plastic

Table 6.4. Effects of packaging material on the quality of yam (*Dioscorea trifida*) after 64 days at 20–29°C and 46–62% rh. Fungal score was 0 = no surface fungal growth, 5 = tuber surface entirely covered with fungi. Necrotic tissue was estimated by cutting the tuber into two length ways and measuring the area of necrosis and expressing it as a percent of the total cut surface.

Package type	Weight loss (%)	Fungal score	Necrotic tissue (%)
Paper bags	26.3	0.2	5
Sealed 0.03 mm thick polyethylene bags with 0.15% of the area as holes	15.7	0.2	7
Sealed 0.03 mm thick polyethylene bags	5.4	0.4	4

film, but mainly by the permeability of the film. Several different plastics are used for this purpose (Box 6.1).

A range of high shrink multilayer high speed machineable shrink films made from polyolefin are available. Polyethylene is also used for shrink film packaging.

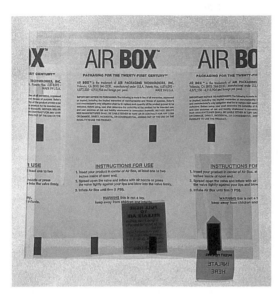

Fig. 21. Double walled plastic film which can provide a modified atmosphere for produce as well as physical protection. (Courtesy of Allan Hilton.)

Box 6.1. A selection of plastic films which have been used for fruit and vegetable packaging.

Cellulose acetate (CA)
Ethylene vinyl acetate copolymers (EVAL)
Ethylene vinyl alcohol copolymers (EVOH)
High density polyethylene (HDPE)
Ionomer
Linear low density polyethylene (LLDPE)
Low density polyethylene (LDPE)
Medium density polyethylene (MDPE)
Polybutylene (PB)
Polyethylene terephthalate (PET)
Polyolefin
Polypropylene (PP)
Polystyrene (PS)
Polyvinyl butyral (PVB)
Polyvinyl chloride (PVC)
Polyvinylidene chloride (PVDC)

FILM PERMEABILITY

Respiring fresh fruits and vegetables sealed in plastic films will cause the atmosphere to change in particular O_2 levels to be depleted and CO_2 levels to be increased. The transmission of CO_2 and O_2 though plastic films will vary with film type, but generally films are four to six times more permeable to CO_2 than to O_2. However, Barmore (1987) indicates that the relationship between CO_2 and O_2 permeability and that of water vapour is not so simple. Variation in transmission of water vapour can therefore be achieved to some extent, independently of transmission of CO_2 and O_2 using such techniques as producing multilayer films by coextrusion or applying adhesives between the layer. This permeability is invariably restricting and often holes are punctured in the bags to improve ventilation (Table 6.5).

The permeability of films to gases (including water vapour) varies with the type of material from which they are made, temperature, in some cases humidity, the accumulation and concentration of the gas and the thickness of the material. Schlimme and Rooney (1994) showed that there were a range of permeabilities which could be obtained from films with the same basic specifications. A selection of these are summarized as follows. Polyethylene film, as indicated above, is available in several forms with different properties.

Low density polyethylene is produced by polymerization of the ethylene monomer which produces a branched chain polymer with a molecular weight of 14,000–1,400,000 and a density ranging from 0.910 to 0.935 g cm^{-1}. Low density polyethylene film has the following specifications: 3900–13,000 O_2 cm^3 m^{-2} day^{-1}, 7700–77,000 CO_2 cm^3 m^{-2} day^{-1}, at 1 atm for 0.0254 mm thick at 22–25°C at various or unreported relative humidity and 6–23.2 water vapour g m^{-2} day^{-1} at 37.8°C and 90% rh (Schlimme and Rooney, 1994).

Table 6.5. Effects of number and size of perforation in 3 lb (1.36 kg) 150 gauge polyethylene film bags of Yellow Globe onions on the relative humidity in the bags, rooting of the bulbs and weight loss after 14 days at 24°C (Hardenburg, 1955).

Perforations		% rh in bag	Onions rooted (%)	Weight loss (%)
Number	Size (mm)			
0	–	98	71	0.5
36	1.6	88	59	0.7
40	3.2	84	40	1.4
8	6.4	–	24	1.8
16	6.4	54	17	2.5
32	6.4	51	4	2.5
0*	–	54	0	3.4

*Kraft paper with film window.

High density polyethylene has 75–90% crystalline structure with an ordered linear arrangement of the molecules with little branching and a molecular weight of 90,000–175,000. It has a typical density range of 0.995–0.970 g cm^{-1} and has greater tensile strength, stiffness and hardness than low density polyethylene. High density polyethylene film has the following specifications: 520–4000 O_2 cm^3 m^{-2} day^{-1}, 3900–10,000 CO_2 cm^3 m^{-2} day^{-1}, at 1 atm for 0.0254 mm thick at 22–25°C at various or unreported relative humidity and 4–10 water vapour g m^{-2} day^{-1} at 37.8°C and 90% rh (Schlimme and Rooney, 1994).

Medium density polyethylene has a density in the range of 0.926 and 0.940, 2600–8293 O_2 cm^3 m^{-2} day^{-1}, 7700 to 38,750 CO_2 cm^3 m^{-2} day^{-1}, at 1 atm for 0.0254 mm thick at 22–25°C at various or unreported relative humidity and 8–15 water vapour g m^{-2} day^{-1} at 37.8°C and 90% rh (Schlimme and Rooney, 1994).

Linear low density polyethylene combines the properties of low density polyethylene film and high density polyethylene film giving a more crystalline structure than low density polyethylene film but with a controlled number of branches which makes it tougher and suitable for heat sealing. It is made from ethylene with butene, hexene or octene; with the latter two co-monomers giving enhanced impact resistance and tear strength. Permeability was given as 7000–9300 O_2 cm^3 m^{-2} day^{-1}, at 1 atm for 0.0254 mm thick at 22–25°C at various or unreported relative humidity and 16–31 water vapour g m^{-2} day^{-1} at 37.8°C and 90% rh (Schlimme and Rooney, 1994).

Where the permeability of the film is too low the crop may be damaged because of the accumulation of water or CO_2 or depletion of O_2. In those cases holes can be punched in the film to improve ventilation (see Table 6.3).

Film permeability to gases is by active diffusion where the gas molecules dissolve in the film matrix and diffuse through in response to the concentration gradient (Kester and Fennema, 1986). A formula to describe film permeability was given by Crank (1975) as follows:

$$P = \frac{Jx}{A(p_1 - p_2)}$$

where J = volumetric rate of gas flow through the film at steady state, x = thickness of film, A = area of permeable surface, p_1 = gas partial pressure on side 1 of the film, p_2 = gas partial pressure on side 2 of the film ($p_1 > p_2$).

GAS FLUSHING IN MODIFIED ATMOSPHERE PACKAGING

The levels of CO_2 and O_2 can take some time to change inside the modified atmosphere pack. In order to speed up this process the pack can be flushed with nitrogen to reduce the O_2 rapidly or the atmosphere can be flushed with an appropriate mixture of CO_2, O_2 and nitrogen. In other cases the pack can be connected to a vacuum pump to remove the air so that the respiratory gases can change within the pack more quickly. Gas flushing is more important for non-respiring products such as meat or fish, but it can profitably be used with fresh fruits and vegetables. In work described by Aharoni *et al.* (1973), yellowing and decay of leaves were reduced when the lettuce cultivar Hazera Yellow was pre-packed in closed polyethylene bags in which the O_2 concentration was reduced by flushing with nitrogen. Similar but less effective results were obtained when the lettuce was pre-packaged in closed polyethylene bags not flushed with nitrogen, or when open bags were placed in polyethylene-lined cartons. Andre *et al.* (1980a) showed that fungal development during storage could be prevented in asparagus spears by packing them in polyethylene bags with silicon elastomer windows and flushing with 30% CO_2 for 24 h or by maintaining a CO_2 concentration of 5–10%.

The changes in the atmosphere inside the sealed plastic film bag depend on the characteristics of the material used to make the package, the environment inside and outside the package as well as the respiration of the produce it contains. Changes in gas content may take some time to reach equilibrium, but the speed at which the atmosphere is changed can be accelerated by gas flushing. This was illustrated in experiments described by Zagory (1990) which showed the effects of flushing chilli peppers stored in plastic film with a mixture containing 10% CO_2 and 1% O_2 compared to no gas flushing (Fig. 22).

MODELLING

Ben-Yehoshua *et al.* (1995) reviewed modified atmosphere packaging under the following headings: modified humidity packaging with an example of the effects on bell peppers, interactive and microporous films, effects of perforations in the film, and modelling of packaging. Day (1994) also reviewed the concept of mathematical modelling of modified atmosphere packaging for minimally processed fruit and vegetables. Evelo (1995) showed that the package volume did not affect the steady state gas concentrations inside a modified atmosphere package but did affect the non-steady state conditions. A modified atmosphere packaging model was developed using a systems-oriented approach which allowed the selection of a respiration model according to the available data. Temperature dependence was explicitly incorporated into the model. The model was suitable for assessing optimal modified atmosphere packaging in realistic distribution chains. O_2 and CO_2 concentrations inside plastic film packages containing mango cultivar Nam Dok Mai fruits were modelled by Boon-Long *et al.* (1994). Parameters for polyvinyl chloride, polyethylene and polypropylene films were included. A method of determining respiration rate from the time

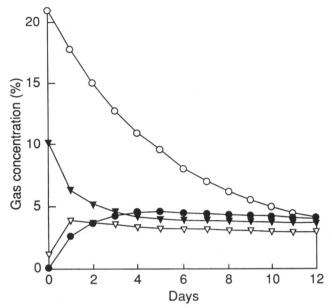

Fig. 22.   A comparison of passive modified atmosphere packaging and active modified atmosphere packaging on the rate of change of CO_2 and O_2 in Anaheim chilli pepper fruit packed in Cryovac SSD-310 film (Zagory, 1990). ○, O_2 passive; ●, CO_2 passive; ▽, O_2 active; ▼, CO_2 active.

history of the gas concentrations inside the film package instead of from direct measurements was devised. Satisfactory results were obtained in practical experiments carried out to verify the model.

Many other mathematical models have been developed and published (e.g. Cameron *et al.*, 1989; Lee *et al.*, 1991; Lopez-Briones *et al.*, 1993) which can help to predict the atmosphere around fresh produce sealed in plastic film bags. The latter suggested the following model:

$$\frac{1}{x}=\frac{\hat{A}m}{x_0 S}\cdot\frac{1}{K}+\frac{1}{x_0}$$

where x = O_2 concentration in the pouches (%), x_0 = initial O_2 concentration within the pouches (%), $\hat{A}$ = proportionality between respiration rate and O_2 concentration including the effect of temperature (ml g^{-1} day^{-1} atm^{-1}), K = O_2 diffusion coefficient of the film (includes effects of temperature) (ml m^{-2} day^{-1} atm^{-1}), S = surface area for gas exchange (m^2), and m = weight of plant tissue (g) (source Lopez-Briones *et al.*, 1993).

Lee *et al.* (1991) obtained a set of differential equations representing the mathematical model for the modified atmosphere system. In this case, the rate of reaction (r) is equal to:

$$r=V_m\,C_1\,/\left[K_M+C_1(1+C_2/K_1)\right]$$

where C_1 is the substrate concentration (which is O_2 concentration in the case of respiration rate), V_m and K_M are parameters of the classical Michaelis–Menten kinetics, V_m being the maximal rate of enzymatic reaction, and K_M is the Michaelis constant, C_2 is the inhibitor concentration (CO_2 concentration in the case of respiration), and K_1 is the constant of equilibrium between the enzyme-substrate-inhibitor complex and free inhibitor.

Combining this equation with Fick's law for O_2 and CO_2 permeation, Lee *et al.* (1991) estimated parameters of this model (V_m, K_M, and K_1) from experimental data and then performed numerical calculation of the equations.

Strawberries (cultivars Pajaro and Selva) were used to test a model describing gas transport through microperforated polypropylene films and fruit respiration involved in modified atmosphere packaging by Renault *et al.* (1994). Some experiments were conducted with empty packs initially filled with either 100% nitrogen or 100% O_2. Simulations agreed very well with experiments only if the cross-sectional area of the microperforations was replaced by areas of approximately half the actual areas in order to account for the resistance of air around the perforations. It was also possible to fit the model to gas concentration changes in packs filled with strawberries, although deviations were encountered due to contamination of strawberries by fungi. The model was used to quantify the consequences of the variability of pack properties (number of microperforations per pack and cross-sectional area of these perforations) on equilibrium gas concentrations and to define minimum homogeneity requirements for modified atmosphere packaging.

EFFECT OF PRODUCT ON GAS CONTENT

The quantity of produce inside the sealed plastic film bag has been shown to affect the equilibrium gas content (Table 6.6), but the levels of CO_2 and O_2 do not always follow what would be predicted from permeability data and respiration load of the crop. Zagory (1990) showed that there was a negative linear relationship between weight of fresh chillies in a sealed Cryovac SSD-310 film package between CO_2 and O_2 levels and CO_2 levels were reduced with increasing product in the bag while O_2 levels were proportionately higher.

The number of fruit packed in each plastic bag can alter the effect of modified atmosphere packaging. An example of this is that plantains packed with six fruits in a bag ripened in 14.6 days compared to 18.5 days when fruits were packed individually when stored at 26–34°C (Thompson *et al.*, 1972).

Zagory (1990) also demonstrated the relationship between the weight of produce in a package (20 × 30 cm) and its O_2 and CO_2 content for one type of plastic film. The relationship between O_2 and CO_2 levels was linear

Table 6.6. Effects of the amount of asparagus spears sealed inside different film type on the equilibrium CO_2 and O_2 content at 20°C (Lill and Corrigan, 1996).

Plastic film type	Permeability at 20°C in ml m^{-3} atm^{-1} day^{-1}	Product load (g)	% CO_2	% O_2
W R Grace RD 106	23,200 CO_2, 10,200 O_2	100	2.5	9.7
W R Grace RD 106	23,200 CO_2, 10,200 O_2	150	3.2	6.2
W R Grace RD 106	23,200 CO_2, 10,200 O_2	200	3.5	4.1
Van Leer Packaging Ltd	33,200 CO_2, 13,300 O_2	100	3.6	4.5
Van Leer Packaging Ltd	33,200 CO_2, 13,300 O_2	200	4.2	3.6
Van Leer Packaging Ltd	33,200 CO_2, 13,300 O_2	250	4.6	2.0
Chequer Systems Ltd	45,300 CO_2, 16,400 O_2	100	3.4	15.4
Chequer Systems Ltd	45,300 CO_2, 16,400 O_2	250	3.9	16.4
Chequer Systems Ltd	45,300 CO_2, 16,400 O_2	3000	6.1	11.3

(Fig. 23), but varied considerably with a fourfold variation in produce weight, which illustrates the importance of varying just one factor.

PERFORATION

Punching holes in the plastic can maintain a high humidity around the produce, but it may be less effective in delaying fruit ripening (Table 6.2) because it does not have the same effect on the CO_2 and O_2 content of the atmosphere inside the bag. The holes may be very small and in these cases they are commonly referred to as microperforations.

Various studies have been conducted on the effects of packing various produce at different temperatures on the internal gas content. Fruits of the pear cultivars Okusankichi and Imamuraaki decayed most if placed in unsealed 0.05 mm plastic bags and least in sealed bags with five pinholes (bags with ten holes had more decay). Weight loss of fruit after 7 months of storage was 8–9% and less than 1% in unsealed and sealed bags, respectively. In sealed bags, CO_2 concentration reached 1.9% after 2 months of storage (Son *et al.*, 1983).

Hughes *et al.* (1981) showed that capsicums sealed in various plastic films had a higher percentage of marketable fruit than those stored in air (Table 2.4).

ABSORBENTS IN MODIFIED ATMOSPHERE PACKAGING

'Active packaging' of fresh produce has been carried out for many years. This system usually involves the inclusion of a desiccant or O_2 absorber within or as part of the packaging material. These are mainly used to control

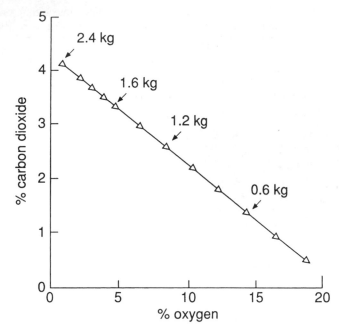

Fig. 23. Equilibrium gas concentration in a modified atmosphere package as a function of weight for Anaheim chilli pepper fruit packed in Cryovac SSD-310 film (Zagory, 1990).

insect damage, mould growth, rancidity and discoloration in a range of perishable food products such as meat, herbs, grains, beans, spices and dairy products. One such product is marketed as Ageless and uses iron reactions to absorb O_2 from the atmosphere (Abe, 1990).

Ascorbic acid based sachets (which also generate CO_2) and cathecol based sachets are also used as O_2 absorbers. The former are marketed as Ageless G or Toppan C or as Vitalon GMA (when combined with iron), and the latter as Tamotsu. Mineral powders are incorporated into some films for fresh produce, particularly in Japan (Table 6.7).

It was claimed that among other things these films can remove or absorb ethylene, excess CO_2 and water and be used for broccoli, cucumber, lotus root, kiwifruit, tomato, sweetcorn and cut flowers (Industrial Material Research 1989, quoted by Abe, 1990). Eun *et al.* (1997) described silver-coated ceramic coatings used on LDPE film for the storage of enoki mushrooms. Coated films extended the storage life of the mushrooms from 10 days for those stored in uncoated LDPE to 14 days for those in silver-coated ceramic films.

Ethylene absorbents can be included inside modified atmosphere packages and the following examples illustrate some of their uses. A major source of deterioration of limes during marketing is their rapid weight loss which can give the skins a hard texture and an unattractive appearance.

Packing limes in sealed polyethylene film bags inside cartons resulted in a weight loss of only 1.3% in 5 days, but all the fruits degreening more rapidly than those which were packed just in cartons where the weight loss was 13.8% (Thompson *et al.*, 1974b). However, this degreening effect could be countered by including an ethylene absorbent in the bags (Fig. 24).

Chemicals may be placed inside the bags along with the fruits or vegetables which can absorb gases. Scott and Wills (1974) described experiments in which compounds that absorb ethylene, CO_2 and water were packed with cultivar William's Bon Chrétien (Bartlett) pears held in sealed polyethylene bags. After storage at $-1°C$ for 18 weeks in air or in bags with a CO_2 absorbent (calcium hydroxide) fruit was in poor condition externally but free of brown heart. All other fruit was stored in the presence of about 5% CO_2 and was externally in excellent condition but was affected by brown heart. The ethylene absorbent potassium permanganate reduced the mean level of ethylene from 395 to 1.5 $\mu l \ l^{-1}$ and reduced brown heart from

Table 6.7. Commercially available films in Japan (Source: Abe, 1990).

Trade name	Manufacturer	Compound	Application
FH film	Thermo	Ohya stone/PE	Broccoli
BF film	BF distribution research	Coral sand/PE	Home
Shupack V	Asahi-kasei	Synthetic-zealite	–
Uniace	Idemitsu pet chem	Silicagel mineral	Sweetcorn
Nack fresh	Nippon unicar	Cristobalite	Broccoli and sweetcorn
Zeomic	Shinanen	Ag-zeolite/PE	–

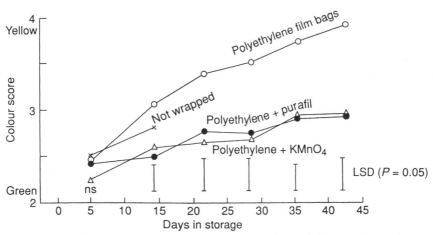

Fig. 24. Effects of modified atmosphere packaging in polyethylene film and an ethylene absorbent included within the pack on the rate of degreening of limes stored in tropical ambient conditions of 31–34°C and 29–57% rh (Thompson *et al.*, 1974).

68 to 36%. The presence of $CaCl_2$ tended to increase brown heart. The storage treatments also affected the level of six other organic volatiles in the fruit.

Proprietary products such as Ethysorb and Purafil are available which can be placed inside the crop store or even inside the actual package containing the crop. They are made by impregnating an active alumina carrier (Al_2O_3) with a saturated solution of potassium permanganate and then drying it. The carrier is usually formed into small granules, the smaller the granules the larger the surface area and therefore the quicker their absorbing characteristics. Any molecule of ethylene in the package atmosphere which comes into contact with the granule will be oxidized so the larger surface area is an advantage. The oxidizing reaction is not reversible and the granules change colour from purple to brown which indicates that they need replacing. Strop (1992) studied the effects of Ethysorb sachets in polyethylene film bags containing broccoli with and without Ethysorb. She found that the ethylene content in the bags after 10 days at 0°C was 0.423 ppm for those without Ethysorb and 0.198 for those with. The rate of ethylene removal from packages using this material is affected by humidity (Lidster et al., 1985). At the high humidities found in modified atmosphere packages the rate of ethylene removal of potassium permanganate was shown to be reduced.

Kiwifruit of the cultivar Bruno were stored at –1°C in sealed polyethylene bags of varying thickness, with and without an ethylene absorbent (Ethysorb). The gaseous composition of the modified atmospheres generated by the respiration of the fruit was dictated by the thickness of the polyethylene and the presence of the absorbent. The ripening rate of the fruit and its keeping quality during 6 months of storage and 7 days of subsequent shelf life at 20°C were compared with those of fruit stored in unsealed 0.02 mm polyethylene bags. Without ethylene removal from the storage atmosphere, there was no significant effect of the modified atmospheres in the sealed bags on the rate of fruit ripening and no improvement in its keeping quality. However, when ethylene was absorbed from the storage atmosphere, post-storage fruit ripening was retarded by increasing the thickness of the polyethylene bag and, as a result, the CO_2 level within it. This was indicated by a significant negative correlation between the CO_2 level in the bags and the increase in total soluble solids and a significantly positive correlation with fruit firmness. Retarded fruit softening resulted in improved keeping quality, and the storage life was extended to 6 months by storing the fruit at –1°C in sealed 0.04–0.05 mm thick polyethylene bags containing the absorbent. The average composition of the atmosphere in these bags was 3–4% CO_2, 15–16% O_2 and less than 0.01 μl l^{-1} C_2H_4 (Ben Arie and Sonego, 1985).

Carbon dioxide absorbents or a permeable window can be used with this technique to prevent atmospheres developing which could damage the crop. The concentration of the various respiratory gases, O_2, CO_2 and ethylene, within the crop is governed by various factors including the gas exchange equation which is a modification of Fick's law:

$$-ds/dt = (C_{in} - C_{out})DR$$

where $-ds/dt$ = rate gas transport out of the crop, C_{in} = the concentration of gas within the crop, C_{out} = the concentration of gas outside the crop, D = gaseous diffusion coefficient in air, and R = a constant specific to that particular crop.

Companies are marketing films which they claim can greatly increase the storage life of packed fruit and vegetables. These are marketed with names such as Maxifresh, Gelpack and Xtend.

Where an ethylene absorbent (potassium permanganate) was included in the tube the increase in storage life was three to four times compared to unwrapped fruit. Fruits sealed in polyethylene tubes with ethylene absorbents could be stored for 6 weeks at 20°C or 28°C and 16 weeks at 13°C (Satyan et al., 1992).

The investigations on the effects of modified atmosphere conditions were taken by Ali Azizan (1988 quoted by Abdullah and Pantastico, 1990). He observed that the total soluble solids, total titratable acidity and pH in Pisang Mas bananas did not change during storage in modified atmospheres at ambient temperature. However, Wills (1990) mentioned an unsuitable selection in packaging materials can still accelerate the ripening of fruits or enhance CO_2 injury when the ethylene accumulated over a certain period.

POLYETHYLENE BAGS WITH AN ADJUSTABLE DIFFUSION LEAK

A simple method of controlled atmosphere storage was developed for strawberries (Anon, 1920). The gaseous atmospheres containing reduced amounts of O_2 and moderate amounts of CO_2 were obtained by keeping the fruit in a closed vessel fitted with an adjustable diffusion leak.

Marcellin (1973) described the use of polyethylene bags with silicone rubber panels which allow a certain amount of gas exchange which have been used for storing vegetables. Good atmosphere control within the bags was obtained for globe artichokes and asparagus at 0°C and green peppers at 12–13°C when the optimum size of the bag and of the silicon gas-exchange panel had been determined. Gas exchange for carrots and turnips at 1–2°C was hampered by condensation on the inner surface of the panel. The storage life of all the vegetables was extended compared with storage in air, but losses due to rotting were high for green peppers which seemed unsuitable for this type of storage. Rukavishnikov et al. (1984) described controlled atmosphere storage under a 150–300 μ polyethylene film, with windows of membranes selective for O_2 and CO_2 permeability made of polyvinyl trimethyl silane or silicon organic polymers. Apples and pears stored at 1–4°C under the covers had 93 and 94% sound fruit, respectively, after 6–7 months, whereas in ordinary storage at similar temperatures over 50% losses occurred after 5 months.

Hong *et al.* (1983) described the storage behaviour of fruits of *Citrus unshiu* at 3–8°C in plastic films with or without a silicone window. After 110–115 days 80.8–81.9% of fruits stored with the silicone window were healthy with good coloration of the peel and excellent flavour, while those without a silicone window were 59.4–76.8% with poor coloration of the peel and poor flavour. Controls stored for 90 days had 67% healthy fruit with poor quality and shrivelling of the peel, calyx browning and a high rate of moisture loss. The size of the silicone window was 20–25 cm^2 kg^{-1} fruit giving less than 3% CO_2 with > 10% O_2.

The suitability of the silicone membrane system for controlled atmosphere storage of Winter Green cabbage was studied by Gariepy *et al.* (1984) using small experimental chambers. Three different controlled atmosphere starting techniques were then evaluated using a cabbage cultivar imported from Florida. In each case, there was a control with the product stored under regular atmosphere at the same temperature. The silicone membrane system maintained controlled atmosphere storage conditions of 3.5–5% CO_2 and 1.5–3% O_2, where 5 and 3%, respectively, were expected. After 198 days of storage, total mass loss was 14% under controlled atmosphere storage compared to 40% under regular atmosphere. The three methods used to achieve the controlled atmosphere storage conditions did not have any significant effect on the storability of cabbage, but all maintained 3–4% CO_2, and 2–3% O_2, while 5 and 3%, respectively, were expected. In both experiments, cabbage stored under controlled atmosphere storage showed better retention of colour, fresher appearance, firmer texture and lower total mass losses compared to those stored under regular atmosphere.

Raghavan *et al.* (1982) used a silicone membrane system for storing carrots, celery, rutabagas and cabbage and this was evaluated. Carrots were stored for 52 weeks and celery, rutabagas and cabbage for 16 weeks. The required higher CO_2 levels were achieved within the membrane system compared to normal air composition in a standard cold room. Design calculations for selection of the membrane area are also presented in the paper.

SAFETY

This subject has recently been dealt with in detail in a review by Church and Parsons (1995). There is considerable legislation related to food sold for human consumption. In the UK for example, the Food Safety Act 1990 states that it is an offence to sell or supply food for human consumption if it does not meet food safety requirements. The use of modified atmosphere packaging has health and safety implications. One factor that should be taken into account is that the gases in the atmosphere could possibly have a stimulating effect on microorganisms. Farber (1991) stated that while modified atmosphere packaged foods have become increasingly more common in North America, research on the microbiological safety of these foods was still lacking. The growth of aerobic microorganisms is generally optimum at about 21% O_2 and falls off sharply with reduced O_2 levels while generally for anaerobic microorganisms their optimum growth is at 0% O_2 and falls as the O_2 level increases (Day, 1996). With many modified atmospheres containing increased levels of CO_2, the aerobic spoilage organisms, which usually warn consumers of spoilage, are inhibited, while the growth of pathogens may be allowed or even stimulated which raises safety issues (Farber, 1991). Hotchkiss and Banco (1992) stated that extending shelf life of refrigerated foods might increase microbial risks in modified atmosphere packaged produce in at least three ways:

• increasing the time in which food remains edible increases the time in which even slow growing pathogens can develop or produce toxin,
• retarding the development of competing spoilage organisms,
• packaging of respiring produce could alter the atmosphere so that pathogen growth is stimulated.

The ability of *Escherichia coli* 0157:H7 to grow on raw salad vegetables which have been subjected to processing and storage conditions simulating those routinely used in commercial practice has been demonstrated by Abdul Raouf *et al.* (1993). The influence of modified atmosphere packaging, storage temperature, and time on survival and growth of *E. coli* 0157:H7 inoculated onto shredded lettuce, sliced cucumber and shredded carrot was determined. Packaging in an atmosphere containing 3% O_2 and 97% nitrogen had no apparent effect on populations of *E. coli* 0157:H7, psychrotrophs or mesophiles. Populations of viable *E. coli* 0157:H7 declined on vegetables stored at 5°C and increased on vegetables stored at 12 and 21°C for up to 14 days. The most rapid increases in populations of *E. coli* 0157:H7 occurred on lettuce and cucumbers stored at 21°C. These results suggest that an unknown factor or factors associated with carrots may inhibit the growth of *E. coli* 0157:H7. The reduction in pH of vegetables was correlated with initial increases in populations of *E. coli* 0157:H7 and naturally occurring microfloras. Eventual decreases in *E. coli* 0157:H7 in samples stored at 21°C, were attributed to the toxic effect of accumulated acids. Changes in visual appearance of vegetables were not influenced substantially by growth of *E. coli* 0157:H7.

Church and Parsons (1995) mentioned that there was a theoretical potentially fatal toxigenesis through infections by *Clostridium botulinum* in the depleted O_2 atmospheres in modified atmosphere packed fresh vegetables. It was claimed that this toxigenesis has not been demonstrated in vegetable products without some sensory indication (Zagory and Kader, 1988 quoted by Church and Parsons, 1995). Roy *et al.* (1995) showed that the optimum in-package O_2 concentration for suppressing cap opening of fresh mushrooms was 6% and that lower O_2 concentrations in storage are not recommended because they could promote growth and toxin production by *C. botulinum*. Betts (1996) in a review of hazards related to modified atmosphere packaging of food indicated that vacuum packing of shredded lettuce had been implicated in a botulinum poisoning outbreak.

It is not part of this work to discuss the effects of controlled atmosphere storage or modified atmosphere packaging on meat and fish. However, there is considerable literature on the subject; including work on safety aspects. A couple of fairly recent examples are cited here. Gill and DeLacy (1991) described experiments where a strain each of *Escherichia coli* and *Salmonella typhimurium* were inoculated onto samples of high pH (> 6.0) beef. Samples were packaged under vacuum or CO_2 and stored at 8, 10, 12, 15, 20 or 30°C. In vacuum packs, *E. coli* and *S. typhimurium* grew at all storage temperatures. At temperatures between 8 and 12°C inclusive, both organisms grew at rates less or no more than those of the spoilage flora after significant lag periods. At temperatures of 15°C or above, growth rates were equal to or greater than those of the spoilage flora, and lag periods were insignificant. In CO_2 packs, neither organism grew at 82°C and *S. typhimurium* did not grow at 10°C. Subsequent tests showed that *E. coli* did not grow at 9°C, nor *S. typhimurium* at 11°C. At 12°C, both organisms grew, after prolonged lags, at rates markedly slower than that of the spoilage flora. At 15°C, their growth rates were similar to that of the spoilage flora. At higher temperatures, both organisms grew without significant lags at rates greater than that of the spoilage flora. Gill and Reichel (1989) experimented with strains of the cold tolerant pathogens *Yersinia enterocolitica*, *Aeromonas hydrophila* and *Listeria monocytogenes* which were inoculated onto samples of beef at pH above 6.0. Samples were packaged under vacuum or under CO_2 and stored at −2, 0, 2, 5 or 10°C. In vacuum packs, *Y. enterocolitica* and *A. hydrophila* grew at all storage temperatures at rates similar to or faster than those of the spoilage flora. The lags before growth of *Y. enterocolitica* at 0, 2 and 5°C, and of *A. hydrophila* at 10°C were not significant relative to the proliferation of the spoilage flora. *Listeria monocytogenes* did not grow in vacuum pack at −2°C, and at higher temperatures generally grew at rates less than that of the spoilage flora after relatively long lag periods. In CO_2 packs, all organisms grew at 10°C, but only *Y. enterocolitica* grew at 5°C and then slowly after a prolonged lag. No organisms grew in CO_2 packs at 2°C or at lower temperatures.

At 25°C the Gram-positive bacteria isolated from black drum fish were more resistant to CO_2 than the Gram-negative organisms, while at 4°C none of the organisms grew in 25% CO_2, the lowest concentration tested. When exposed to air after being incubated in CO_2 enriched environments of 25–100%, the organisms in most instances grew at normal rates indicating limited residual effect of CO_2. The effect of temperature on relative CO_2 inhibition was investigated in detail for *Micrococcus* spp. and *Pseudomonas fluorescens*. In an atmosphere containing 25% CO_2 in air at 20°C both *Pseudomonas fluorescens* and *Micrococcus* spp. showed approximately 25%

inhibition as compared with growth in air at the same temperature while at 10°C *P. fluorescens* was completely inhibited and *Micrococcus* spp. showed 95% inhibition (Lannelongue and Finne, 1986). Channel catfish (*Ictalurus punctatus*) fillet strips were packaged air, 25% CO_2 and 80% CO_2 environments and stored at 2 and 8°C for 4 weeks. *Listeria monocytogenes* was isolated from fish under 25% CO_2 and air at 8°C in the second and third weeks. *Salmonella* spp. was present in fish stored in all storage atmospheres at 8°C, but only in fish packed in air at 2°C. The best treatment was fish packed in 80% CO_2 and stored at 2°C (Silva and White, 1994).

There are other possible dangers in using high concentrations of CO_2 and O_2 in storage. Low O_2 and high CO_2 can have a direct lethal effect on human beings working in those atmospheres. Great care needs to be taken when using modified atmosphere packaging containing 70 or 80% O_2, or even higher levels, because of potential explosions.

Another safety issue is the possibility of the films being used in modified atmosphere packaging being toxic. Schlimme and Rooney (1994) reviewed the possibility of constituents of the polymeric film used in modified atmosphere packaging migrating to the food that they contain. They showed that it is unlikely that the polymerized constituents would be transferred to the food because of their high molecular weight and insolubility in water. All films used can contain some non-polymerized constituents which could be transferred to the food and in the USA the Food and Drugs Administration and also the European Community have regulations related to these 'indirect additives'. The film manufacturer must therefore establish the toxicity and extraction behaviour of the constituents with specified food simulants.

HUMIDITY

The gas permeability of some plastics used for film packaging are sensitive to environmental humidity. Roberts (1990) showed that gas transmission of polyamides (nylons) can increase by about three times when the relative humidity is increased from 0 to 100% and with ethyl vinyl alcohol copolymers the increase can be as high as 100 times over the same range.

Moisture given out by the produce can condense on the inside of the pack. This is especially a problem where there are large fluctuations in external temperatures because the humidity is high within the pack and easily reaches dew-point where the film surface is cooler than the pack air. Antifogging chemicals can be added during the manufacture of plastic films. These do not affect the quantity of moisture inside the packs, but it causes the moisture which has condensed on the inside of the pack to form sheets rather than discrete drops. They can eventually form puddles at the bottom of the pack.

The high humidity provided by individual film wrapping was beneficial particularly to crops such as citrus fruits and cauliflowers. Zong (1989) showed that water stress (in non-wrapped samples or film wrapped samples containing silica gel desiccant) initiated metabolic changes in citrus fruits, but resulted only in desiccation with cauliflowers. Individual film wrapping delayed broccoli senescence by altering the pack atmosphere (O_2 and CO_2 levels). Individual film wrapping delayed senescence of cucumbers and Chinese cabbage by increasing humidity and altering the pack atmosphere. The results indicate that physiological responses to individual film wrapping vary with commodity, and that the respiratory pattern of the commodity cannot be used as a basis for predicting commodity responses to individual film wrapping.

INCREASE IN POSTHARVEST LIFE

Partially ripe tomatoes, packed in suitably permeable plastic film giving 4–6% CO_2 and 4–6% O_2 with about 90% rh, can be expected to have 7 days longer shelf life at ambient temperature than those stored without wrapping (Geeson, 1989). Under these conditions the eating quality should not be impaired but film packaging that results in higher CO_2 and lower O_2 levels may prevent ripening and result in tainted fruit when the fruit are ripened after removal from the packs.

Modified atmosphere storage in polyethylene bags has been shown to extend the storage life of bananas (Scott *et al.*, 1971). Their storage life was therefore extended for 6 days compared to non-wrapped fruit at 20°C. In a subsequent trial, bananas were packed in 13.6 kg commercial packs inside polyethylene film bags. During 48 h transport in Australia the fruit could be kept in good conditions at ambient temperatures and then be held in a commercial ripening room for several days, and still had a low weight loss, good flavour and nice appearance when subsequently ripened (Scott *et al.*, 1971). Satyan *et al.* (1992) stored Williams bananas in 0.1 mm thick polyethylene tubes at 13, 20 and 28°C and found that their storage life was increased two to three times compared to unwrapped fruit.

Latundan bananas (*Musa* AAB) can be stored in 0.08 mm thick polyethylene bags at ambient temperature of 26–30°C for up to 13 days and Lakatan bananas (*Musa* AA) on the other hand, shown a lag of 3 days before colour break down upon exposure to air (Abdullah and Tirtosoekotjo, 1989; Abdullah and Pantastico, 1990). Tiangco *et al.* (1987) also observed that green life of Saba bananas (*Musa* BBB) held in modified atmosphere conditions at ambient temperature had a considerable extension of storage life. The pre-climacteric period of bananas can be further extended by combining both modified atmosphere and refrigeration. Tongdee (1988) found that the green life of Kluai Khai banana fruits (*Musa* AA) could be maintained for more than 45 days in polyethylene bags at 13°C. This was longer by more than 20 days compared to the fruits stored at 25°C under the same treatment. A more practical method was to wrap the fruits in polyethylene film, which reduced fruit weight loss by evapotranspiration and slowed down banana respiration rate, by modifying the atmosphere in the packs, this allowed fruits to be stored for several weeks (Marchal and Nolin, 1990).

Plantains stored in modified atmospheres had a considerably longer postharvest life than those stored unwrapped (Fig. 25). Plantains stored in polyethylene which had holes in them ripened at the same speed as unwrapped fruit although the speed of ripening was more variable in fruits packed in perforated polyethylene (Thompson *et al.*, 1972).

VACUUM PACKING

Nair and Tung (1988) reported that Pisang Mas had an extension of 4–6 weeks at 17°C when it was stored in evacuated collapsed polyethylene bags by applying vacuum not exceeding 300 mmHg.

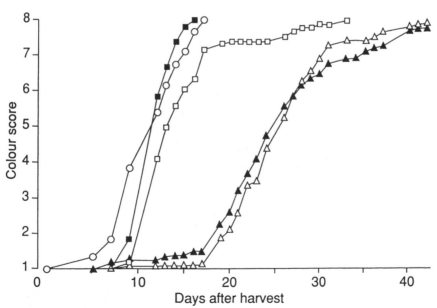

Fig. 25. Effects of plastic film wraps on the ripening of plantain fruits as measured by changes in skin colour during storage in tropical ambient conditions of some 20°C (Thompson *et al.*, 1974). ○, Unwrapped; △, polyethylene; ▲, polyethylene (evacuated); □, perforated polyethylene (individual fruits); ■, perforated polyethylene (six fruits).

CONCLUSION

Modified atmosphere packaging, using various plastic films, has been known for several decades to have great potential in extending the post-harvest life of fruit and vegetables, but it has never reached its full potential. It can be clearly demonstrated experimentally that it can have similar effects to controlled atmosphere storage on the postharvest life of crops and the mechanism of the effects is doubtless the same. Its limited uptake by the industry probably reflects the limited control of the gases around and within the fruit and vegetables achieved by this technology. This can result in unpredictable effects on postharvest life and quality of the commodity packaged in this way. Some recent concerns about the safety of products in modified atmosphere packaging also militate against its use.

Chapter 7

Recommended Controlled Atmosphere Storage Conditions for Selected Crops

Recommended storage conditions, especially where a storage period is included, may well be referring to the maximum time that the fruit or vegetable could possibly be stored. They usually include the assumption that the crop is in excellent condition before being placed in storage and may also assume that pre-storage treatments such as fungicides and precooling have been applied and that there is no delay between harvesting the crop and placing it in storage. In terms of climacteric fruits there would also be an assumption that the fruit had been harvested at an appropriate stage of maturity.

Apple, *Malus domestica*

Various techniques are used in commercial storage of apples, one of which can be described in the following example. In a commercial store in the UK (East Kent Packers, 1994, personal communication) where ultra low O_2 storage is used, controlled atmospheres are achieved by sealing the store directly after loading and reducing the temperature to below 10°C, bags of hydrated lime having been placed into the store prior to sealing as a method of removing respiratory CO_2. The temperature is then reduced further and when it is below 5°C, nitrogen is injected to reduce the O_2 level to about 3% for Cox's Orange Pippin. The O_2 level is allowed to decline to about 2% for 7 days through fruit respiration and then progressively to 1.2% over the next 7 days.

The recommended storage conditions for apples varies between cultivars, country of production and the year in which the research was carried out. This latter factor reflects the available technology for controlling the gas

117

concentrations in the store and recommendations have changed with time. It is currently possible to have precise control of gas concentrations within the store due to accurate gas analysis and the use of computers to control store conditions, which were not available when much of the research into controlled atmosphere storage conditions was carried out. Burton (1974a) pointed out that the differences in the O_2 concentrations recommended for different varieties of apple are not readily explicable. Varietal differences in susceptibility to CO_2 injury could possibly result from anatomical, rather than biochemical differences. Variability of plant material prevents precise control of intercellular atmosphere; recommended atmospheres can be designed only to avoid completely anaerobic conditions and a harmful level of CO_2 in the centre of the least permeable individual fruit or vegetable.

General storage recommendations include 1–2% CO_2 and 2–3% O_2 for non-chilling sensitive cultivars and 2–3% CO_2 with 2–3% O_2 for chilling sensitive cultivars (SeaLand, 1991). Lawton (1996) gave a general recommendation of 0–2°C, 95% relative humidity (rh) with 1–5% CO_2 and 2–3% O_2. Blythman (1996) mentioned that controlled atmosphere storage of apples was carried out in 2% CO_2 with 3% O_2 and 95% nitrogen. Another general recommendation was storage at 0–2°C with 1–2% CO_2 and 2–3% O_2 (Kader, 1985). However, Kader (1992) revised these recommendations to 1–5% CO_2 with 1–3% O_2. In northern Italy there was a general recommendation of 1–2% O_2 (Chapon and Trillot, 1992). Recommendations for specific cultivars have been given in the following.

Allington Pippin

Fidler (1970) recommended 3–3.5°C with 3% CO_2 and 3% O_2.

Aroma

Storage trials of Aroma apples at 2°C and 88% rh in air, or 2% O_2 with 2% CO_2 or 3% O_2 with 3% CO_2 for 115, 142 and 169 days were described by Haffner (1993). One set of samples analysed immediately and a second set of samples analysed after 1 week in a cold store and a third set after 1 week in a cold store followed by 1 week in ambient conditions of 18–20°C, 25–30% rh to simulate shelf life. Weight loss was 0.6% per month for controlled atmosphere storage compared with 1.5% in cold storage. Controlled atmosphere storage controlled *Gloeosporium album* (*Pezicula alba*) and *G. perennans* (*P. malicorticis*). Fruits undergoing simulated shelf life experiments showed a severe loss in quality after storage with damage from *Gloeosporium* rot, wilting and over-ripeness. Controlled atmosphere storage delayed the maturity process, controlled weight loss and resulted in better eating quality during shelf life.

Boskoop

Meheriuk (1993) recommended 3–5°C with 0.5–2% CO_2 and 1.2–2.3% O_2 for 5–7 months. Koelet (1992) advocated 3–3.5°C with 0.5–1.5% CO_2 and 2–2.2% O_2. Hansen (1977) suggested a temperature of not lower than 4°C with 3% CO_2 and 3% O_2. Schaik (1994) described experiments where fruits of the cultivar Schone van Boskoop (also called Belle de Boskoop) from several orchards were held in controlled atmosphere storage with a combined scrubber/separator or with a traditional lime scrubber, in 0.7% CO_2 and 1.2% O_2, at 4.5°C, followed by holding at 15°C for 1 week or at 20°C for 2 weeks. Ethylene concentration during storage with a scrubber/separator rose to only 20 ppm, compared with up to 1000 ppm in a cell without a scrubber. The average weight loss, rots (mostly *Gloeosporium* spp.), flesh browning and core flush were also slightly lower in the scrubber/separator cell, but the incidence of bitter pit was slightly greater. Using a scrubber/separator did not affect fruit firmness and colour. M. Herregods (1993, personal communication) and Meheriuk (1993) summarized general recommendations from a variety of countries as follows:

	°C	CO_2	O_2
Belgium	3–3.5	< 1	2–2.2
Denmark	3.5	1.5–2	3–3.5
France	3–5	0.5–0.8	1.5
Germany (Saxony)	3	1.1–1.3	1.3–1.5
Germany (Westphalia)	4–5	2	2
Holland	4–5	Much less than 1	1.2
Switzerland	4	2–3	2–3

Bowden's Seedling

Fidler (1970) recommended 3.5°C with 5% CO_2 and 3% O_2.

Bramley's Seedling

Fidler and Mann (1972) recommended 3–4°C with 8–10% CO_2 and about 11–13% O_2. Wilkinson and Sharples (1973) recommended 3.3–4.4°C with 8–10% CO_2 and no scrubber for 8 months. Sharples and Stow (1986) recommended 4–4.5°C with 8–10% CO_2 where no scrubber is used and 4–4.5°C with 6% CO_2 and 2% O_2 where a scrubber is used. The following conditions were said to be used in their commercial controlled atmosphere stores by East Kent Packers Limited: 3.5–4°C with 5% CO_2 and 1% O_2 (1994, personal communication). Johnson (1994 and personal communication 1997) recommended the following:

Temperature °C	CO_2	O_2	Storage time in weeks
4–4.5	8–10	11–13	39
4–4.5	6	2	39
4–4.5	5	1	44

Cacanska Pozna

Jankovic and Drobnjak (1994) stored Cacanska Pozna at 1°C and 85–90% rh, either in a controlled atmosphere (7% CO_2 with 7% O_2) or under normal atmospheric conditions. Weight losses and decay were negligible in controlled atmospheres (0–0.78%). Fruits stored in controlled atmosphere stores exhibited no physiological disorders.

Charles Ross

Fidler (1970) recommended 3.5°C with 5% CO_2 and 3% O_2. Sharples and Stow (1986) recommended 3.5–4°C and 5% CO_2 for fruit stored without a scrubber and 5% CO_2 with 3% O_2 for those stored with a scrubber.

Cortland

Anon (1968) recommended 3.3°C with 2% then 5% CO_2 and 3% O_2 for 5–6 months. Meheriuk (1993) recommended 0–3°C with 5% CO_2 and 2–3.5% O_2 for 4–6 months. M. Herregods (1993, personal communication) and Meheriuk (1993) summarized general recommendations from a variety of countries as follows:

	Temperature °C	CO_2	O_2
Canada (Nova Scotia)	3	4.5	2.5
Canada (Nova Scotia)	3	1.5	1.5
Canada (Quebec)	3	5	2.5–3
USA (Mass.)	0	5	3

Cox's Orange Pippin

Fidler and Mann (1972) and Wilkinson and Sharples (1973) recommended 3.3–3.9°C with 5% CO_2 and 3% O_2 for 5 months, but where core flush is a problem then 0–1% CO_2 with 1.8–2.5% O_2 for 6 months was recommended.

Stoll (1972) suggested 4°C with 1–2% CO_2 and 3% O_2. Sharples and Stow (1986) advocated 3.5–4°C with 5% CO_2 where no scrubber is used and 3.5–4°C with 5% CO_2 and 3% O_2 (for storage to mid-February), less than 1% CO_2 with 2% O_2 (for storage to late March), less than 1% CO_2 with 1.25% O_2 (for storage to late April) and less than 1% CO_2 with 1% O_2 (for storage to early May) where a scrubber is used. East Kent Packers Limited used 3.5°C with 1% CO_2 and 1.2% O_2 for Cox's Orange Pippin in their commercial stores (1994, personal communication). Koelet (1992) recommended 3–3.5°C with 0.5–1.5% CO_2 and 2–2.2% O_2. Meheriuk (1993) recommended 3–4.5°C with much less than 1–2% CO_2 (*sic*) and 1–3.5% O_2 for 5–6 months. Johnson (1994, and personal communication 1997) recommended the following:

Temperature °C	CO_2	O_2	Storage time in weeks
3.5–4	5	3	20
3.5–4	< 1	2	23
3.5–4	< 1	1.2	28
3.5–4	< 1	1	30

M. Herregods (1993, personal communication) and Meheriuk (1993) summarized general recommendations from a variety of countries as follows:

	Temperature °C	CO_2	O_2
Belgium	3–3.5	< 1	2–2.2
Denmark	3.5	1.5–2	2.5–3.5
England	4–4.5	< 1	1.25
France	3–4	< 1	1.2–1.5
Germany (Saxony)	3	1.7–1.9	1.3–1.5
Germany (Westphalia)	4	2	1–2
Holland	4	Much less than 1	1.2
New Zealand	3	2	2
Switzerland	4	2–3	2–3

Crispin

Sharples and Stow (1986) recommended 3.5–4°C with 8% CO_2 where no scrubber is used, but that there was a risk of the physiological disorder of core flush at 8% CO_2.

Delicious

See also the sections on Red Delicious, Starking and Golden Delicious. Drake (1993) described the effects of a 10-day delay in harvesting date and/or a 5-day, 10-day or 15-day delay in the start of controlled atmosphere storage (1% O_2 and 1% CO_2) on Delicious fruits (Bisbee, Red Chief and Oregon Spur strains). Delayed harvest increased red colour of the skin at harvest in Bisbee and Red Chief, but not Oregon Spur; soluble solids content and size also increased, but up to 12% of firmness was lost, depending on strain. Immediate establishment of controlled atmosphere conditions after harvest resulted in good quality fruits after 9 months of storage, but reduced quality was evident when controlled atmosphere establishment was delayed by 5 days (the interim period being spent in refrigerated storage at 1°C). Longer delays did not result in greater quality loss. Oregon Spur had more red colour at harvest and after storage than the other two strains. Sensory panel profiles were unable to distinguish between strains, harvest dates or delays in the time of controlled atmosphere establishment.

Mattheis *et al.* (1991) stored Delicious fruit in 0.5% O_2 plus 0.2% CO_2 at 1°C for 30 days and found that they developed high concentrations of ethanol and acetaldehyde. Scald susceptibility in Delicious apples was found to be strain dependent. While storage in 0.7% O_2 effectively reduced scald in Starking and Harrold Red fruits picked over a wide range of maturity stages, it did not adequately reduce scald in Starkrimson fruits after 8 months of storage (Lau and Yastremski, 1993).

Discovery

Discovery has been reported to have poor storage qualities and Sharples and Stow (1986) recommended 3–3.5°C with less than 1% CO_2 and 2% O_2 for 7 weeks.

Edward VII

Fidler (1970) recommended 3.5–4°C with 8–10% CO_2 with information supplied only for stores with no scrubber and Wilkinson and Sharples (1973) recommended 3.3–4.4°C with 8–10% CO_2 and no scrubber for 8 months. Sharples and Stow (1986) also recommended 3.5–4°C and 8% CO_2 for fruit stored without a scrubber.

Egremont Russet

Wilkinson and Sharples (1973) recommended 3.3°C with 5% CO_2 and 3% O_2 for 5 months. Fidler (1970) recommended $3.3–4.4^\circ$C with 7–8% CO_2 with information supplied only for stores with no scrubber.

Ellison's Orange

Fidler (1970) and Sharples and Stow (1986) recommended $4–4.5^\circ$C and 5% CO_2 with 3% O_2 for fruit stored with a scrubber or less than 1% CO_2 with 2% O_2.

Elstar

Sharples and Stow (1986) recommended $1–1.5^\circ$C and less than 1% CO_2 with 2% O_2 for those stored with a scrubber. Meheriuk (1993) recommended $1–3^\circ$C with 1–3% CO_2 and 1.2–3.5% O_2 for 5–6 months. In a trial with Elstar apples stored in 1 or 3% O_2, the ethylene concentration in the lower O_2 concentration remained less than 1 ppm after 8 months. After holding at 20°C for 1 week fruit firmness declined only slightly and taste was considered good. In the higher O_2 concentration, however, the ethylene concentration rose rapidly to almost 100 ppm and fruit hardness declined markedly after holding at 20°C (Schaik and Van-Schaik, 1994). Elstar fruits were stored in controlled atmospheres of 1 or 5% O_2 plus 0.5% CO_2, or 3% O_2 plus 0.5 or 3.0% CO_2 at 1.5°C. An ethylene scrubber was used continuously in the low ethylene cabinets and at intervals in other cabinets to remove excess ethylene produced by the fruits. Fruit firmness was measured 1 week after removal from storage and holding at 20°C. Low O_2 concentrations in addition to low temperature further decreased ethylene sensitivity in apples. Increased CO_2 concentrations did not affect fruit firmness when fruits were stored under low ethylene but were clearly beneficial when ethylene was present in the storage atmosphere (Woltering *et al.*, 1994). Schouten (1997) compared storage at $1–2^\circ$C in either 2.5% CO_2 with 1.2% O_2 or less than 0.5% CO_2 with 0.3–0.7% O_2. The fruit stored in 0.5% CO_2 with 0.3–0.7% O_2 were shown to be of better quality both directly after storage and after a 10 day shelf life period in air at 18°C. M. Herregods (1993, personal communication) and Meheriuk (1993) summarized general recommendations from a variety of countries as follows:

	Temperature °C	CO_2	O_2
Belgium	1	2	2
Canada (Nova Scotia)	0	4.5	2.5
Canada (Nova Scotia)	0	1.5	1.5
Denmark	2–3	2.5–3	3–3.5
England	1–1.5	< 1	2
France	1	1–2	1.5
Germany (Saxony)	2	1.7–1.9	1.3–1.5
Germany (Westphalia)	2–3	3	2
Holland	1–2	2.5	1.2
Slovenia	1	3	1.5

Empire

Meheriuk (1993) recommended 0–2°C with 1–2.5% CO_2 and 1–2.5% O_2 for 5–7 months. M. Herregods (1993, personal communication) and Meheriuk (1993) summarized general recommendations from a variety of countries as follows:

	Temperature °C	CO_2	O_2
Canada (British Columbia)	0	1.5	1.5
Canada (Ontario)	0	2.5	2.5
Canada (Ontario)	0	1	1
USA (Michigan)	0	3	1.5
USA (New York)	0	2–2.5	1.8–2
USA (Pennsylvania)	−0.5–+0.5	0–2.5	1.3–1.5

Fiesta

Sharples and Stow (1986) recommended 1–1.5°C and less than 1% CO_2 with 2% O_2 for those stored with a scrubber.

Fortune

Fidler (1970) recommended 3.5–4°C with 7–8% CO_2 for stores with no scrubber or either 5% CO_2 and 3% O_2 or less than 1% CO_2 and 2% O_2 for stores fitted with a scrubber.

Fuji

Meheriuk (1993) recommended 0–2°C with 0.7–2% CO_2 and 1–2.5% O_2 for 7–8 months. Fuji were stored at 0°C in a controlled atmosphere (1.5% O_2, with less than 0.5% CO_2). Fruits were removed from cold storage after 3, 5, 7 and 9 months and assessed for storage life and sensory acceptability on return to air at 20°C after 0, 4, 11, 18 and 25 days. Harvest maturity was not critical for Fuji apples, which maintained a good level of consumer acceptability for up to 9 months in storage and 25 days at 20°C in air (Jobling et al., 1993). Work in Australia showed that fruits store well in 2% O_2 and 1% CO_2 at 0°C (Tugwell and Chvyl, 1995). M. Herregods (1993, personal communication) and Meheriuk (1993) summarized general recommendations from a variety of countries as follows:

	Temperature °C	CO_2	O_2
Australia (South)	0	1	2
Australia (Victoria)	0	2	2–2.5
Brazil	1.5–2	0.7–1.2	1.5–2
France	0–1	1–2	2–2.5
Japan	0	1	2
USA (Washington)	0	1–2	1–2

Gala

Gala were stored at 0°C in 1.5% O_2 with less than 0.5% CO_2 for up to 9 months (Jobling et al., 1993). Johnson (1994) recommended 3.5–4°C in less than 1% CO_2 with 2% O_2 for 23 weeks. However, more recent work has shown that Gala grown in the UK can be stored at 1.5°C in O_2 concentrations as low as 1% with CO_2 concentrations of 2.5–5% (Stow, 1996). Meheriuk (1993) recommended –0.5–3°C with 1–5% CO_2 and 1–2.5% O_2 for 5–6 months. Work in Australia showed that fruits of Royal Gala can be stored for up to 5 months in 2% O_2 and 1% CO_2 at 0°C (Tugwell and Chvyl, 1995), but Stow (1996) in the UK showed that there was a large loss in flavour after 3 months storage. M. Herregods (1993, personal communication) and Meheriuk (1993) summarized general recommendations from a variety of countries as follows:

	Temperature °C	CO_2	O_2
Australia (South)	0	1	2
Australia (Victoria)	0	1	1.5–2
Brazil	1–2	2.5–3	1.6–2
Canada (British Columbia)	0	1.5	1.2
Canada (Nova Scotia)	0	4.5	2.5
Canada (Nova Scotia)	0	1.5	1.5
Canada (Ontario)	0	2.5	2.5
France	0–2	1.5–2	1–2
Germany (Westphalia)	1–2	3–5	1–2
Holland	1	1	1.2
Israel	0	2	0.8–1
New Zealand	0.5	2	2
South Africa	–0.5	2	2
Switzerland	0	2	2
USA (Washington)	0	1–2	1–2

Gloster

Sharples and Stow (1986) recommended for Gloster 69, 1.5–2°C with less than 1% CO_2 and 2% O_2. Hansen (1977) advocated 0–1°C with 3% CO_2 and 2.5% O_2. Meheriuk (1993) suggested 0.5–2°C with 1–3% CO_2 and 1–3% O_2 for 6–8 months. M. Herregods (1993, personal communication) and Meheriuk (1993) summarized general recommendations from a variety of countries as follows:

	Temperature °C	CO_2	O_2
Belgium	0.8	2	2
Canada (Nova Scotia)	0	4.5	2.5
Canada (Nova Scotia)	0	1.5	1.5
Denmark	0	2.5–3	3–3.5
France	1–2	1–1.5	1.5
Germany (Saxony)	2	1.7–1.9	1.3–1.5
Germany (Westphalia)	1–2	2	2
Holland	1	3	1.2
Slovenia	1	3	1
Slovenia	1	3	3
Spain	0.5	2	3
Switzerland	2–4	3–4	2–3

Golden Delicious

Stoll (1972) recommended 2.5°C with 5% CO_2 and 3% O_2. Koelet (1992) recommended 0.5–1°C with 1–3% CO_2 and 2 to 2.2% O_2. Meheriuk (1993) recommended –0.5–2°C with 1.5–4% CO_2 and 1–2.5% O_2 for 8–10 months. Golden Delicious were stored at 3% CO_2 with 21% O_2, 3% CO_2 with 3% O_2, 3% CO_2 with 1% O_2, 1% CO_2 with 3% O_2, 1% CO_2 with 1% O_2 and 1% CO_2 with 21% O_2. All controlled atmosphere storage combinations suppressed volatile production except for fruits stored in 1% CO_2 with 3% O_2 in which volatile production was high (Harb et al., 1994). The effect of ultra low O_2 (3% CO_2 with 1% O_2) and low ethylene (1 ppm C_2H_4) storage on the production of ethylene, CO_2, volatile aroma compounds and fatty acids of Golden Delicious apples harvested at pre-climacteric and climacteric stages were evaluated during 8 months of storage and during the following 10 days at 20°C in air. 1-aminocyclopropane 1-carboxylic acid oxidase activity, increase in ethylene and CO_2 and production of aroma volatiles and fatty acids were lower during the 10 days after storage following ultra low O_2 storage than following non-controlled atmosphere, cold storage. The reduction of all these parameters was more pronounced in fruits harvested at the pre-climacteric stage than in those harvested when climacteric. Increasing CO_2 concentrations from 1 to 3% under ultra low O_2 conditions (1% O_2) intensified the inhibition of respiration and the production of ethylene, aroma and fatty acids. On the other hand, removal of ethylene from the store atmosphere only slightly affected CO_2 and aroma production by apples (Brackmann et al., 1995). Van der Merwe (1996) recommended –0.5°C and 1.5% CO_2 with 1.5% O_2 for 9 months. Sharples and Stow (1986) recommended 1.5–3.5°C with 5% CO_2 and 3% O_2.

In experiments conducted over 4 years at 1% O_2 with 1% CO_2 or 1% O_2 with 6% CO_2, fruit rot, scald and flesh browning were prevented to a large extent (Strempfl et al., 1991). Starking Delicious and Golden Delicious can be stored for up to 9 months without appreciable loss of quality at 0°C in 3% CO_2 with 3% O_2 (Ertan et al., 1992).

Golden Delicious in storage at 0°C had a threshold for peel injury at 15% CO_2 with 3% O_2, with little difference between 3 and 5% O_2 levels at 20% CO_2. The development of superficial scald (which did not occur at 3% O_2) diminished at 15% O_2 with increasing CO_2 and was completely inhibited by 20% CO_2. Fruit softening was inhibited with increasing CO_2 levels in a similar manner at both 3 and 5% O_2 levels, even though the higher O_2 level enhanced softening. No softening occurred at 20% CO_2 regardless of O_2 level both during storage and subsequent ripening. This could be related to the total inhibition of ethylene evolution, which occurred at 20% CO_2 at both O_2 levels. Chlorophyll degradation was predominantly inhibited by reducing the O_2 level. Titratable acidity was highest when CO_2 was increased to 10% at 15% O_2 and to 5% at 3% O_2 (Ben Arie et al., 1993).

Controlled atmosphere storage treatments suppressed aroma production compared with storage at 1°C. The greatest reduction was found under
ultra low O_2 (1% O_2) and high CO_2 (3%) conditions. A partial recovery of
aroma production was observed when controlled atmosphere stored fruits
were subsequently stored for 14 days under cold storage conditions (Brackmann *et al.*, 1993). Anon (1968) recommended −1.1–0°C with 1–2% CO_2 and
2–3% O_2. M. Herregods (1993, personal communication) and Meheriuk
(1993) summarized general recommendations from a variety of countries as
follows:

	Temperature °C	CO_2	O_2
Australia (South)	0	2	2
Australia (Victoria)	0	1	1.5
Australia (Victoria)	0	5	2
Belgium	0.5	2	2
Brazil	1–1.5	3–4.5	1.5–2.5
Canada (British Columbia)	0	1.5	1–1.2
Canada (Ontario)	0	2.5	2.5
China	5	4–8	2–4
France	0–2	2–3	1–1.5
Germany (Saxony)	2	1.7–1.9	1.3–1.5
Germany (Westphalia)	1–2	3–5	1–2
Holland	1	4	1.2
Israel	−0.5	2	1–1.5
Italy	0.5	2	1.5
Slovenia	1	3	1
Slovenia	0	3	1
South Africa	−0.5	1.5	1.5
Spain	0.5	2–4	3
Switzerland	2	5	2–3
Switzerland	2	4	2–3
USA (New York)	0	2–3	1.8–2
USA (New York)	0	2–3	1.5
USA (Penn.)	−0.5–+0.5	0–0.3	1.3–2.3
USA (Washington)	1	< 3	1–1.5

Storing Golden Delicious infested with eggs of *Rhagoletis pomonella* in
100% N_2 atmospheres at 20 ± 1°C resulted in 100% mortality after 8 days but
apples developed detectable off-flavour and it was suggested that the treatment may be useful for cultivars less susceptible to anoxia (Ali Niazee *et al.*,
1989).

Granny Smith

Meheriuk (1993) recommended $-0.5-2°C$ with 0.8–5% CO_2 and 0.8–2.5% O_2 for 8–10 months. Van der Merwe (1996) suggested $-0.5°C$ and 0–1% CO_2 with 1.5% O_2 for 7 months. Granny Smith apples were stored at $-0.5°$ for 6 months in normal atmosphere, for 9 months in 1.5% O_2 and 0% CO_2. After storage the fruits were ripened at 20°C for 7 days before evaluation for superficial scald. In a normal atmosphere, all control fruits developed scald. In controlled atmosphere storage only a few apples developed scald (Eeden *et al.*, 1992).

In South Tyrol (Italy) controlled atmosphere storage using ultra low O_2 regimes (0.9–1.1% O_2 compared with conventional levels of 2.5–3.0% O_2) controlled scald on Granny Smith fruits for about 150 days. No accumulated alcoholic flavour or any external or internal visual symptoms of low O_2 injury, either during or after storage, were observed (Nardin, 1994).

Granny Smith apples were kept at 46°C for 12 h, 42°C for 24 h, or 38°C for 72 or 96 h before storage at 0°C for 8 months in 2–3% O_2 with 5% CO_2. Heat treated fruit were firmer at the end of storage and had a higher soluble solids to acid ratio and a lower incidence of superficial scald than fruits not heat treated. Pre-storage regimes of 46°C for 24 h or 42°C for 48 h resulted in fruit damage after storage (Klein and Lurie, 1992).

After 6.5 months of storage, at $-0.5°C$ the average percentage of Beurre d'Anjou pears and Granny Smith apples with symptoms of scald ranged from zero in fruits stored in 1.0% O_2 with 0% CO_2 or 1.5% O_2 with 0.5% CO_2 to 2% in fruits stored in 2.5% O_2 with 0.8% CO_2 and 100% in control fruits stored in air. M. Herregods (1993, personal communication) and Meheriuk (1993) summarized general recommendations from a variety of countries as follows:

	Temperature °C	CO_2	O_2
Australia (South)	0	1	2
France	0–2	0.8–1	0.8–1.2
Germany (Westphalia)	1–2	3	1–2
Israel	–0.5	5	1–1.5
Italy	0–2	1–3	2–3
New Zealand	0.5	2	2
Slovenia	1	3	1
South Africa	0–0.5	0–1	1.5
Spain	1	4–5	2.5
USA (Penn.)	–0.5–+0.5	0–4	1.3–2
USA (Washington)	1	< 1	1

Howgate Wonder

Sharples and Stow (1986) recommended 3–4°C with 8–10% CO_2 with data only available for controlled atmosphere storage with no scrubbing.

Idared

Idared was stored at 1°C and 85–90% rh, either in a controlled atmosphere (7% CO_2, 7% O_2) or under normal atmospheric conditions. Weight loss and decay were negligible in controlled atmospheres (0.0–0.78%). Fruits in controlled atmosphere storage exhibited no physiological disorder (Jankovic and Drobnjak, 1994). Meheriuk (1993) recommended 0–4°C with less than 1–3% CO_2 and 1–2.5% O_2 for 5–7 months. Hansen (1977) recommended a temperature of not lower than 3°C with 3% CO_2 and 3% O_2. Sharples and Stow (1986) recommended 3.5–4.5°C with 5% CO_2 and 3% O_2 or less than 1% CO_2 and 1.25% O_2 for 37 weeks. In order to maintain fruit at a satisfactory level of firmness Stow (1996a) has suggested storage in 1% O_2 with less than 1% CO_2 for no more than 13 weeks. M. Herregods (1993, personal communication) and Meheriuk (1993) summarized general recommendations from a variety of countries as follows:

	Temperature °C	CO_2	O_2
Belgium	1	2	2
Canada (Ontario)	0	2.5	2.5
England	3.5–4	8	–
England	3.5–4	< 1	1.25
France	2–4	1.8–2.2	1.4–1.6
Germany (Westphalia)	3–4	3	1–2
Slovenia	1	3	3
Spain	2	3	3
Switzerland	4	4	2–3
USA (Michigan)	0	3	1.5
USA (New York)	0	2–3	1.8–2
USA (New York)	0	2–3	1.5
USA (Penn.)	−0.5–+0.5	0–4	2–3

Ingrid Marie

Fidler (1970) recommended 1.5–3.5°C with 6–8% CO_2 for stores with no scrubber or less than 1% CO_2 and 3% O_2 for stores fitted with a scrubber.

James Grieve

Fidler (1970) and Sharples and Stow (1986) recommended 3.5–4°C with 6% CO_2 and 5% O_2 or less than 1% CO_2 and 3% O_2.

Jonagold

Koelet (1992) recommended, for what he refers to as 'Jonah Gold', 0.5–1°C with 1–3% CO_2 and 2–2.2% O_2. Meheriuk (1993) recommended 0–2°C with 1–3% CO_2 and 1–3% O_2 for 5–7 months. In experiments with Jonagold stored at 5°C and 95% rh for 9 months it was shown that the lower the O_2 concentration (over the range of 17.0% O_2 with 4.0% CO_2, 1.0% O_2 with 1.0% CO_2 or 0.7% O_2 with 0.7% CO_2) the slower the decrease in firmness and titratable acidity and the slower the change of skin colour. No alcohol flavour or physiological disorders developed in fruits stored at 0.7% O_2 and this concentration was recommended (Goffings et al., 1994). Hansen (1977) recommended 0–1°C with 6% CO_2 and 2.5% O_2. Awad et al. (1993) recommended 1°C for 6 months. Sharples and Stow (1986) recommended 1.5–2°C with less than 1% CO_2 and 2% O_2 or less than 1% CO_2 and 1.25% O_2. Johnson (1994) showed that a slight difference in temperature at the same controlled atmosphere conditions could have an effect on maximum storage life as shown below:

Temperature °C	CO_2	O_2	Storage time in weeks
1.5–2	< 1	1.5–2	32
1.25	< 1	1.5–2	36

Jonagold were stored at 1°C and 85–90% rh, either in a controlled atmosphere (7% CO_2, 7% O_2) or under normal atmospheric conditions. Weight losses and decay were negligible in controlled atmospheres (0.0–0.78%). Fruits stored in controlled atmosphere storage exhibited no physiological disorder (Jankovic and Drobnjak, 1994). After storage at 3% O_2 with 1% CO_2 (high O_2 low CO_2) followed by 3 weeks of shelf life at 20°C, per cent scald was 23%. At 1.5% O_2 with 1% CO_2 (low O_2 high CO_2) and 3% O_2 with 5% CO_2 (high O_2 low CO_2), per cent scald was 2 and 6%, respectively, while at 1.5% O_2 with 5% CO_2 (low O_2 high CO_2) no scald was observed. This shows that in Jonagold both O_2 and CO_2 affect the occurrence of scald. At high O_2 low CO_2 the level of scald (39%) was significantly higher in early harvested apples compared with apples from a normal harvest date (13% scald) (Awad et al., 1993). Jonagold apples were stored at 0°C in air or in a controlled atmosphere. Controlled atmosphere storage at 0°C 1.5% O_2 plus 1.5% CO_2 for 6 months significantly reduced the loss of acidity and firmness, decreased production of volatile compounds by half, but did not influence total soluble solids content (Girard and Lau, 1995). M. Herregods (1993, personal communication) and Meheriuk (1993) summarized general recommendations from a variety of countries as follows:

	Temperature °C	CO_2	O_2
Belgium	0.8	1	2
Canada (British Columbia)	0	1.5	1.2
Canada (Nova Scotia)	0	4.5	2.5
Canada (Nova Scotia)	0	1.5	1.5
Denmark	1.5–2	< 1	2
England	1.5–2	8	–
England	1.5–2	< 1	2
France	0 to 1	2.5–3	1.5
Germany (Saxony)	2	1.7–1.9	1.3–1.5
Germany (Westphalia)	1–2	3–6	1–2
Holland	1	4	1–2
Slovenia	1	3	3
Slovenia	1	6	3
Slovenia	1	3	1.2
Spain	2	3–4	3
Switzerland	2	4	2
USA (New York)	0	2–3	1.8–2
USA (New York)	0	2–3	1.5

Jonathan

Anon (1968) recommended 0°C with 3–5% CO_2 and 3% O_2 for 5–6 months. Stoll (1972) advocated 4°C with 3–4% CO_2 and 3–4% O_2. Meheriuk (1993) suggested –0.5–4°C with 1–3% CO_2 and 1–3% O_2 for 5–7 months. Fidler and Mann (1972), Fidler (1970) and Sharples and Stow (1986) recommended 4–4.5°C with 6% CO_2 and 3% O_2. With Jonathan harvested at the optimum time, the loss of hardness after storage until March or June was less in 0.9% O_2 and 4.5% CO_2 than 1.2% O_2 and 4.5% CO_2 directly after storage or after holding for 1 week at 20°C but not after 2 weeks. With later-harvested fruits, O_2 concentration did not affect fruit hardness. However, flesh browning was greater in fruits stored in 0.9% O_2 than in 1.2% O_2 (Schaik et al., 1994). Jonathan was shown to be sensitive to CO_2 injury, which appeared both externally and internally. The threshold for peel injury was 10% CO_2 and it occurred more severely and earlier at 15% O_2 than at 3% O_2 with little difference between the two O_2 levels at 20% CO_2. The development of superficial scald (which did not occur at 3% O_2) diminished at 15% O_2 with increasing CO_2 and was completely inhibited by 20% CO_2. Fruit softening was inhibited with increasing CO_2 levels in a similar manner at both O_2 levels, even though the higher O_2 level enhanced softening. No softening occurred at 20% CO_2 regardless of O_2 level both during storage and subsequent ripening. This could be related to the total inhibition of ethylene evolution which occurred at 20% CO_2 at both O_2 levels. Chlorophyll degradation was predominantly inhibited by increasing the level of CO_2 (Ben Arie *et al.*, 1993). M. Herregods (1993, personal communication) and Meheriuk (1993) summarized general recommendations from a variety of countries as follows:

	Temperature °C	CO_2	O_2
Australia (South)	0	3	3
Australia (South)	0	1	3
Australia (Victoria)	2	1	1.5
China	2	2–3	7–10
Germany (Westphalia)	3–4	3	1–2
Israel	–0.5	5	1–1.5
Slovenia	3	3	3
Slovenia	3	3	1.5
Switzerland	3	3–4	2–3
USA (Michigan)	0	3	1.5

Jupiter

Sharples and Stow (1986) recommended 3–3.5°C with less than 1% CO_2 and 2% O_2.

Karmijn

Hansen (1977) recommended a temperature of not lower than 4°C with 3% CO_2 and 3% O_2.

Katy (Katja)

Sharples and Stow (1986) recommended 3–3.5°C with less than 1% CO_2 and 2% O_2.

Kent

Sharples and Stow (1986) recommended 3.5–4°C with 8–10% CO_2 and about 11–13% O_2 or less than 1% CO_2 and 2% O_2.

King Edward VII

Fidler and Mann (1972) recommended 3–4°C with 8–10% CO_2 and about 11–13% O_2.

Lady Williams

Lady Williams were stored at 0°C in a controlled atmosphere of 1.5% O_2 with less than 0.5% CO_2. Fruits were removed from storage after 3, 5, 7 and 9 months and assessed for storage life and sensory acceptability on return to air at 20°C after 0, 4, 11, 18 and 25 days. Lady Williams is a late-maturing cultivar and the level of acidity (which decreased with time) is an important quality parameter for fruits of this cultivar (Jobling *et al.*, 1993). Work in Australia showed that fruits can be stored for 6–8 months in 2% O_2 and 1% CO_2 at 0°C (Tugwell and Chvyl, 1995).

Laxton's Fortune

Sharples and Stow (1986) recommended 3–3.5°C with 5% CO_2 and 3% O_2 or less than 1% CO_2 and 2% O_2 using a scrubber and 7–8% CO_2 for stores without a scrubber.

Laxton's Superb

Fidler (1970), Wilkinson and Sharples (1973) and Sharples and Stow (1986) recommended 3.5–4.5°C with 7–8% CO_2 and 3% O_2 for 6 months.

Lord Derby

Sharples and Stow (1986) recommended 3.5–4°C with 8–10% CO_2 with data only available for stores without a scrubber.

Lord Lambourne

Sharples and Stow (1986) recommended 3.5–4°C with 8% CO_2 with data only available for stores without a scrubber.

McIntosh

Anon (1968) recommended 3.3°C with 2–5% CO_2 with 3% O_2 for 6–8 months. Meheriuk (1993) recommended 2–4°C with 1–5% CO_2 and 1.5–3% O_2 for 5–7 months. Sharples and Stow (1986) recommended 3.5–4°C with 8–10% CO_2, with data only available for stores without a scrubber. Sharples and Stow (1986) recommended 3.5–4°C with less than 1% CO_2 and 2% O_2. Hansen (1977) recommended for McIntosh-Rogers a temperature not lower than 3°C with 6% CO_2 and 2.5% O_2. M. Herregods (1993, personal communication) and Meheriuk (1993) summarized general recommendations from a variety of countries as follows:

	Temperature °C	CO_2	O_2
Canada (British Columbia)	3	4.5	2.5
Canada (British Columbia)	3	5	2.5
Canada (British Columbia)	1.7	5	2.5
Canada (Nova Scotia)	3	4.5	2.5
Canada (Nova Scotia)	3	1.5	1.5
Canada (Ontario)	3	5	2.5–3
Canada (Ontario)	3	1	1.5
Canada (Quebec)	3	5	2.5
Germany (Westphalia)	3–4	3–5	2
USA (Mass.)	1–2	5	3
USA (Michigan)	3	3	1.5
USA (New York)	2	3–5	2–2.5
USA (New York)	2	3–5	4
USA (Penn.)	1.1–1.7	0–4	3–4

Melrose

Melrose were stored at 1°C and 85–90% rh, either in a controlled atmosphere (7% CO_2 with 7% O_2) or under normal atmospheric conditions. Weight losses and decay were negligible in controlled atmospheres (0–0.78%). Fruits stored in controlled atmospheres exhibited no physiological disorders whereas bitter pit was observed under normal atmospheric conditions (Jankovic and Drobnjak, 1994). Hansen (1977) recommended 0–1°C with 6% CO_2 and 2.5% O_2. Meheriuk (1993) recommended 0–3°C with 2–5% CO_2 and 1.2–3% O_2 for 5–7 months. M. Herregods (1993, personal communication) and Meheriuk (1993) summarized general recommendations from a variety of countries as follows:

	Temperature °C	CO_2	O_2
Belgium	2	2	2–2.2
France	0–3	3–5	2–3
Germany (Westphalia)	2–3	3	1–2
Slovenia	1	3	3
Slovenia	1	3	1.2

Merton Worcester

Fidler (1970) recommended 3–3.5°C with 7–8% CO_2 for stores with no scrubber or 5% CO_2 and 3% O_2 for stores with a scrubber.

Michaelmas Red

Fidler (1970) recommended 1–3.5°C with 6–8% CO_2 for stores with no scrubber or 5% CO_2 and 3% O_2 for stores with a scrubber.

Monarch

Fidler (1970) recommended 0.5–1°C with 7–8% CO_2 for stores with no scrubber or 5% CO_2 and 3% O_2 for stores with a scrubber.

Morgenduft

In Italy controlled atmosphere storage using ultra low O_2 regimes (0.9–1.1% O_2 compared with conventional levels of 2.5–3.0% O_2) controlled scald on Morgenduft fruits for 210 days. Fruits did not develop any alcoholic flavour or any external or internal visual symptoms of low O_2 injury, either during or after storage (Nardin, 1994).

Mutsu

Fidler (1970) recommended 1°C with 8% CO_2 with data supplied only for stores not fitted with a scrubber. Hansen (1977) recommended 0–1°C with 6% CO_2 and 2.5% O_2. Meheriuk (1993) recommended 0–2°C with 1–3% CO_2 and 1–3% O_2 for 6–8 months. M. Herregods (1993, personal communication) and Meheriuk (1993) summarized general recommendations from a variety of countries as follows:

	Temperature °C	CO_2	O_2
Australia (Victoria)	1	1	1.5
Australia (Victoria)	0	3	3
Denmark	0–2	3–5	3
Germany (Westphalia)	1–2	3–5	1–2
Japan	0	1	2
Slovenia	1	3	1
USA (Michigan)	0	3	1.5
USA (Pennsylvania)	–0.5–+0.5	0–2.5	1.3–1.5

Newton Wonder

Controlled atmosphere storage was not recommended, only air storage at 1.1°C for 6 months (Fidler, 1970; Wilkinson and Sharples, 1973).

Norfolk Royal

Fidler (1970) recommended 3.5°C with 5% CO_2 and 3% O_2.

Northern Spy

Anon (1968) recommended 0°C with 2–3% CO_2 and 3% O_2.

Pink Lady

Work in Australia showed that fruits can be stored for up to 9 months in 2% O_2 and 1% CO_2 at 0°C without developing scald (Tugwell and Chvyl, 1995).

Red Delicious

In South Tyrol (Italy) controlled atmosphere storage, using ultra low O_2 regimes (0.9–1.1% O_2 compared with conventional levels of 2.5–3.0% O_2), controlled scald on Red Delicious fruits for about 150 days without the development of alcoholic flavour or any external or internal visual symptoms of low O_2 injury, either during or after storage (Nardin, 1994). Red Delicious fruits stored for up to 7 months in 1 to 5% O_2 and 2–6% CO_2 had superficial scald levels of less than 3% (Xue et al., 1991). Van der Merwe (1996) recommended –0.5°C and 1.5% CO_2 with 1.5% O_2 for 9 months. Fidler (1970) and Sharples and Stow (1986) recommended 0–1°C with either

5% CO_2 and 3% O_2 or less than 1% CO_2 and 3% O_2. Meheriuk (1993) recommended –0.5 to 1°C with 1–3% CO_2 and 1–2.5% O_2 for 8–10 months. M. Herregods (1993, personal communication) and Meheriuk (1993) summarized general recommendations from a variety of countries as follows:

	Temperature °C	CO_2	O_2
Australia (South)	0	1	2
Australia (Victoria)	0	1.5	1.5–2
Canada (British Columbia)	0–1	1	1.2–1.5
Canada (British Columbia)	0–1	1	0.7
Canada (Ontario)	0	2.5	2.5
Canada (Ontario)	0	1	1
France	0–1	1.8–2.2	1.5
Germany (Westphalia)	1–2	3	1–2
Israel	–0.5	2	1–1.5
Italy	0.5	1.5	1.5
New Zealand	0.5	2	2
South Africa	–0.5	1.5	1.5
Spain	0	2–4	3
USA (Mass.)	0	5	3
USA (New York)	0	2–3	1.8–2
USA (Penn.)	–0.5–+0.5	0–0.3	1.3–2.3
USA (Washington)	0	< 2	1–1.5

Ribstone Pippin

Fidler (1970) recommended 4–4.5°C with 5% CO_2 and 3% O_2.

Rome Beauty

Anon (1968) recommended –1.1–0°C with 2–3% CO_2 and 3% O_2.

Spartan

Fidler (1970) recommended –1–0°C with 5–8% CO_2 for stores with no scrubber or less than 1% CO_2 and 2% O_2 for stores with a scrubber. Sharples and Stow (1986) recommended 1.5–2°C with 6% CO_2 and 2% O_2 for 42 weeks. Meheriuk (1993) recommended 0–2°C with 1–3% CO_2 and 1.5–3% O_2 for 6–8 months. M. Herregods (1993, personal communication) and Meheriuk (1993) summarized general recommendations from a variety of countries as follows:

	Temperature °C	CO_2	O_2
Canada (British Columbia)	0	1.5	1.2–1.5
Canada (Ontario)	0	2.5	2.5
Canada (Quebec)	0	5	2.5–3
Denmark	0–2	2–3	2–3
England	1.5–2	< 1	2
England	1.5–2	6	2
Switzerland	2	3	2–3
USA (New York)	0	2–3	1.8–2
USA (Penn.)	−0.5–+0.5	0–1	2 to 3

Starking

Starking were stored at −0.5°C for 4 months in normal atmosphere, followed by 7 months in 1.5% O_2 and 2% CO_2. After storage the fruits were ripened at 20° for 7 days before evaluation for superficial scald. Starking fruits stored in controlled atmospheres were similar during 1988 and 1989 (Eeden et al., 1992). Starking Delicious could be stored for up to 9 months without appreciable loss of quality at 0°C in 3% CO_2 with 3% O_2 (Ertan et al., 1992). Sfakiotakis et al. (1993) stored Starking Delicious fruit under various conditions including 2.5% O_2 with 2.5% CO_2 and ultra low O_2 (with 1% O_2 and 1% CO_2) storage all at 0.5°C and 90% rh. They showed that these conditions generally retarded fruit softening and markedly reduced scald development. 1% O_2 with 1% CO_2 was the most promising treatment for reducing fruit softening. Low O_2 without CO_2 was equally effective in extending fruit storage life, while 2.5% O_2 with 2.5% CO_2 storage gave moderate results. Sensory evaluation of fruits stored for 7–8 months in 2.5% O_2 with 2.5% CO_2 and 1% O_2 with 1% CO_2 gave high scores without any low O_2 injury or alcoholic taste. During storage, the ethylene scrubber used (Ethysorb) was effective in reducing the ethylene concentration in the gas phase below 0.3 ppm in the 2.5% O_2 with 2.5% CO_2 and ultra low O_2 treatments and below 1.6 ppm in air storage. However, internal fruit ethylene concentrations were found to be above the physiological levels capable of inducing ripening, even in the ultra low O_2 treatments.

Stayman

Anon (1968) recommended −1.1–0°C with 2–3% CO_2 and 3% O_2.

Sundowner

Work in Australia showed that fruits can be stored in 2% O_2 and 1% CO_2 at 0°C with no adverse effects developing when fruit were stored below 1°C (Tugwell and Chvyl, 1995).

Sunset

Sharples and Stow (1986) recommended 3–3.5°C with 8% CO_2, with data only presented for stores without a scrubber.

Suntan

Sharples and Stow (1986) recommended 3–3.5°C with either 5% CO_2 and 3% O_2 or less than 1% CO_2 and 2% O_2.

Tydeman's Late Orange

Sharples and Stow (1986) recommended 3.5–4°C with either 5% CO_2 and 3% O_2 or less than 1% CO_2 and 2% O_2.

Undine

Hansen (1977) recommended a temperature of not under 2°C with 3% CO_2 and 2.5% O_2.

Virginia Gold

Kamath et al. (1992) recommended 2.2°C with 2.5% O_2 and 2% CO_2 for up to 8 months. Under these conditions fruits were firmer than those held in normal cold storage, they did not shrivel even when stored without polyethylene box liners and soft scald was eliminated.

Winston

Sharples and Stow (1986) recommended 1.5–3.5°C with 6–8% CO_2, with data only presented for stores without a scrubber.

Worcester Pearmain

Fidler (1970) recommended 0.5–1°C with 7–8% CO_2 for stores with no scrubber or 5% CO_2 and 3% O_2 for stores with a scrubber. Wilkinson and Sharples (1973) suggested 0.6–1.1°C with 5% CO_2 and 3% O_2 for 6 months. Sharples and Stow (1986) advocated 0.5–1°C and 7–8% CO_2 for fruit stored without a scrubber and 5% CO_2 with 3% O_2 for those stored with a scrubber.

Apricot, *Prunus armeniaca*

SeaLand (1991) recommended 2–3% CO_2 with 2–3% O_2, and typical storage conditions were given by Bishop (1996) as 0–5°C with 2–3% CO_2 and 2–3% O_2. Controlled atmospheres were reported to have a fair effect on storage at 0–5°C with 2–3% CO_2 and 2–3% O_2 but were not used commercially (Kader, 1985, 1989). It was reported that controlled atmosphere storage can lead to an increase in internal browning in some cultivars (Hardenburg *et al.*, 1990).

Fruit of the cultivars Rouge de Roussillon and Canino were picked half ripe or mature then subjected to pre-treatment with high levels of CO_2 (10–30%) for different periods (24, 48 or 72 h) before storage in air, and compared with fruits stored directly in air or in controlled atmosphere (5% O_2 and 5% CO_2). Under ripe fruits exposed to 20% CO_2 for 24 or 48 h, or controlled atmosphere storage, remained firm for longer than those stored in air. CO_2 pre-treatment also seemed to reduce the incidence of brown rot (*Monilinia* spp.). Titratable acidity decreased continuously during cold storage and after 24 days was the same for CO_2 pre-treated and air stored fruits, but remained high for the controlled atmosphere stored fruits. Respiration of apricots stored in air showed a typical climacteric peak, which was delayed by the CO_2 pre-treatments and was completely inhibited by controlled atmosphere storage. The best CO_2 pre-treatments involved exposure to 20% CO_2 for 24 to 48 h. Higher concentrations or a longer duration of exposure resulted in fermentation. The benefits of CO_2 pre-treatments were most noticeable at the beginning of cold storage but gradually declined, and, once the temperature was increased, they disappeared within about 24 h (Chambroy *et al.*, 1991).

Fruits of two cultivars were stored at 0–0.5°C in four different controlled atmospheres with 0, 2.5, 5 or 7.5% CO_2 and 5.0% O_2. Both the cultivars ICAPI 17 COL but particularly ICAPI 30 COL, proved suitable for controlled atmosphere storage for up to 3 weeks. Only in the cultivar ICAPI 30 COL did the development of some ripening parameters seem to be related to CO_2 concentration (Andrich and Fiorentini, 1986).

Fruits of the cultivar Reale d'Imola were harvested and stored at 0 or 6°C, either in air or in an atmosphere of 0.3% O_2 for up to 37 days. Low O_2 concentration significantly increased the ethanol and acetaldehyde content of both fruits especially at 6°C, whereas the methanol content of fruits was not significantly affected by low O_2. Low O_2 reduced total soluble solids when fruits were stored at 6°C and increased fruit firmness at both storage temperatures. Physiological storage disorders, such as flesh discoloration and browning, appeared after 5 days of controlled atmosphere storage at 6°C and 12 days of storage at 0°C (Folchi *et al.*, 1995). Folchi *et al.* (1994) stored the cultivar Reale d'Imola in less than 0.1% CO_2 with 0.3% O_2 and showed an increase in aldehyde and ethanol content of fruit with increased storage time. Truter *et al.* (1994) showed that at –0.5°C, storage in 1.5% CO_2 with 1.5% O_2 of 5% CO_2 with 2% O_2 had a reduced mass loss compared to those stored in air.

For canning apricots Van der Merwe (1996) recommended –0.5°C with 1.5% CO_2 with 1.5% O_2 for up to 2 weeks. Hardenburg *et al.* (1990) recommended 2.5–3% CO_2 with 2–3% O_2 at 0°C for the cultivar Blenheim (Royal) which retained their flavour better than air stored fruit when they were subsequently canned.

Apricot, Japanese *Prunus mume*

Mature green fruits of the Japanese apricot cultivar Ohshuku were stored in 2–3% O_2 with 3, 8, 13 or 18% CO_2 at 20°C and 100% rh. Surface yellowing and flesh softening were delayed at higher CO_2 levels. Pitting injury occurred in air stored fruits, but no injury was observed in controlled atmosphere stored fruits for up to 23 days in 3–13% CO_2. The results indicated that controlled atmosphere stored fruits have a shelf life of 15, 15, 19 and 12 days using 3, 8, 13 and 18% CO_2, respectively (Kaji *et al.*, 1991). Mature green Japanese apricot fruits of four cultivars (Gojiro, Nankou, Hakuoukoume and Shirakaga) were held for 7 days at 25°C under various controlled atmospheres. Anaerobic respiration occurred at O_2 concentrations of up to 2% at 25°C. Fruits held at O_2 concentrations of up to 1% developed browning injury and produced ethyl alcohol at a high rate. The percentage of fruits with water core injury was high in conditions of low O_2 and no CO_2. Ethylene production rates were suppressed at high CO_2 concentrations or at O_2 levels less than 5%. These results indicate that storing Japanese apricot fruits in approximately 10% CO_2 and 3–4% O_2 may help to retain optimum quality at 25°C, but that such controlled atmosphere storage conditions delay the respiratory climacteric rises slightly and increase ethyl alcohol production after the third day of storage (Koyakumaru *et al.*, 1995). The cultivar Gojiro was stored for 3 days at 25°C under several controlled atmosphere conditions. Compared with fruits in ambient air, O_2 uptake and ethylene production decreased when fruits were exposed to a mixture of 19.8% CO_2 and 21% O_2; they decreased even more when the controlled atmosphere storage gas contained 5 or 2% O_2. However, at 2% O_2 and no CO_2, the percentage of physiologically injured fruits increased considerably in or after storage. When fruits were exposed to a mixture of high CO_2 and low O_2, the rates of O_2 uptake and production of CO_2, acetaldehyde, ethyl alcohol, and the percentages of injured brown fruits increased as the volume of the gas mixture was increased. These results suggest that removing ethylene from the storage atmosphere and maintaining the storage atmosphere at about 8% CO_2 and low O_2 (at least 2%) are important to retain fruit quality (Koyakumaru et al., 1994).

Artichoke, Globe artichoke, *Cynara scolymus*

Typical storage conditions were given as 0–5°C with 2–3% CO_2 and 2–3% O_2 by Bishop (1996). Monzini and Gorini (1974) recommended 0.5–1.5°C with 0–2.5% CO_2 and 5% O_2 for 20–30 days. Storage at 3–5% CO_2 with 2–3% O_2 was recommended by SeaLand (1991) and 3% CO_2 with 3% O_2 in cold storage (unspecified) for 1 month to reduce browning (Pantastico, 1975). Saltveit (1989) and Kader (1992) recommended 0–5°C with 2–3% CO_2 and 2–3% O_2 but claimed that controlled atmosphere storage only had a slight effect on storage life. Kader (1985) recommended 0–5°C with 3–5% CO_2 and 2–3% O_2 but indicated that it was not used commercially. Poma Treccani and Anoni (1969) recommended 3% CO_2 combined with 3% O_2 for up to 1 month. They showed that it reduced browning discoloration of the bracts. Miccolis and Saltveit (1988) described storage of large, mature artichokes in humidified air at 7°C for 1 week, 2.5°C for 2 weeks, or 0°C for 3 weeks. All storage conditions resulted in significant quality loss, and there was little or no beneficial effect of lowered O_2 (1, 2.5 or 5%) and elevated CO_2 (2.5 or 5%) in storage at 0°C. Bracts and receptacles blackened in all treatments after 4 weeks of storage. The degree of blackening was insignificant in air, but became severe as the O_2 level decreased from 5 to 1% and CO_2 increased from 2.5 to 5%. Controlled atmosphere storage of large artichokes was therefore not recommended.

The cultivar Violeta was stored for 15–28 days at 1°C and 90–95% rh, with 1–6% O_2 and 2–8% CO_2. The best results, for physical, chemical and organoleptic properties, and a shelf life of 3–4 days, were obtained with storage for 28 days in 2% O_2 and 6% CO_2 (Artes-Calero et al., 1981).

Storage trials were carried out by Mencarelli (1987) at 1 ± 0.5°C and 90–95% rh in a range of controlled atmospheres (5 and 10% O_2 and 2–6% CO_2) for up to 45 days, or in air for up to 35 days after heat sealing in perforated plastic films or films of differing permeability. Artichokes stored in 5 and 10% O_2 without CO_2 remained turgid with good organoleptic quality. Addition of 2% CO_2, however, controlled the development of superficial mould, but 4% CO_2 had adverse effects. Perforated heat sealed films gave poor results as did low permeability films but films sufficiently permeable to allow gaseous exchange while maintaining high humidity gave good results with high turgidity and good flavour and appearance. Storage in crates lined with perforated polyethylene film was recommended by Lutz and Harden-burg (1968). Five storage trials carried out by Andre *et al.* (1980) on the cultivar Violet de Provence showed that their storage life in air was 1 week at room temperature and 3–4 weeks at 1°C. This was extended to 2 months by a combination of vacuum cooling (pre-refrigeration), wrapping in poly-ethylene bags and storage at 1°C in a range of controlled atmospheres.

Asian pear, Nashi, *Pyrus pyrifolia, P. ussuriensis* variety *sinensis*

Zagory *et al.* (1989) and A.A. Kader (quoted by Zagory *et al.*, 1989) reviewed work on several cultivars and found the following. Storage experi-ments at 2°C in 1, 2 or 3% O_2 of the cultivars Early Gold and Shinko showed no clear benefit compared to storage in air. Early Gold fruits appeared to be of reasonably good quality after up to 6 months of storage at 2°C. Shinko became prone to internal browning after the third month in storage perhaps due to CO_2 injury. Low O_2 atmospheres reduced yellow colour develop-ment, compared with air. The cultivar 20th Century was shown to be sensi-tive to CO_2 concentrations above 1% when exposed for longer than 4 months or to 5% CO_2 when exposed for more than 1 month. In other work Kader (1989) recommended 0–5°C with 0–1% CO_2 and 2–4% O_2 but claimed that it had limited commercial use. Meheriuk (1993) indicated that con-trolled atmosphere storage of Nashi was still in an experimental stage but gave provisional recommendations of 2°C with 1–5% CO_2 and 3% O_2 for about 3–5 months and that longer storage periods may result in internal browning. Meheriuk also stated that some cultivars are subject to CO_2 injury but indicates that this is when they are stored in concentrations of over 4%.

Asparagus, *Asparagus officinalis*

Storage at 0°C with 12% CO_2 or at 5°C or just above at 7% CO_2 retarded decay and toughening (Hardenburg *et al.*, 1990). Lawton (1996) recom-mended 0°C, 95% rh, 9% CO_2 and 5% O_2. SeaLand (1991) recommended 5–10% CO_2 and 20% O_2. Fellows (1988) recommended 5°C with a maximum of 10% CO_2 and a minimum of 10% O_2. Monzini and Gorini (1974) recommended 1°C with 10% CO_2 and 5–10% O_2. Kader (1985, 1992) recommended 0–5°C with 5–10% CO_2 and 21% O_2, which had a good effect but was of limited commercial use. Saltveit (1989) recommended 1–5°C with 10–14% CO_2 and 21% O_2 which had only a slight effect. Typical storage conditions were given as 0–3°C with 10–14% CO_2 by Bishop (1996). Lill and Corrigan (1996) showed that during storage at 20°C, spears stored in 5–10% O_2 combined with 5–15% CO_2 had an increased postharvest life from 2.6 days in air to about 4.5 days. Lipton (1968) showed that levels of 10% CO_2 could be injurious, causing pitting, at storage temperatures of 43°F (6.1°C), but at 35°F (1.7°C) no CO_2 injury was detected. In atmospheres containing 15% CO_2 spears were injured at both temperatures. Lipton also showed that storage in 10% CO_2 reduced rots due to *Phytophthora*, but in atmospheres containing 5% CO_2 there were no effects. Other work has also shown that fungal development during storage could be prevented by flushing with CO_2 (30% for 24 h) or by maintaining a CO_2 concentration of 5–10% (Andre *et al.*, 1980). Hardenburg *et al.* (1990) indicated that brief (unspecified) exposure to 20% CO_2 can reduce soft rot at the butt end of spears.

In several trials, the vacuum precooling of asparagus spears in a controlled atmosphere (in polyethylene bags with silicon elastomer windows) within 6–8 h of harvest, followed by cold storage at 1°C was the best long-term method of storage (20–35 days) (Andre *et al.*, 1980b). Lill and Corrigan (1996) experimented with different modified atmosphere packs and found they all significantly extended the shelf life of the spears by between 83 and 178% depending on film type. The least permeable film (W.R. Grace RD 106 and Van Leer Packaging Ltd) gave the longer shelf life extensions, the former film had 2.5–3.5% CO_2 and 4.1–9.7% O_2 equilibrium gas content at 20°C, and the latter 3.6–4.6% CO_2 and 2.0–4.5% O_2.

Aubergine, Eggplant, *Solanum melongena*

High CO_2 in the storage atmosphere can cause surface scald browning, pitting and excessive decay, and these symptoms are similar to those caused by chilling injury (Wardlaw, 1937). CO_2 concentrations of 5, 8 or 12% during storage all resulted in CO_2 injury, characterized by external browning without tissue softening (Mencarelli *et al.*, 1989). Storage in 0% CO_2 with 3–5% O_2 was recommended by SeaLand (1991). Viraktamath *et al.* (1963, quoted by Pantastico, 1975) claimed that atmospheres containing CO_2 levels of 7% or higher were injurious, but Pantastico could not confirm these results.

Avocado, *Persea americana*

Both Overholster (1928) and Wardlaw (1937) showed that the cultivar Fuerte stored at 45°F (7.2°C) had a 2-month storage life in 3 or 4–5% CO_2 with 3 or 4–5% O_2. This was more than a month longer than they could be stored at the same temperature in air. Biale (1950) also reported similar results. Stahl and Cain (1940) recommended 3% CO_2 with 10% O_2. Storage at 3–5% CO_2 and 3–5% O_2 delayed fruit softening and 9% CO_2 with 1% O_2 at 10°C kept the cultivar Lula in an acceptable eating quality and appearance for 60 days (Hardenburg *et al.*, 1990). Earlier work had also shown that Lula can be stored for 60 days at 10°C with 9% CO_2 and 1% O_2 or for 40 days at 45°F (7.2°C) with 10% CO_2 and 1% O_2 (Pantastico, 1975). Spalding and

Reeder (1972) showed that Lula can be stored for 8 weeks at 4–7°C and 98–100% rh with 10% CO_2 and 2% O_2. Pantastico (1975) showed similar results in that Fuerte stored at 5–6.7°C had double or triple the storage life in 3–5% CO_2 with 3–5% O_2 than in air. In other work 3–10% CO_2 with 2–5% O_2 was recommended for both Californian and tropical avocados (SeaLand, 1991). 5°C, 85% rh, 9% CO_2 and 2% O_2 was a general recommendation for avocados by Lawton (1996). Typical storage conditions were given as 10–13°C with 3–10% CO_2 and 2–5% O_2 by Bishop (1996). A general comment was that storage at 5–13°C with 3–10% CO_2 and 2–5% O_2 had a good effect and was of some commercial use (Kader, 1985, 1992). Kader (1993) in his review of storage conditions for tropical and subtropical fruit recommended 3–10% CO_2 combined with 2–5% O_2 at 10°C, with a range of 5–13°C, and that the benefits of reduced O_2 were good and those of increased CO_2 were also good. If the CO_2 level is too high skin browning can occur and off-flavour can develop, and if O_2 levels are too low internal flesh browning can occur and also off-flavour. Bleinroth *et al.* (1977) recommended 10% CO_2 with 6% O_2 at 7°C which gave a storage life for the cultivars Fuerte and Anaheim of 38 days compared to only 12 days in air at the same temperature. Fellows (1988) recommended a maximum of 5% CO_2 and a minimum of 3% O_2. Van der Merwe (1996) recommended 5.5°C with 10% CO_2 with 2% O_2 for up to 4 weeks. Lizana *et al.* (1993) described storage of the cultivar Gwen at 6°C and 90% rh in controlled atmosphere conditions of either 5% CO_2 with 2% O_2, 5% CO_2 with 5% O_2, 10% CO_2 with 2% O_2, 5% CO_2 with 2% O_2, 10% CO_2 with 5% O_2, or 0.03% CO_2 with 21% O_2 (control) for 35 days followed by 5 days at 6°C in normal air and 5 days at 18°C to simulate shelf life. All fruit in controlled atmosphere storage retained their firmness and were in excellent quality regardless of treatment while the control fruit were very soft even after 35 days storage at 6°C. Lizana and Figuero (1997) compared several controlled atmosphere storage conditions on the cultivar Hass and found that 6°C with 5% CO_2 and 2% O_2 was optimum. Under these conditions they could be stored for at least 35 days and then kept at the same temperature in air for a further 15 days and still be in good condition. Corrales-Garcia (1997) found that Hass avocados stored at 2 or 5°C for 30 days in air, 5% CO_2 with 5% O_2 or 15% CO_2 with 2% O_2 had higher chilling injury for fruits stored in air that the fruits in controlled atmosphere storage, especially those stored in 15% CO_2 with 2% O_2.

High CO_2 treatment (20%) applied for 7 days continuously at the beginning of storage or for 1 day a week throughout the storage period at either 2 or 5°C maintained fruit texture and delayed ripening compared to fruit stored throughout in air (Saucedo Veloz *et al.*, 1991). Storage in total nitrogen or total CO_2, however caused irreversible injury to the fruit (Stahl and Cain, 1940).

Controlled atmosphere storage can reduce chilling injury symptoms. There were less chilling injury symptoms in the cultivars Booth 8, Lula and Taylor after refrigerated storage in controlled atmosphere storage than in air

(Haard and Salunkhe, 1975; Pantastico, 1975). Spalding and Reeder (1975) showed that storage at 0% CO_2 with 2% O_2 or 10% CO_2 with 21% O_2 fruit had less chilling injury and less anthracnose during storage at 7°C than fruits stored in air. Intermittent exposure of the cultivar Hass to 20% CO_2 increased their storage life at 12°C and reduced chilling injury during storage at 4°C compared to those stored in air at the same temperatures (Marcellin and Chevez, 1983).

In experiments described by Meir et al. (1993) at 5°C, increasing CO_2 levels over the range of 0.5, 1, 3 or 8% and O_2 levels of 3 or 21% inhibited ripening (on the basis of fruit firmness and peel colour change) and reduced the incidence of chilling injury. The most effective CO_2 concentration was 8%, with either 3 or 21% oxygen; with a combination of 3% O_2 and 8% CO_2 fruits could be stored for 9 weeks. After controlled atmosphere storage, fruits ripened normally and underwent typical peel colour changes. Some injury to the fruit peel was observed, probably attributable to low O_2 concentrations. Peel damage was seen in 10% of fruits held in 3% O_2 with either 3 or 8% CO_2 but it was too slight to affect their marketability.

Storage life was extended by 3–8 days at various temperatures by sealing individual fruit in polyethylene film bags (Haard and Salunkhe, 1975). In earlier work it had been shown that Fuerte fruit sealed individually in 0.025 mm thick polyethylene film bags for 23 days at 14–17°C ripened normally on subsequent removal to higher temperatures. Levels of gases inside the bags after 23 days storage were 8% CO_2 and 5% O_2 (Aharoni et al., 1968). Thompson et al. (1971) showed that sealing various seedling varieties of West Indian avocados in 125 gauge polyethylene film bags greatly reduced fruit softening during storage at various temperatures (Table 7.1).

Table 7.1.   Effects of storage temperature and wrapping treatment (125 gauge polyethylene film bags compared to unwrapped fruit) on the number of days to softening of avocados (Thompson et al., 1971).

	7°C	13°C	27°C
Not wrapped	32	19	8
Sealed film bags	38	27	11

Banana, *Musa*

The application of controlled atmosphere storage has a considerable significance in the proper shipment, storage and ripening of bananas. Bananas are shipped commercially from some Latin American countries either in controlled atmosphere containers or in reefer ships where equipment has been installed in the refrigerated holds which enable the levels of O_2 and CO_2 to be controlled. Experimental reports published over several decades have indicated various gas combinations that have been suitable for extending the storage life of bananas. The benefits of reduced O_2 were very good and those of increased CO_2 were also very good in that they both delayed ripening and high CO_2 also reduced chilling injury symptoms. CO_2, above certain concentrations, can be toxic to bananas. Symptoms of CO_2 toxicity are softening of green fruit, an undesirable texture and flavour.

Many of the reports in the literature do not include which type or variety of banana is being referred to. Typical storage conditions for unspecified bananas were given as 12–16°C with 2–5% CO_2 and 2–5% O_2 by Bishop (1996). Smock *et al.* (1967) showed that at 15°C with O_2 levels of 2% and CO_2 of 8%, bananas can be stored for 3 weeks compared to the fruits which had green life for only 2 days at room temperature. Hardenburg *et al.* (1990) recommended 5% CO_2 with 4% O_2 which can extend the shelf life by two to three times. SeaLand (1991) recommended 2–5% CO_2 with 2–5% O_2. Fellows (1988) recommended a maximum of 5% CO_2. Anon (1978) reported that in experiments with 25 controlled atmosphere storage combinations 0.5, 1.5, 4.5, 13.5 and 21% O_2 combined with 0, 1, 2, 4 and 8% CO_2 all at 20°C the optimum controlled atmosphere storage conditions appeared to be 1.5–2.5% O_2 combined with possibly 7–10% CO_2. Under these conditions there was an extension of the green life of the fruit of about six times. Van der Merwe (1996) recommended 11.5°C with 7% CO_2 with 2% O_2 for 6–8 weeks for South Africa. Kader (1985, 1992) recommended storage at 12–15°C with 2–5% CO_2 and 2–5% O_2. Kader (1989) also recommended 12–16°C, 2–5% CO_2 and 2–5% O_2 and subsequently Kader (1993) in his review of storage conditions for tropical and subtropical fruit recommended 2–5% CO_2 combined with 2–5% O_2 at 14°C, with a range of 12–16°C.

In a study in South-East Asia it was found that banana cultivars differed in their response to controlled atmosphere storage conditions. Generally, the respiration rate of fruit was slowed down with increased CO_2 and the decreased O_2 in the surrounding atmosphere. As a result of reduced respiratory activity, the processes associated with ripening were slowed (Abdullah and Pantastico, 1990). Abdullah and Pantastico (1990) also observed that the Lakatan cultivar (*Musa* AA) can tolerate O_2 levels of as low as 2% at 15°C, if the CO_2 level was around 8%. A similar response to controlled atmosphere storage was shown by Bungulan bananas (*Musa* AAA) which were stored at 5–8% CO_2 and 3% O_2 (Calara, 1969; Pantastico, 1976). Castillo *et al.* (1967) recommended that Latundan (*Musa* AAB) should be stored at 10–13.5% O_2. The general recommendation for storage of the Dwarf Cavendish cultivar (*Musa* AAA) was 2% O_2 and 6–8% CO_2 at 15°C (Pantastico and Akamine, 1975). Abdul Rahman *et al.* (1997) indicated that anaerobic respiration occurred in the cultivar Berangan at 12°C in 0.5% O_2 but not in 2% O_2. They also showed that after storage for 6 weeks in both 2% O_2 and 5% O_2 fruit had a lower respiration rate at 25°C than those which had been stored for the same period in air.

Sarananda and Wilson Wijeratnam (1997) found that fruit stored at 14°C in 1–5% O_2 had lower levels of crown rot than those stored in air. This fungistatic effect continued even during ripening in air at 25°C. They also found that CO_2 levels of 5 and 10% actually increased rotting levels during storage at 14°C.

Woodruff (1969) recommended 5% or lower levels of CO_2 in combination with 2% O_2. In Australia, McGlasson and Wills (1972) ventilated bananas in 3% O_2 and 5% CO_2 for 182 days at 20°C and the fruit still ripened normally when removed to air. Lawton (1996) recommended 14–16°C, 95% rh, 2–5% CO_2 and 2–5% O_2. Some preliminary work in Australia on Williams, a Cavendish type (*Musa* AAA), suggested that concentrations around 1.5–2.5% O_2 and 7–10% CO_2 at 20°C was the optimum controlled atmosphere storage condition. Under these conditions, an extension of pre-climacteric period of about six times was achieved compared to that in normal air (Sandy Trout Food Preservation Laboratory, 1978). Dillon *et al.* (1989) found that fruits stored in 5% O_2 had lower activities of some enzymes during storage and ripened more slowly than fruits stored in air.

Another type of controlled atmosphere storage treatment was reported by Wills *et al.* (1990). Cavendish bananas were stored in a total nitrogen atmosphere at 20°C for 3 days soon after harvest. These fruit took about 27 days to ripen in subsequent air storage, compared with untreated fruit which ripened after about 19 days. Parsons *et al.* (1964) found that bananas could be stored satisfactorily for several days at 15.6°C in an atmosphere of 99–100% nitrogen. However, this treatment should only be applied to high quality fruit without serious skin damage and with short periods of treatment, otherwise fruits failed to develop a full yellow colour when subsequently ripened even though the flesh softened normally. Parsons *et al.* (1964) observed that Cavendish (*Musa* AAA) fruit stored in less than 1% O_2 failed to ripen normally, developed off-flavour, a dull yellow to brown skin with a 'flaky' grey pulp. Also Wilson (1976) mentioned that fermentation takes place when bananas are ripened at 15.5°C in an atmosphere containing 1% O_2.

Modified atmosphere packaging of fruits in polyethylene film bags is commonly used in international transport (Fig. 26). Besides retaining a desirable high humidity around the fruit it has been shown that packing in non-perforated bags prolonged the pre-climacteric life of Mas banana fruits (Tan *et al.*, 1990). According to Shorter *et al.* (1987) the storage life of banana can be increased five times when they are stored in plastic film (where the gas content stabilized at about 2% O_2 and 5% CO_2) with an ethylene scrubber compared to fruit stored without wraps. If the film was insufficiently permeable, packaging Apple bananas (*Musa* AA) in polyethylene film and storage at 13–14°C accumulated levels of CO_2 which could prove toxic to the fruit (Wei and Thompson, 1993). In this latter study, the symptoms of CO_2 injury were observed when CO_2 levels were between about 5 and 14%. These levels occurred with some hands of fruit in 150

Fig. 26. Modified atmosphere packaging of bananas for international transport in 1996.

gauge polyethylene bags and all the fruit in 200 gauge polyethylene bags, but only after 3 weeks, and not at all in 100 gauge bags. Injury was characterized by darkening of the skin and softening of the outer pulp while the inner pulp (core) remained hard and astringent. Often there would be a distinct irregular ring of dark brown tissue in the outer cross section of the pulp. In some of the fingers the pulp developed a tough texture. Weight loss of fruits in polyethylene bags was low. After 4 weeks storage weight losses were 1.5, 1.8 and 2.1% respectively for fruit stored in 100, 150 and 200 gauge bags and 12.2% for fruit stored without wrapping. A patented modified atmosphere packaging method is called Banavac (Badran, 1969).

Controlled atmosphere storage can also affect the shelf life of fruit after they have begun to ripen. Bananas which have been initiated to ripen by exposure to exogenous ethylene then immediately stored in 1% O_2 at 14°C remained firm and green for a 28 day storage period but ripened almost immediately when transferred to air at 21°C (Liu, 1976a, 1976b).

Some work, which was carried out in the Cameroon, was not relevant to the international trade but shows an interesting effect. Fruits of the cultivar Poyo (*Musa* AAA) held in 50% O_2 concentrations during cooling periods of 12 h at 20°C had reduced high temperature damage during subsequent storage at 30–40°C (Dick and Marcellin, 1985).

Beet, *Beta vulgaris*

SeaLand (1991) reported that unspecified controlled atmosphere storage conditions had only a slight to no effect on beet storage. Shipway (1968) showed that atmospheres containing over 5% CO_2 can damage beet. Monzini and Gorini (1974) recommended 0°C with 3% CO_2 and 10% O_2 for 1 month.

Berries, *Rubus* spp.

Berries retain their flavour well if they are cooled rapidly and stored in an atmosphere containing up to 40% CO_2 for 2 days (Hulme, 1971). Kader (1989) recommended 0–5°C with 10–15% CO_2 and 5–10% O_2 for optimum storage. Agar *et al.* (1994a) showed that the blackberry cultivar Thornefree can be stored at 0–2°C with 20–30% CO_2 and 2% O_2 for 6 days and had up to 3 days subsequent shelf life in air at 20°C. Storage with 20–40% CO_2 can be used to maintain the quality of machine harvested blackberries for processing during short-term storage at 20°C (Hardenburg *et al.*, 1990).

Bitter melon, *Momordica charantia*

Fruits of bitter melon were harvested at horticultural maturity and stored for 2 weeks in various conditions. Fruit quality was similar after storage at 15°C in controlled atmospheres (21, 5 or 2.5% oxygen with 0, 2.5, 5 or 10% carbon dioxide) or in air. Fruits stored for 3 weeks in 2.5% oxygen with 2.5 or 5% carbon dioxide showed greater retention of green colour and had less decay (unspecified) and splitting than air-stored fruits (Zong *et al.*, 1995).

Black sapote, *Diospyros ebenaster*

No data can be found on controlled atmosphere storage. The factor limiting storage was reported to be softening, and fruits wrapped in plastic film had 40–50% longer storage life at 20°C than fruits not wrapped (Nerd and Mizrahi, 1993).

Blackcurrant, *Ribes nigrum*

Stoll (1972) recommended 2–4°C with 40–50% CO_2 and 5–6% O_2. After storage with CO_2 levels of 40% at 18.3°C for 5 days the fruit had a subsequent shelf life of 2 days with 2–3% of the fruit being unmarketable due to rotting (Wilkinson, 1972). Skrzynski (1990) described experiments where fruits of the cultivar Roodknop were held at 6–8°C for 24 h and then transferred to 2°C for storage for 4 weeks in one of the following:

- 20% CO_2 with 3% O_2
- 20% CO_2 with 3% O_2 for 14 days then 5% CO_2 with 3% O_2
- 10% CO_2 with 3% O_2
- 5% CO_2 with 3% O_2
- ambient air.

The best retention of total ascorbic acid content was obtained in both treatments with 20% CO_2. In years with favourable weather preceding the harvest, storage in 20% CO_2 completely controlled the occurrence of moulds caused mainly by *Botrytis*, *Mucor* and *Rhizopus* species. Smith (1957) recommended storage of fruit at 2°C with 50% CO_2 for 7 days followed by a further 3 weeks at 2°C with 25% CO_2 for juice manufacture. With longer storage periods there was an accumulation of alcohol and acetaldehyde but the juice quality was not affected. The cultivar Rosenthal was stored at 1°C in either a controlled atmosphere of 10, 20 or 30% CO_2 all with 2% O_2 or a high CO_2 environment of 10, 20 or 30% CO_2 all with over 15% O_2. The optimum CO_2 concentration was found to be 20%. Ethanol accumulation was higher under controlled atmosphere storage than high CO_2 environment conditions. Fruits could be stored for 3–4 weeks under controlled atmosphere storage or high CO_2 environment conditions, compared with 1 week, for fruits stored at 1°C in a normal atmosphere (Agar *et al.*, 1991).

Blueberry, Bilberry, Whortleberry, *Vaccinium corymbosum*

The storage of fresh blueberries for 7–14 days at 2°C in an atmosphere in 15% CO_2 delayed their decay by 3 days after they had been returned to ambient temperature compared to storage in air at the same temperature (Ceponis and Cappellini, 1983). They also showed that storage in 2% O_2 had no added effect over the CO_2 treatment. Kader (1989) recommended 0–5°C with 15–20% CO_2 and 5–10% O_2 for optimum storage. Ellis (1995) recommended 0.5°C and 90–95% rh with 10% O_2 and 10% CO_2 for 'medium term' storage.

Breadfruit, *Artocarpus altilis, A. communis*

No work on controlled atmosphere storage was located in the literature but storage at 12.5°C sealed in 150 gauge polyethylene bags kept the fruit in good condition for up to 13 days compared to 80% of the unwrapped fruit being unmarketable after only 2 days (Thompson *et al.*, 1974a, 1974b). At 28°C fruit could be stored for 5 days in sealed 100 gauge polyethylene film bags (Maharaj and Sankat, 1990a), while at 12 or 16°C in sealed 100 gauge polyethylene film bags they could be stored for 14 days (Maharaj and Sankat, 1990a). Worrel and Carrington (1994) also showed that storing breadfruit sealed in plastic film maintained their green colour and fruit quality. They used 40 μm high density polyethylene film or low density polyethylene film and showed that fruit quality could be maintained for 2 weeks at 13°C.

Broccoli, Sprouting broccoli, *Brassica oleracea* variety *italica*

Storage at 5–10% CO_2 with 1–2% O_2 was recommended by SeaLand (1991). Studies that were conducted by Lipton and Harris (1974), McDonald (1985) and Deschene *et al.* (1991) indicated that concentrations of up to 10% CO_2 in combination with 1% O_2 to be optimal for controlled atmosphere storage. Shelf life extension at 10°C could be achieved with storage in 2–3% O_2 with 4–6% CO_2 (Ballantyne *et al.*, 1988). Kader (1985, 1992) recommended storage at 0–5°C with 5–10% CO_2 and 1–2% O_2 which had a good effect but was of limited commercial use. Saltveit (1989) also recommended 0–5°C with 5–10% CO_2 and 1–2% O_2 but showed that this combination had a 'high level of effect'. Typical storage conditions were given as 0–5°C with 5–10% CO_2 and 1–2% O_2 by Bishop (1996). Fellows (1988) recommended a maximum of 15% CO_2 and a minimum of 1% O_2. Izumi *et al.* (1996a, 1996b) found that the best storage conditions for cultivar Marathon florets were given as 0.5% O_2 with 10% CO_2 at 0 and 5°C, and 1% O_2 and 10% CO_2 at 10°C. Controlled atmosphere storage in levels of CO_2 ranging from 1.8 to 10% with 2% O_2 can prolong storage life to 3–4 weeks (Gorini, 1988).

Broccoli was held at 5°C in a stream of humidified air (control) or of a humidified mixture of 14% O_2 with 10% CO_2 or the spears wrapped with a non-perforated flexible polyvinylchloride film for 3 weeks. Weight loss was reduced by 17% in the controlled atmosphere treatment and by 50% by wrapping. Broccoli stored in controlled atmosphere or wrapped in poly-vinylchloride film retained their market quality significantly better than the unwrapped controls. Atmosphere composition within packages reached equilibrium within the first 24 h and did not change significantly over the following 3 weeks. O_2 consumption and CO_2 production by samples held in controlled atmosphere or in films were reduced by 30–40% in relation to the controls (Forney *et al.*, 1989).

The chlorophyll degradation was effectively delayed by storage in 5 and 10% CO_2. The samples stored at less than 1.5% O_2 and in cold storage (0–1°C) showed similar effects on chlorophyll degradation. Total carotenoid contents remained almost constant during 8 weeks of storage in 0.5% CO_2 with 0.8–3.0% O_2 and under normal atmospheres, but increased in samples stored under 10% CO_2 with 3% O_2 (Yang and Henze, 1988).

Minimum water loss from broccoli in storage was found in an atmosphere containing 5% CO_2 and 3% O_2. A reduction of O_2 to 1.5 or 0.8% and

0% CO_2 caused the most water loss compared with other modified atmospheres tested. Compared with normal atmosphere controls, atmospheres containing 5–10% CO_2 and 3% O_2 caused a small reduction in decay while 0.8 and 1.5% O_2 caused more rapid decay. CO_2 at 5 and 10% and O_2 at 3% were, by visual assessment, the most effective in preserving chlorophyll. Most controlled atmosphere combinations had no effect on flavour, but the samples stored at 10% CO_2 and 3% O_2 had an undesirable flavour after 8 weeks. CO_2 and O_2 had little effect on broccoli firmness (Yang and Henze, 1987).

Storage for 13 weeks in 15% O_2 with 6% CO_2 resulted in a modest 3.5% retention of chlorophyll and a 30% reduction of trim loss as compared to air. Storage of Green Valient in 8% CO_2 plus 10% O_2, 6% CO_2 plus 2% O_2 or 8% CO_2 plus 1% O_2 reduced chlorophyll loss by 13, 9 and 32% and reduced trim loss 40, 45 and 41% and their respiration rate as compared to air (McDonald 1985). An optimum storage life of about 8 weeks was attained at 0°C and 100% rh but close control of these conditions was required. Storage life could be further extended by reducing the O_2 concentration to less than 1%. Increasing the CO_2 concentration to 6% caused injury after 4–5 weeks. For storage at 20°C, CO_2 concentrations of 6–14% extended post-storage shelf life (Klieber and Wills, 1991).

When freshly cut heads of broccoli (cultivars Commander and Green Duke) were stored in air at 23 or 10°C, the florets rapidly senesced. Chlorophyll levels declined by 80–90% within 4 days at 23°C and within 10 days at 10°C. Storage at 5 or 10°C in 5% CO_2 with 3% O_2 at approximately 80% rh strongly inhibited loss of chlorophyll (Deschene et al., 1991). 60% CO_2 with 20% O_2 reduced the respiration rate in broccoli (Kubo et al., 1989a, 1989b). The cultivar Stolto was stored for 6 weeks at 1°C under the following percentages of CO_2 with O_2: 0% with 20%, 10% with 20%, 6% with 2.5%, 10% with 2.5%, and 15% with 2.5%. There was delayed development of soft rot and mould and colour and chlorophyll retention was better under controlled atmospheres than in air mainly due to increased CO_2 concentration. However, after 6 weeks of storage under an atmosphere containing 10% or more CO_2, the rate of respiration increased simultaneously with the development of undesirable odours and physiological injury. Among the atmospheres tested, 6% CO_2 and 2.5% O_2 was the best for long term (> 3 weeks) maintenance of broccoli quality while avoiding physiological injury (Makhlouf et al., 1989b).

Storage in 2% O_2 with 10% CO_2 at 15°C was recommended by Saijo (1990). 11% O_2 with 10% CO_2 at 4°C storage significantly reduced the growth of microorganisms on broccoli and extended the length of time they were subjectively considered acceptable for consumption. However, colour and texture was not significantly influenced by a controlled atmosphere during refrigerated storage compared to those stored at the same temperature in air (Berrang et al., 1990). The effects of controlled atmosphere storage on the survival and growth of Aeromonas hydrophila on fresh asparagus, broccoli and cauliflower were examined. Two lots of each vegetable were inoculated with *A. hydrophila* 1653 or K144. A third lot served as an uninoculated control. Following inoculation, vegetables were stored at 4 or 15°C under a controlled atmosphere storage system previously shown to extend the shelf life of each commodity or under ambient air. Populations of *A. hydrophila* were enumerated on the initial day of inoculation and at various intervals for 10 days (15°C) or 21 days (4°C) of storage. Direct plating of samples with selective media was used to enumerate *A. hydrophila*. The organism was detected on most lots of vegetables as they were received from a commercial produce supplier. Without exception, the controlled atmosphere storage system lengthened the time vegetables were subjectively considered acceptable for consumption. However, controlled atmosphere storage did not significantly affect populations of *A. hydrophila* which survived or grew on inoculated vegetables (Berrang *et al.*, 1989).

Buds of the cultivar Green Valiant were stored in the dark at 25°C in a continuous stream of nitrogen containing the following percentages of CO_2 and O_2: 0 and about 20 (air), 0 and 2.5, 6 and 20, and 10 and 20. Under high CO_2, respiration was reduced, but ethylene production and ethylene forming enzyme activity were temporarily stimulated early in this treatment. Under low O_2, respiration, ethylene production, ethylene forming enzyme activity, and 1-aminocyclopropane 1-carboxylic acid content were reduced, and the climacteric and chlorophyll losses were delayed to a greater extent than under high CO_2 (Makhlouf *et al.*, 1989a). Broccoli florets (cultivar Marathon) were held for 4 days at 10°C in containers with air flow (20.5% O_2 and less than 0.5% CO_2), restricted air flow (down to 17.2% O_2 and up to 3.7% CO_2), no flow (down to 1.3% O_2 and up to 30% CO_2) and nitrogen flow (less than 0.01% O_2 and less than 0.25% CO_2). Sensory analysis of cooked broccoli indicated a preference for the freshly harvested and air stored broccoli with samples stored under no flow conditions showing the opposite results (Hansen *et al.*, 1993).

Experiments were carried out to compare the rates of respiration and ethylene production by intact broccoli (cultivar Green Valiant) heads and florets and to determine the optimal atmosphere for preservation of florets at 4°C. Minimally processing broccoli heads into florets increased the rate of respiration throughout storage at 4°C in air, in response to wounding stress. Ethylene production was also stimulated after 10 days. Atmospheres for optimal preservation of the florets were evaluated using continuous streams of the following defined atmospheres: 0% CO_2 with 20% O_2 (air control), 6% CO_2 with 1% O_2, 6% CO_2 with 2% O_2, 6% CO_2 with 3% O_2, 3% CO_2 with 2% O_2 and 9% CO_2 with 2% O_2. The atmosphere consisting of 6% CO_2 with 2% O_2 resulted in extended storage of broccoli florets from 5 weeks in air to 7 weeks. This was demonstrated by delayed yellowing, prolonged chlorophyll retention, reduced development of mould and offensive odours, and better water retention. These beneficial effects were especially noticeable when the florets were returned from controlled atmospheres at 4°C to air at

Fig. 27. Modified atmosphere packaging of broccoli on sale in a UK supermarket in 1997.

20°C. It was concluded that minimal processing had little influence on optimal storage atmosphere, suggesting that recommendations for intact produce are useful guidelines for modified atmosphere packaging of minimally processed vegetables (Bastrash *et al.*, 1993). Modified atmosphere packaging is commonly used in spears prepared for sale in supermarkets (Fig. 27).

Brussels sprouts, *Brassica oleracea* variety *gemmifera*

Storage at levels of 5–7.5% CO_2 and 2.5–5% O_2 helped to maintain quality at 5 or 10°C, but at 0°C with O_2 levels below 1% internal discoloration can occur (Hardenburg *et al.*, 1990). SeaLand (1991) recommended 5–7% CO_2 with 1–2% O_2. Beneficial effects have been reported with storage in atmospheres containing 2.5, 5, and 7% CO_2 (Pantastico, 1975). Kader (1985) and Saltveit (1989) recommended 0–5°C with 5–7% CO_2 and 1–2% O_2 which had a good effect on storage but was of no commercial use. Kader (1989) recommended 0–5°C with 5–10% CO_2 and 1–2% O_2 which had an excellent potential benefit, but limited commercial use. Typical storage conditions were given as 0–5°C with 5–7% CO_2 and 1–2% O_2 by Bishop (1996). Brussels sprouts could be kept in controlled atmosphere storage at 0–1°C in 6% CO_2 with 15% O_2 or 6% CO_2 with 3% O_2 for up to 4 months (Damen, 1985) or for 80 days in 7% O_2 and 8% CO_2 (Niedzielski, 1984). High quality sprouts could be maintained for 10 weeks in cold storage at 1.5–2.0°C, which could be extended to 12 weeks in controlled atmosphere storage containing 5% CO_2 (Peters and Seidel, 1987).

Sprouts of the cultivars Lunette, Rampart and Valiant were stored for 2, 3 or 4 weeks at 2.5, 5 or 7.5°C with 0.5, 1, 2, 4 or 21% O_2, or in the following percentage combinations of O_2 and CO_2 : 1 with 10, 2 with 10 or 20 with 10. Storage was followed by 2 or 3 days of aeration at 10°C. Low O_2 levels reduced the rate of CO_2 production relative to that in air, but rates were similar among the low O_2 levels. Ethylene production was low at 2.5 and 5°C in all atmospheres, but at 7.5°C it was 20–170% higher in air than in low O_2. Ethylene production was virtually stopped during exposure to high CO_2, but increased markedly during aeration. Since low O_2 levels retarded yellowing and 10% CO_2 retarded decay development, the combination of low O_2 with high CO_2 effectively extended the storage life of sprouts at 5 and 7.5°C. The beneficial effect of controlled atmosphere storage was still evident after return of the samples to normal air. The sprouts retained good appearance for 4 weeks at 2.5°C, whether stored in controlled atmosphere or in air. Storage in 0.5% O_2 occasionally induced a reddish tan discoloration of the heart leaves and frequently an extremely bitter flavour in the non green portion of the sprouts. None of the atmosphere modifications appreciably affected either the tissue pH or texture of the sprouts (Lipton and Mackey, 1987).

The cultivar Rampart was stored at 2–3°C and more than 75% rh loose, and on the stem sprouts at 1°C and more than 95% rh. Loose sprouts became severely discoloured at the stalk end after only 1 week in air and after 2 weeks in 6% CO_2 and 3% O_2. Sprouts on the stem remained fresh in 6% CO_2 and 3% O_2 for 9 weeks at 2 to 3°C, and for 16 weeks at 1°C (Pelleboer, 1983).

Butter bean, Lima bean, *Phaseolus lunatus*

Storage in CO_2 concentrations of 25–35% inhibited fungal and bacterial growth without adversely affecting the beans (Brooks and McColloch, 1938).

Cabbage, *Brassica oleracea* variety *capitata*

Cabbage is perhaps the most common vegetable to be stored commercially in controlled atmosphere storage conditions. Stoll (1972) recommended 0°C with 3% CO_2 and 3% O_2 for red and Savoy cabbage and 0–3% CO_2 with 3% O_2 for white cabbage. Storage of Danish cultivars at 0°C with 2.5–5% CO_2 and 5% O_2 was successfully carried out for 5 months (Isenburg and Sayle, 1969). Hardenburg *et al.* (1990) showed that the optimum conditions for controlled atmosphere storage was 2.5–5% CO_2 with 2.5–5% O_2 at 0°C, while SeaLand (1991) recommended 5–7% CO_2 with 3–5% O_2 for green, red or Savoy. Kader (1985) recommended 0–5°C with 5–7% CO_2 and

3–5% O_2 or 0–5°C with 3–6% CO_2 and 2–3% O_2 (Kader, 1992) which was claimed to have had a good effect and was of some commercial use for long-term storage of certain cultivars. Saltveit (1989) recommended 0–5°C with 3–6% CO_2 and 2–3% O_2 which had a high level of effect. Typical storage conditions were given as 0°C with 5% CO_2 and 3% O_2 by Bishop (1996) for white cabbage. After storage of several white cabbage cultivars at 0–1°C for 39 weeks the percentage recovery of marketable cabbage after trimming was 92% in 5% CO_2, 3% O_2 and only 70% for those which had been stored in air (Geeson, 1984). Cabbage stored for 159 days in air had a 39% total mass loss and under controlled atmosphere storage of 3–4% CO_2 with 2–3% O_2 for 265 days the total mass loss was only 17%. The cabbage in controlled atmosphere storage showed better retention of green colour, fresh appearance and texture than those in air (Gariepy et al., 1985).

The cultivars Lennox and Bartolo were stored in air, 3% O_2 with 5% CO_2 or 2.5% O_2 with 3% CO_2. Disease incidence was lower at both the controlled atmosphere storage conditions and there were no trim losses or yield reductions and helped to retain green colour in Lennox (Prange and Lidster, 1991). Abscissic acid concentration increased during storage at 0°C in air. This increase was reduced in a 1% O_2 atmosphere which was shown also to delay the yellowing of the outer laminae and maintained a higher chlorophyll content (Wang and Ji, 1989).

Berard (1985) described experiments in which 25 cultivars were placed at 1°C and 92% rh either in air or rooms sealed to obtain 2.5% O_2 with 5% CO_2 for up to 213 days. Controlled atmosphere storage usually reduced or eliminated grey speck disease and reduced the incidence and severity of vein streaking, but not in every case, and considerably delayed degreening, and eliminated abscission and loss of dormancy during the first 122 days of storage. Two non-parasitic disorders (grey speck and vein streaking) of cabbage stored for 24 weeks were shown to be either eliminated or reduced in controlled atmosphere storage of 2.5% O_2 with 5% CO_2 compared to air storage (Berard et al., 1985). Black midrib and necrotic spot were both absent at harvest but developed on several of the 25 cultivars examined after 175 or 122 days. In comparison with storage in air those stored in 2.5% O_2 with 5% CO_2 had increased incidence of black midrib and it also favoured the development of inner head symptoms on susceptible cultivars. Controlled atmosphere storage similarly increased the incidence of necrotic spot in the core of the heads of cultivar Quick Green Storage, this effect being most evident in a season when senescence of cabbage was most rapid. Even though both disorders were initiated in the parenchyma cells, black midrib and necrotic spot had a distinct histological evolution and affected different cultivars under similar conditions of growth and storage (Berard et al., 1986).

Storage at 0.5–1.5°C, and 60–75% rh with 3% O_2 and 4.5–5% CO_2 lengthened the period for which cabbage can be stored by a least 2 months and improved quality compared to storage at 2–7°C with 60–75% rh in air (Zanon and Schragl, 1988).

Storage of the cultivar Tip Top at 5°C in 5% CO_2 with 5% O_2 gave the slowest rate of deterioration and at 2.5°C a CO_2 concentration > 2.5% was detrimental. The cultivar Treasure Island had a longer storage life than Tip Top in all CO_2 to O_2 combinations. At 2.5°C in 5% CO_2 with 20.5% O_2, 97% of Treasure Island was saleable after 120 days (Schouten, 1985). Controlled atmosphere storage at −0.5–0°C in 5–8% CO_2 prevented the spread of *Botrytis cinerea* and total storage losses were lower under controlled atmosphere storage than air (Nuske and Muller, 1984). Gariepy *et al.* (1984) showed that controlled atmosphere storage of cabbage at 3.5–5% CO_2 and 1.5–3% O_2 for 198 days had a total mass loss of 14% compared to 40% in air and better retention of colour, fresher appearance and firmer texture. The cultivar Winter Green was stored at 1.3°C with 5–6% CO_2, 2–3% O_2, 92% nitrogen and traces of other gases for 32 weeks compared to air storage at 0.3°C. The average trimming losses were less than 10% for the controlled atmosphere stored cabbage while those in air exceeded 30% and controlled atmosphere stored cabbage retained their colour, flavour and texture better (Raghavan *et al.*, 1984). Less disease (mainly *Botrytis cinerea* rot) and better head colour were observed with controlled atmosphere storage treatments (5% CO_2, 2.5% O_2 and the remainder as nitrogen) than with air during storage at 1°C and 75%, 85% or 100% rh (Pendergrass and Isenberg, 1974).

The need for fresh air ventilation at regular intervals to maintain ethylene concentrations at low levels was emphasized by Meinl *et al.* (1988).

The best plastic covering for the cultivar Nagaoka King stored in wooden boxes was polyethylene sheets (160 × 60 × 0.02 mm) with 0.4% perforations wrapped around the top and long sides of the boxes, with the slatted ends of the boxes left open (Peters *et al.*, 1986).

Cactus pear, Prickly pear, *Tuna* spp., *Opuntia ficus indica, O. robusta*

Cantwell (1995) stated that no work had been published on controlled atmosphere storage of cactus pear although she cites the beneficial effects of lining boxes of fruit with polyethylene film especially with paper or other absorbent material to absorb condensation. Testoni and Eccher Zerbini (1993) recommended 5°C with 2 or 5% CO_2 and 2% O_2 for prickly pear which reduced the incidence of chilling injury and rot development.

Capsicum, Sweet pepper, Bell pepper, *Capsicum annum* variety *grossum*

Kader (1985, 1992) recommended 8–12°C with 0% CO_2 and 3–5% O_2 which had a fair effect but was of limited commercial use. Saltveit (1989) recommended 8–12°C with 0% CO_2 and 2–5% O_2 which had only a slight effect. SeaLand (1991) recommended 0–3% CO_2 with 3–5% O_2. Storage with O_2 levels of 3–5% was shown to retard respiration but high CO_2 can reduce loss of green colour but also result in calyx discoloration (Hardenburg et al., 1990). Transport life could be extended in controlled atmospheres of 2–8% CO_2 and 4–8% O_2 and the storage of California Wonder at 8.9°C in air was 22 days, but in 10% CO_2 with 5% O_2 it was 38 days (Pantastico, 1975). Otma (1989) showed that at 8°C and > 97% rh the optimum conditions for controlled atmosphere storage were 2% CO_2 with 4% O_2. Storage of cultivar Jupiter fruits for 5 days at 20°C in 1.5% O_2 resulted in post-storage respiratory suppression of CO_2 production for about 55 h after transfer to air (Rahman et al., 1995).

Fruits of the cultivar Jupiter were held in 1.5% O_2 with 98.5% nitrogen or air for 5 days at 20°C. Fruits held in 1.5% O_2 were then exposed to humidified air for 24 h at 20°C. After 5 days in 1.5% O_2 or in air, samples were removed for isolation of mitochondria. Samples were also removed 24 h after transfer to air. The steady state oxidative capacity of mitochondria isolated from fruits stored in 1.5% O_2 for 5 days was reduced by about 45% compared with mitochondria from air stored fruits. During subsequent holding for 24 h in air, the O_2 uptake capacity recovered and attained values similar to those isolated from fruits held continuously in air. There were no observed differences in the ultrastructural features of mitochondria isolated from fruits held in 1.5% O_2 for 5 days compared with those from freshly harvested fruits. Although CO_2 production and O_2 consumption rates from whole fruit exhibited residual inhibition following low O_2 exposure, the inhibition did not persist in isolated mitochondria. These data suggest that the residual respiratory responses may be due to the suppression of extra-mitochondrial low affinity oxidases (Rahman et al., 1993b). The cultivar Jupiter fruits were stored in 1.5, 5 or 10% O_2, at 20°C for 24 h then returned to air. The lowest O_2 concentration (1.5%) exerted the greatest residual effect on fruit CO_2 production and O_2 uptake measured within 10 min of return to air (53 and about 80% lower, respectively, than for air stored fruits). No ethanol was detected in the headspace gas of fruits stored in 1.5% O_2. CO_2 production continued to be suppressed for about 24 h after transfer from 1.5% O_2 to air. Storage at 5% O_2 resulted in less suppression of CO_2 production and O_2 uptake upon transfer to air than storage at 1.5%, while 10% O_2 exerted no residual effect. Extending the storage period in 1.5% O_2 to 72 h extended the residual effect from 24 to 48 h. In a separate experiment, ethylene production was not affected by storage in 1.5 or 4% O_2 for 24 or 72 h. The residual effect exhibited in whole fruits was not apparent in mitochondria isolated from fruits stored in 1.5 or 4% O_2 (Rahman et al., 1993). Mazurka fruits were stored for 15 days at 8°C in atmospheres containing 3% CO_2 with 3% O_2 or 0% CO_2 with 21% O_2. For each atmosphere, humidity levels of 85, 90, 95 or 100% were compared. Fruits were then stored for 7 days in air at 20°C and 70% rh. The incidence of an unspecified decay during the post-storage period increased as rh increased during storage. Controlled atmosphere storage in 3% CO_2 with 3% O_2 reduced the incidence of post-storage decay compared with storage in 0% CO_2 with 21% O_2. Weight loss and softening increased as humidity during storage decreased but these were unaffected by the gas composition of the storage atmosphere (Polderdijk *et al.*, 1993). Fruits of the cultivar California Wonder were stored in O_2 with nitrogen mixtures with 1, 3, 5, 7 or 21% (air) O_2 for 4 weeks at 10°C. After 2 weeks, 33% of the control air stored fruits had decayed while only 9% of those stored in 1% O_2 had decayed and the benefit of 1% O_2 in preventing decay continued throughout the 4 weeks of storage. The 3 and 5% O_2 atmospheres slightly reduced decay for a short time while the incidence of decay in fruits stored in 7% O_2 was no different to that in air stored fruits. Fruits stored in low O_2 atmospheres had lower internal ethylene contents that air stored fruits and in 1% O_2 fruits had significantly lower internal CO_2 contents than the control air stored fruits and fruits stored in the other low O_2 atmospheres. Colour retention was greater in fruits stored in low O_2 atmospheres than in air stored fruits (Luo and Mikitzel, 1996).

Carambola, Star fruit, *Averrhoa carambola*

Storage at 7°C and 85–95% rh with either 2.2% O_2 and 8.2% CO_2 or 4.2% O_2 and 8% CO_2 resulted in low losses of about 1.2% in 1 month and fruit retained a bright yellow colour with also good retention of firmness, Brix and acidity compared to fruits stored in air (Renel and Thompson, 1994).

Carrot, *Daucus carota* subspecies *sativus*

Controlled atmosphere storage was not recommended since carrots stored in 5–10% CO_2 with 2.5–6% O_2 had more mould growth and rotting than those stored in air (Hardenburg *et al.*, 1990). Increased decay was also reported in atmospheres of 6% CO_2 with 3% O_2 compared to storage in air (Pantastico, 1975). Ethylene or high levels of CO_2 in the storage atmosphere can give the roots a bitter flavour (Fidler, 1963). However, storage in 1–2% O_2 at 2°C for 6 months was previously reported to have been successful (Platenius *et al.*, 1934) and Fellows (1988) recommended controlled atmosphere storage in a maximum of 4% CO_2 and a minimum of 3% O_2.

Storage atmospheres containing 1, 2.5, 5 or 10% O_2 inhibited both sprouting and rooting during storage at 0°C, but increased mould infection. Atmospheres containing 21% or 40% O_2 reduced mould infection, but increased sprouting and rooting of stored carrots (Abdel-Rahman and Isenberg, 1974).

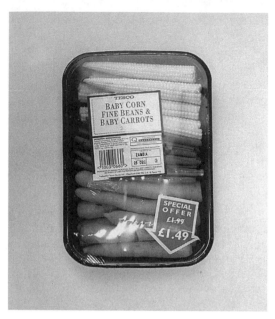

Fig. 28. Modified atmosphere packaging of baby carrots, babycorn and green beans on sale in a UK supermarket in 1997.

Modified atmosphere packaging is used commercially to extend the marketable life of immature carrots (Fig. 28).

Cassava, Tapioca, Manioc, Yuca, *Manihot esculenta*

Plumbley and Rickard (1991) reviewed the factors involved in the post-harvest deterioration of cassava. Roots can deteriorate within a day or so of harvesting due to a physiological disorder called 'vascular streaking' (Thompson and Arango, 1977). Coating roots with paraffin wax kept them in good condition for 1–2 months at room temperature in Bogota in Colombia (Young *et al.*, 1971; Thompson and Arango, 1977) and this coating treatment is commonly applied to roots exported from Costa Rica. Another treatment is dipping the roots in a fungicide (benomyl or thiabendazole) and packing them in plastic directly after harvest (Thompson and Arango, 1977). Oudit (1976) showed that freshly harvested roots could be kept in good condition for up to 4 weeks when stored in polyethylene bags. Waxed cassava cultivar Valencia taken from 12-month-old plants were stored at 25°C for 3 days in either 54–56% rh or 95–98% rh and O_2 levels of either 21% (air) or 1%. Low humidity in air enhanced vascular streaking development with a 46% incidence. Storage in 1% O_2 at low humidity reduced vascular streaking to an incidence of 15%. However, vascular streaking was almost completely inhibited (1.4% incidence) at high humidity irrespective of the O_2 level. It appeared that the occurrence of vascular streaking in cassava tubers was primarily related to water stress in the tissue, while O_2 was secondarily involved in the discoloration reaction (Aracena *et al.*, 1993).

Cauliflower, *Brassica oleracea* variety *botrytis*

Storage at 0°C with 10% CO_2 and 10% O_2 for 5 weeks was recommended by Wardlaw (1937). Storage at 0°C in 10% CO_2 and 11% O_2 (Smith, 1952) kept them in good condition for 5 weeks. Stoll (1972) advocated 0°C with 0–3% CO_2 and 2–3% O_2. Adamicki (1989) described successful storage of autumn cauliflowers at 1°C with 2.5% CO_2 and 1% O_2 for 71–75 days or in the same atmospheres but at 5°C for 45 days. Monzini and Gorini (1974) recommended 0°C with 5–10% CO_2 and 5% O_2 for 50–70 days. SeaLand (1991) recommended 2–5% CO_2 with 2–5% O_2. Fellows (1988) recommended a maximum of 5% CO_2 and a minimum of 2% O_2. Kader (1985) recommended 0–5°C with 2–5% CO_2 and 2–5% O_2 which had a fair effect but, in a later paper suggested that controlled atmosphere storage of cauliflower was of no commercial use (Kader, 1992). Romo Parada *et al.* (1989) showed that curds stored at 1°C and 100% rh in 3% O_2 with either 2.5 or 5% CO_2 were still acceptable after 7 weeks of storage, while 3% O_2 with 10% CO_2 caused softening, yellowing and increased leakage. Curds of cultivar Primura were stored successfully for 6–7 days at 0–1°C and 90–95% rh in circulating air and for 20–25 days in controlled atmospheres of 4–5% CO_2 with 16–27% O_2 was recommended (Saray, 1988). Considerable work has been done in Holland (originally in the Sprenger Institute) since the 1950s. This work has shown very little benefit of controlled atmosphere storage, but storage at 0–1°C and > 95% rh with 5% CO_2 and 3% O_2 gave a better external appearance but had no effect on curd quality (Mertens and Tranggono, 1989). Tataru and Dobreanu (1978) recommended 5–6% CO_2 with 3% O_2.

Summer and autumn grown cultivars Idol and Master were stored at 1 or 5°C and 80–90% rh. The summer crops were stored for 30 days at both temperatures and the autumn crops for 73–75 days at 1°C or 45 days at 5°C. Both summer and autumn crops in controlled atmospheres had better quality (leaf colour, curd colour, firmness and market value) than in normal atmosphere. The best results were obtained in 2.5% CO_2 and 3% O_2 or 5% CO_2 and 3% O_2. Curds of the summer crop stored in controlled atmospheres had higher dry matter, sugar and ascorbic acid contents than when stored in air. Curds of the autumn crop stored in 2% CO_2 and 3% O_2 had a superior composition to curds stored in other controlled atmosphere treatments at both storage temperatures (Adamicki and Elkner, 1985).

In other work controlled atmospheres did not extend their storage life and CO_2 levels of 5% or more or O_2 levels of 2% or less injured the curds (Hardenburg *et al.*, 1990). After storage at either 4.4 or 10°C with levels of

5% CO_2 in the storage atmosphere some injury was evident after the curds were cooked (Ryall and Lipton, 1972). Storage was compared at 0–1°C and less than 95% rh in an atmosphere containing 5% CO_2 and 3% O_2 or in normal air for 4 or 6 weeks with subsequent shelf life studies at 10°C and 85% rh. It was concluded that storage in 5% CO_2 and 3% O_2 had a very small effect, if any, on respiration of the curds (Mertens and Tranggono, 1989). Work on the cultivar Pale Leaf 75 by Tomkins and Sutherland (1989) with curds stored at 1°C for up to 47 days in normal refrigerated air, or in 5% CO_2 with 2% O_2 or 0% CO_2 with 2% O_2 showed that curds stored in 2% O_2 alone suffered severe, irreversible low O_2 injury and were discarded after 27 days of storage. After 47 days storage in 5% CO_2 with 2% O_2 plus the 4-day marketing period, curd quality was acceptable owing to the reduction in incidence of soft rot and black spotting noted in air stored curds. In air, curds were unsaleable after only 27 days of storage plus the 4-day marketing. Cooking quality of the curds was impaired when cooked directly after 47 days in controlled atmosphere storage, but was acceptable if the curds were held for 1.5–3 h in air before cooking. Controlled atmosphere storage at 4°C in 18% O_2 with 3% CO_2 for 21 days had no significant effect on the growth of microorganisms compared to those stored in air at the same temperature (Berrang et al., 1990). Storage for 8 days at 13°C in air or in high CO_2 (15% CO_2, 21% O_2, remainder nitrogen) accelerated the deterioration of microsomal membranes during storage and caused an early loss in lipid phosphate (Voisine et al., 1993). Kaynas et al. (1994) stored the cultivar Iglo in controlled atmosphere storage (3% CO_2 with 3% O_2), polyethylene film bags (30 µm), polyethylene film bags with 5 mm diameter hole per kilogram head, PVC film bag (15 µm) or air, for various times at 1°C and 90–95% rh. They found that the cauliflowers could be stored in controlled atmosphere or PVC film bags for a maximum of 6 weeks with a shelf life of 3 days at 20°C, which was double the storage life of those stored in air unwrapped. The permeability of the PVC film used in the studies was given as 200 g of water vapour m^{-2} day^{-1} and 12,000 ml O_2 m^2 day^{-1} atm^{-1}.

Celeriac, Turnip rooted celery, *Apium graveolens* variety *rapaceum*

Controlled atmosphere storage was not recommended because 5–7% CO_2 and low O_2 increased decay during 5 months of storage (Hardenburg et al., 1990). SeaLand (1991) showed that controlled atmosphere storage had a slight to no effect. Saltveit (1989) recommended 0–5°C with 2–3% CO_2 and 2–4% O_2 which had only a slight effect. Pelleboer (1984) showed that with celeriac storage in air at 0–1°C gave better results than controlled atmosphere storage.

Celery, *Apium graveolens* variety *dulce*

Storage with 5% CO_2 and 3% O_2 at 0°C reduced decay and reduced loss of green colour (Hardenburg *et al.*, 1990). SeaLand (1991) recommended 2–5% CO_2 with 2–4% O_2. Storage at either 0, 4 or 10°C for 7 days with 25% CO_2 resulted in browning at the base of the petioles, reduced flavour and a tendency for petioles to break away more easily (Wardlaw, 1937). 1–4% O_2 in storage was shown to only slightly help to preserve the green colour and although 2.5% CO_2 may be injurious, levels of 9% during 1 month of storage caused no damage (Pantastico, 1975). Kader (1985) recommended 0–5°C with 0% CO_2 and 2–4% O_2 or 0–5°C with 0–5% CO_2 and 1–4% O_2 (Kader, 1992) which had a fair effect but was of limited commercial use in mixed loads with lettuce. Reyes (1989) reviewed work on controlled atmosphere storage and concluded that at 0–3°C with 1–4% CO_2 and 1–17.7% O_2, storage could be prolonged for 7 weeks and he specifically referred to his recent work which showed that at 0–1°C and 2.5–7.5% CO_2 with 1.5% O_2 market quality could be maintained for 11 weeks. Saltveit (1989) recommended 0–5°C with 3–5% CO_2 and 1–4% O_2, but it only had a slight effect. Total weight loss of less than 10% over a 10-week period was achieved by storing celery in atmospheres containing 1% O_2 combined with 2 or 4% CO_2 at 0°C. Significant increases in marketable celery resulted when ethylene was scrubbed from some atmospheres. It is suggested that improved visual colour, appearance and flavour, and increased marketable celery yield, justify the use of 4% CO_2 in celery storage (Smith and Reyes, 1988). The cultivar Utah 52 70, stored at 0–1°C in 1.5% O_2, had better marketable quality after 11 weeks than that stored in air. Marketable level was improved by using 2.5–7.5% CO_2 in the storage atmosphere, but not by 2–4% CO_2 (Reyes and Smith, 1987). Celery was stored in air or 3–4% CO_2 with 2–3% O_2 compositions for 62 and 71 days, respectively. The controlled atmosphere stored celery retained better texture and crispness than celery in air. At the end of storage no real difference in total mass loss was observed between the controlled atmosphere stored and air stored celery (Gariepy *et al.*, 1985).

At 1°C, the growth *in vitro* of *Sclerotinia sclerotiorum* on celery extract agar, compared with normal air, was most suppressed in a storage atmosphere containing 7.5% CO_2 with 1.5% O_2, but only slightly suppressed in 4% CO_2 with 1.5% O_2 or in 1.5% O_2 alone. Watery soft rot caused by *S. sclerotiorum* was severe on celery stored in normal air for 2 weeks at 8°C. A comparable level of severity took 10 weeks to develop at 1°C. At 8°C suppression of this disease was greatest in atmospheres of 7.5–30% CO_2 with 1.5% O_2, but only slightly reduced in 4–16% CO_2 with 1.5% O_2 or in 1.5–6% O_2 alone (Reyes, 1988). A combination of 1 or 2% O_2 and 2 or 4% CO_2 prevented black stem development during storage (Smith and Reyes, 1988). Decay was most severe on celery stored in 21% O_2 compared to controlled atmosphere storage; *Botrytis cinerea* and *S. sclerotiorum* were isolated most frequently from decayed celery (Reyes and Smith, 1987). After storage at 1°C celery grown in Ontario in Canada developed a black discoloration of the stalks which appeared outwardly in a striped pattern along the vascular strands. In cross section the vascular strands were discoloured and appeared blackened. A controlled storage atmosphere of 3% O_2 with 2% CO_2 almost completely eliminated the disorder. Ethylene and pre-storage treatment with sodium hypochlorite had little or no influence on the occurrence of the disorder but there was some indication of difference in cultivar susceptibility (Walsh et al., 1985).

Cherimoya, *Annona cherimola*

Hatton and Spalding (1990) recommended 8°C with 10% CO_2 and 2% O_2. Kader (1993) recommended 10°C, with a range of 8–15°C, combined with 5% CO_2 and 2–5% O_2. De la Plaza et al. (1979) showed that good quality was maintained for 22 days in storage at 8.5°C and 90% rh with 10% CO_2 and 2% O_2. Fruits of the cultivar Fino de Jete were stored in air and controlled atmosphere conditions of 3% O_2 in combination with 0% CO_2, 3% CO_2 or 6% CO_2 at 9°C. Low O_2 caused the greatest reductions in respiration, sugar content and acid content, whereas high CO_2 caused the greatest reductions in ethylene production and softening rate. High CO_2 in combination with low O_2 had an additive reducing effect on ethylene production, softening rate and malic acid content and did not significantly affect sugar and citric acid content. CO_2 at 3% and 6% delayed the softening by 5 and 14 days beyond 3% O_2 with 0% CO_2 and air storage, respectively. This allowed sufficient accumulation of sugars and acids to reach an acceptable quality. The results suggest that 3% O_2 with 3% CO_2 and 3% O_2 with 6% CO_2 atmospheres can extend the storage life of cherimoya fruits stored at 9°C by 2 weeks beyond that of fruits stored in air (Alique and Oliveira, 1994). The cultivar Concha Lisa was harvested in Chile 240 days after pollination. Respiration rates during storage at 10°C in air showed a typical climacteric pattern with a peak after some 15 days. The climacteric was delayed by storage in 15% or 10% O_2 and fruits kept in 5% O_2 did not show a detectable climacteric rise and did not produce ethylene. All fruits ripened normally after being transferred to air storage at 20°C. However, the time needed to reach an edible condition differed with O_2 level and was inversely proportional to O_2 concentration during storage (11, 6 and 3 days to ripen following 30 days of storage at 5, 10 and 20% O_2, respectively). The delay of the climacteric by storage in 5% O_2, coupled with the continued ability to ripen, makes this treatment suitable for controlled or modified atmosphere cherimoya packaging (Palma et al., 1993). Berger et al. (1993) showed that waxed (unspecified) fruit of the cultivar Bronceada could be stored at 10°C, 90% rh with 0% CO_2 and 5% O_2 for 3 weeks without visible change. Fruits of cultivar Fino de Jete were stored for 4 weeks at 10–12°C in chambers supplied with a continuous flow of CO_2. The CO_2 treatment prolonged the storage life of cherimoyas by at least 3 weeks compared to those stored in air (Martinez-Cayuela et al., 1986).

Not all reports on controlled atmosphere storage have been positive. De la Plaza (1980) and Moreno and De la Plaza (1983) showed that fruits of the varieties Fino de Jete and Campa stored in 10% CO_2 and 2% O_2 had a higher respiration rate than fruits stored at the same temperature in air. Also there were some interactions between variety and controlled atmosphere storage treatment.

Pre-treatment of the cultivar Fino de Jete for 3 days at 6°C in 20% CO_2 combined with 20% O_2 maintained fruit firmness and colour compared to those not treated (Escribano et al., 1997).

Cherry, *Prunus avium* (sweet cherry), *P. cerasus* (sour cherry)

Storage with 20–25% CO_2 or 0.5–2% O_2 helped to retain firmness, green stems and bright fruit colour (Hardenburg et al., 1990). Lawton (1996) recommended 0–5°C, 95% rh, in combination with 10–15% CO_2 and 3–10% O_2. SeaLand (1991) recommended 10–12% CO_2 with 3–10% O_2 for sour cherries and 20–25% CO_2 with 10–20% O_2 for sweet cherries. Typical storage conditions were given as 0–5°C with 10–15% CO_2 and 3–10% O_2 by Bishop (1996). Storage with CO_2 levels of up to 30% reduced decay (Haard and Salunkhe, 1975).

Storage for sweet cherry cultivars was recommended at 0–5°C with 10–12% CO_2 and 3–10% O_2 (Kader, 1985) or 0–5°C with 10–15% CO_2 and 3–10% O_2 (Kader, 1989). For the cultivars Napoleon, Stella and Karabodur, Eris et al. (1994) found that the optimum controlled atmosphere conditions of those tested (0% CO_2 with 21% O_2, 5% CO_2 with 5% O_2, 10% CO_2 with 3% O_2, 20% CO_2 with 2% O_2 and 0% CO_2 with 2% O_2 at 0°C and 90–95% rh) were 5% CO_2 with 5% O_2. Ionescu et al. (1978) recommended 0°C 5% CO_2 with 3% O_2 for 30 days with 9% losses for the sweet cultivars Hedelfingen and Germersdorf. For the sour cultivars, Crisana (Paddy) and Mocanesti they also recommended 0°C 5% CO_2 with 3% O_2 but for only 20 days which resulted in about 7% loss.

The treatment times required to completely kill specific insects by O_2 levels at or below 1% suggests that low O_2 atmospheres are potentially useful as postharvest quarantine treatments for some fruits. Fruits of the cultivar Bing, were treated with 0.25% or 0.02% O_2 (balance nitrogen) at 0, 5 or 10°C to study the effects of these insecticidal low O_2 atmospheres on fruit postharvest physiology and quality attributes. Development of alcoholic off-flavour, associated with ethanol accumulation, was the most common and important detrimental effect that limited fruit tolerance to low O_2 (Ke and Kader, 1992b). Chen et al. (1981) compared a range of controlled atmosphere storage treatments on the cultivar Bing at −1.1°C for 35 days and found that the combination of 0.03% CO_2 combined with 0.5–2% O_2 maintained the greenness of the stems, brighter fruit colour and higher acidity than other treatments. Storage in 10% CO_2 had similar effects with the exception of maintaining the stem greenness. Folchi *et al.* (1994) stored the cultivar Nero 1 in less than 0.1% CO_2 with 0.3% O_2 and showed an increase in aldehyde and ethanol content of fruit with increased storage time.

Bertolini (1972) stored the cultivar Durone Neo I at 0°C for up to 20 days in air, 20% CO_2 with 17% O_2 or in 0.05 mm polyethylene film bags. Fruits stored in the polyethylene bags retained their freshness best as well as the colour of both the fruit and the stalks.

Chestnut, *Castanea crenata*

It was reported by Lee *et al.* (1983) that the best storage results for the cultivar Okkwang were obtained when they were sealed in 0.1 mm polyethylene bags (30×40 cm) with nine to 11 pinholes. The total weight loss was less than 20% after 8 months of storage. During storage the CO_2 concentration in the polyethylene bags reached 5–7% and the O_2 concentration in the polyethylene bags was reduced to between 7 and 9%.

Chicory, *Cichorium intybus*

Storage of Witloof chicory in 4–5% CO_2 with 3–4% O_2 at 0°C delayed greening of the tips in light and delayed opening of the heads (Hardenburg *et al.*, 1990). Saltveit (1989) recommended 0–5°C with 4–5% CO_2 and 3–4% O_2 for Witloof chicory which had only a slight effect.

Tindall (1983) recommended 0°C for 20 days in plastic wraps or bags with the tops left open which suggests that modified atmosphere packaging might be detrimental.

Chillies, Chilli peppers, Hot peppers, Cherry peppers, *Capsicum annum,* *C. frutescens*

Storage at 0% CO_2 with 3–5% O_2 was recommended by SeaLand (1991). Kader (1985, 1992) recommended 8–12°C with 0% CO_2 and 3–5% O_2 which he observed had a fair effect but was of no commercial use, but 10–15% CO_2 was beneficial at 5–8°C. Saltveit (1989) recommended 12°C with 0–5% CO_2 and 3–5% O_2 for the fresh market which had only a slight effect, and for processing 5–10°C with 10–20% CO_2 and 3–5% O_2 was recommended which had a only a moderate effect. Fresh chillies are often marketed in modified atmosphere packaging (Fig. 29).

Fig. 29. Modified atmosphere packaging of chillies in a supermarket in Belgium.

Chinese cabbage, *Brassica pekinensis*

Storage recommendations by Hardenburg *et al.* (1990) were 1% O_2 at 0°C. Storage in 1% O_2 was shown to extend its storage life and reduce the decline in ascorbic acid, chlorophyll and sugar content in the outer leaf laminae (Wang and Kramer, 1989). Saltveit (1989) recommended 0–5°C with 0–5% CO_2 and 1–2% O_2 which had only a slight effect. Wang and Kramer (1989) showed that 1% O_2 storage at 0°C extended the storage life of Chinese cabbage compared to those stored in air.

Harvested heads of the cultivar WR 60, which is susceptible to vein browning (necrosis), were stored at 0–1, 2.5 or 4°C in air or in controlled atmosphere storage with 0.5% CO_2 and 2% O_2, or at 0–1°C with various CO_2 and O_2 combinations for 3 or 4 months and then at 15°C for 1 week. After 3 months there was a high incidence of vein necrosis in cabbage that were stored in air at 0–1°C, and in controlled atmosphere storage at 0–1°C with 0.5% CO_2 and 2% O_2. After 4 months in controlled atmosphere storage at 0–1°C the incidence of rot (unspecified) increased as the CO_2 percentage rose to 6%. The recommended storage conditions in air were 0–1°C, with a maximum storage period of 6 weeks (Mertens, 1985).

In trials with the cultivars Tip Top and Treasure Island stored at 2.5 or 5.0°C in 0.5, 2.5 or 5.0% CO_2 combined with 1–20.5% O_2, with Tip Top the storage was best at 5°C with 5% CO_2 with 5% O_2. The results with Treasure Island were inconclusive because the control heads kept very well with an average of 68% saleable produce after 120 days (Apeland, 1985). After 4 months of storage of the Chinese cabbage cultivars Chiko and WR 60 in

2–3°C with 0.5% CO_2 and 3% O_2 or less the average percentage of healthy heads was 72%, and after 5 months, 60% or more were healthy. After subsequent storage at 15°C and 85% rh (to simulate shelf life) the corresponding percentages were 58 and 43. Following storage in CO_2 with O_2 concentrations of 6 with 3 and 6 with 15% there was a rapid fall in quality during holding, confirming that high CO_2 levels were harmful (Pelleboer and Schouten, 1984).

Citrus hybrids, *Citrus* spp.

Chase (1969) stored Temple oranges (*C. sinensis* × *C. reticulata*) and Orlando Tangelo (*C. sinensis* × *C. paradisi*) at 1°C in atmospheres containing 5% CO_2 with 10% O_2. There was little benefit from controlled atmosphere storage compared to air storage for Orlando Tangelos but when Temple oranges were stored for 5 weeks in controlled atmosphere storage followed by 1 week at 21°C in air their flavour was superior to those stored in air throughout.

Cranberry, American cranberry, *Vaccinium macrocarpon*, European cranberry, *V. oxycoccus*

Controlled atmosphere storage was not successful in increasing their storage life (Hardenburg *et al.*, 1990), but Kader (1989) recommended 2–5°C with 0–5% CO_2 and 1–2% O_2.

Cucumber, *Cucumis sativus*

Stoll (1972) recommended 14°C with 5% CO_2 and 5% O_2. Storage at 5–7% CO_2 with 3–5% O_2 was recommended by SeaLand (1991). Fellows (1988) recommended a maximum of 10% CO_2 and a minimum of 3% O_2. Storage in 5% O_2 was shown to retard yellowing (Lutz and Hardenburg, 1968). Storage at 8.3°C with 3–5% O_2 gave a slight extension in storage life (Pantastico, 1975) also storage in 5% CO_2 with 5% O_2 extended their storage life (Ryall and Lipton, 1972; Pantastico, 1975). Kader (1985, 1992) recommended 8–12°C with 0% CO_2 and 3–5% O_2 which had a fair effect but was of no commercial use. Saltveit (1989) recommended 12°C with 0% CO_2 and 1–4% O_2 for the fresh market, and 4°C with 3–5% CO_2 and 3–5% for pickling cucumbers both of which had only a slight effect. A storage temperature of 12.5°C with 5% O_2 and 5% CO_2 levels were recommended and not stored with ethylene producing fruits or vegetables, since chlorophyll will be lost rapidly (Schales, 1985). Reyes (1989) showed that the virulence of *Mucor mucedo* (Mucor rot) and *Botrytis cinerea* (grey mould) were suppressed

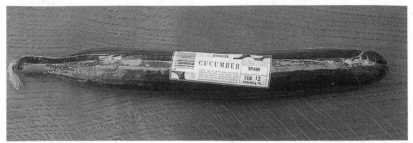

Fig. 30.   Shrink film wrapping of cucumbers giving modified atmosphere packaging.

most in 7.5% CO_2 with 1.5% O_2. Mercer and Smittle (1992) stored fruits of the cultivar Gemini II in atmospheres containing 0, 5 or 10% CO_2 and 5 or 20% O_2 at 5 or 6°C for 2, 4 or 6 days in 1989 and at 5°C for 5 days or at 3°C for 10 days in 1990. Chilling injury symptoms were assessed after 2–4 days at 25°C. High CO_2 concentration and low O_2 concentration delayed the onset of chilling injury symptoms, but did not prevent their development. Chilling injury symptoms increased with longer exposure to chilling temperatures and were associated with solubilization of cell wall polysaccharides.

In Europe fruit are commonly marketed in a shrink wrapped plastic sleeve to help retain their texture and colour which provides a modified atmosphere (Fig. 30).

Durian, *Durio zibethinus*

Kader (1993) recommended 10.5°C, with a range of 12–20°C, combined with 5–20% CO_2 and 3–5% O_2. Storage of 85% mature Mon Tong durians at 22°C showed that O_2 levels of 5 or 7.5% inhibited ripening, but fruit ripened normally when returned to air. With 2% O_2 ripening was inhibited and the fruit failed to ripen when removed and stored in air while CO_2 levels of up to 20% in air did not affect the speed of ripening or the quality of the ripe fruit (Tongdee *et al.*, 1990). Fruits of the cultivars Chanee, Kan Yao and Mon Tong were harvested at three maturity stages and stored at 22°C. The ripening fruits exhibited a very marked climacteric rise in respiration and a parallel increase in ethylene evolution, immediately upon harvest. The rates of CO_2 and ethylene production at harvest and the peak climacteric respiratory value were higher in fruits harvested at a more advanced stage. The number of days between harvest and the climacteric peak was reduced as the maturity stage at harvest advanced. Ripening of fruits stored in low O_2 and/or high CO_2 atmosphere was studied. Low O_2 (10%) caused a significant reduction in CO_2 and ethylene production. However, the onset of ripening and ripe fruit quality were not affected. Ripening was inhibited in fruits stored in O_2 at 5–7.5% but recovered when fresh air was subsequently introduced. O_2 at this level did not affect ripening in fruits harvested at an advanced stage of maturity. Fruits stored in 2% O_2 failed to resume ripening when removed to air. Fruits stored in CO_2 at 10 or 20% were either not affected or showed a slight reduction in ethylene production. Thus high levels of CO_2 alone did not influence the onset of ripening or other quality attributes of ripe fruits. High CO_2 (5, 10, 15 or 20%) in combination with low O_2 (10%) in the storage atmosphere had a greater effect on the condition of the aril than either high CO_2 or low O_2 alone. The aril remained hard in the less mature fruits stored in 10% O_2 and 15 or 20% CO_2 (Tongdee *et al.*, 1990).

Endive, Escarole, Witloof chicory, *Cichorium endiva*

Storage in 25% CO_2 can cause the central leaves to turn brown (Wardlaw, 1937). Monzini and Gorini (1974) recommended 0°C with 1–5% CO_2 and 1–5% O_2 for 45–50 days.

Lips *et al.* (1989) described work where Witloof chicory roots (cultivars Zoom and Toner) were harvested at the end of October, defoliated and placed in airtight or non airtight containers in a cold store (1°C). The airtight containers were aerated once a week or once every 2 or 3 weeks. Roots were removed for forcing at regular intervals between 5 and 19 weeks of storage (cultivar Zoom) or between 20 and 32 weeks of storage (cultivar Toner). Roots stored in airtight containers aerated once every 3 weeks did not produce any chicons after 12 or more weeks of storage. Net productivity (chicon weight to root weight ratio) remained stable throughout the storage period in roots stored in airtight containers aerated once a week or once every 2 weeks, whereas it declined in control roots kept in non-airtight containers. Chicon quality remained relatively constant in cultivar Zoom throughout the storage period in roots stored in airtight containers aerated once a week or once every 2 weeks, but declined in control roots of both cultivars stored in non-airtight containers and in cultivar Toner roots stored in airtight containers aerated once a week or once every 2 weeks. Average weight losses of 2 and 11% were recorded for cultivar Zoom roots kept in airtight and non-airtight containers, respectively, for 19 weeks, and 4 and 14% for cultivar Toner roots kept in airtight and non-airtight containers, respectively, for 32 weeks.

Fig, *Ficus carica*

Storage in high CO_2 atmospheres reduced mould growth without affecting the flavour of the fruit (Wardlaw, 1937). Hardenburg *et al.* (1990) also showed that storage with enriched CO_2 was a useful supplement to refrigeration. Storage at 0–5°C with 15% CO_2 and 5% O_2 was recommended by Kader (1985) as was 0–5°C with 15–20% CO_2 and 5–10% O_2 by Kader (1989, 1992). SeaLand (1991) also recommended 15% CO_2 with 5% O_2. Colelli *et al.* (1991) showed that good quality of fresh Mission figs were maintained for up to 4 weeks when kept at 0, 2.2 or 5°C in atmospheres enriched with 15 or 20% CO_2. The visible benefits of exposure to high CO_2 levels were a reduction of the incidence of decay and the maintenance of a bright external appearance. Ethylene production was lower, and fruit softening was slower in figs stored at high CO_2 concentrations compared to those kept in air. Ethanol content of the fruits stored in 15 or 20% CO_2 increased slightly during the first 3 weeks and moderately during the fourth week, while acetaldehyde concentration increased during the first week, then decreased. It was concluded that the postharvest life of fresh Mission figs can be extended by 2–3 weeks by holding them at 0–5°C in atmospheres enriched with 15–20% CO_2 but off-flavours could be a problem by the fourth week of storage.

Garlic, *Allium sativum*

Storage was recommended with 0% CO_2 and 1–2% O_2 (SeaLand, 1991). Monzini and Gorini (1974) recommended 3°C with 5% CO_2 and 3% O_2 for 6 months.

Garlic sprouts were stored at 0°C in ten combinations of O_2 (3–6.5%) and CO_2 (0–12%). Optimum storage was in 3% O_2 with 8% CO_2, which maintained levels of chlorophyll, reducing and total sugars and freshness of the flower buds after 235 days but there was some spoilage and rotting. Rotting decreased with decreasing O_2 concentration, while high CO_2 reduced mould growth (Zhou *et al.*, 1992b). The use of a carbon molecular sieve nitrogen generator in controlled atmosphere storage of sprouted garlic was investigated. The storage chambers were flushed with nitrogen produced by the generator to reduce the O_2 content to 1–5% and CO_2 was allowed to increase to 27% by respiration. After 240–270 days of storage, the quality of sprouted garlic remained high (Zhou *et al.*, 1992b).

Grape, *Vitus vinifera*

Storage recommendations were 1–3% CO_2 with 3–5% O_2 (SeaLand, 1991). Kader (1989, 1992) recommended 0–5°C with 1–3% CO_2 and 2–5% O_2 but showed that controlled atmosphere storage was incompatible with sulphur dioxide fumigation. Magomedov (1987) showed that different cultivars required different controlled atmosphere storage conditions. The cultivars Agadai and Dol'chatyi stored best in 3% CO_2 with 5% O_2 whereas for the cultivar Muskat Derbentskii 5% CO_2 with 5% O_2 or 3% CO_2 with 2% O_2 were more suitable. Muskat Derbentskii, Dol'chatyi and Agadai had a storage life of up to 7, 6 and 5 months, respectively, in controlled atmosphere storage.

The best results for storage of the cultivar Moldova were in controlled atmospheres with either 8 or 10% CO_2 with 2–3% O_2. Under these conditions, 89, 80 and 75% of first grade grape were obtained after 5, 6.5 and 7.5 months of storage, respectively. In another trial with the cultivars Muscat of Hamburg and Italia, grape stored in air or in controlled atmosphere storage for 3–7 months was assessed in relation to profitability. The best results were again obtained in controlled atmosphere storage of 8% CO_2 with 2–3% O_2 for Muscat of Hamburg and of 5% CO_2 with 2–3% O_2 for Italia (Khitron and Lyublinskaya, 1991). Turbin and Voloshin (1984) showed that for storage for > 5 months, 8% CO_2 with 3–5% O_2 was most suitable for Muskat Gamburgskii (Hamburg Muscat), 5–8% CO_2 with 3–5% O_2 for Italia and 8% CO_2 with 5–8% O_2 for Galan.

Waltham Cross and Barlinka table grape cultivars were stored in controlled atmospheres at −0.5°C for 4 weeks. The percentage of loose berries in Waltham Cross was highest (> 5%) in 21% O_2 and 5% CO_2 than in lower O_2 levels (Laszlo, 1985).

Berry and Aked (1997) working on Thompson Seedless stored at 0–1°C for up to 12 weeks showed that atmospheres containing 15–25% CO_2 inhibited infection with *Botrytis cinerea* by between 95 and 100% without detrimentally affecting their flavour. It can be concluded that controlled atmosphere storage could be an alternative to sulphur dioxide fumigation. Mitcham *et al.* (1997) showed that high CO_2 levels in the storage atmosphere could be used for controlling insect pests of grapes. In trials where four cultivars were exposed to 0°C with 45% CO_2 and 11.5% O_2 complete control of *Platynota stultana*, *Tetranychus pacificus* and *Frankliniella occidentalis* was achieved without injury to the table grapes.

The cultivar Agiorgitiko was stored at 23–27°C for 10 days either in 100% CO_2 or in air. Fruits from both treatments were held at 0°C for 20 h before analysis. In the case of volatile compounds, 114 compounds were identified in the CO_2 treated fruits, compared with only 60 in the fruits stored in air (Dourtoglou *et al.*, 1994).

Ben-Arie (1996) described a method of modified atmosphere packaging for sea freight transport. This involved whole pallet loads of half a tonne of fruit in cartons being wrapped in polyethylene film. The advantage of this compared to individual cartons being lined with polyethylene film was that any condensation remained on the outside of the box and precooling was more rapid since the boxes were precooled before the plastic was applied.

Grapefruit, Pummelo, *Citrus pardisi*

Storage experiments have given inconclusive results with some indication that fruit stored in 10% CO_2 at 4.5°C for 3 weeks had less pitting than those stored in air. Pre-treatment with 20–40% CO_2 for 3 or 7 days at 21°C reduced physiological disorders on fruit stored at 4.5°C for up to 12 weeks (Hardenburg *et al.*, 1990). SeaLand (1991) recommended 5–10% CO_2 with 3–10% O_2 for grapefruit from California, Arizona, Florida, Texas and Mexico. Storage at 10–15°C with 5–10% CO_2 and 3–10% O_2 had a fair effect on storage but controlled atmosphere storage was not used commercially (Kader, 1985, 1992). Controlled atmosphere storage was not considered beneficial but typical storage conditions were given as 10–15°C with 5–10% CO_2 and 3–10% O_2 by Bishop (1996).

Hatton *et al.* (1975) and Hatton and Cubbedge (1982) showed that exposing grapefruit to 40% CO_2 at 21°C reduced chilling injury symptoms (brown staining and rind pitting) during subsequent storage.

Guava, *Psidium guajava*

Short-term exposure to controlled atmospheres containing 10, 20 or 30% CO_2 had no effect on respiration in guavas, but evolution of ethylene was substantially reduced by all levels of CO_2 (Pal and Buescher, 1993). Mature-green fruit were exposed to either air or 5% CO_2 with 10% O_2 for 24 h before storage in air. Temperatures of either 4 or 10°C were used throughout (2 weeks) and then transferred to 20–23°C for 3 days of shelf life studies. The colour development of the controlled atmosphere treated fruit developed slower than those kept in air throughout and were considered to be of better quality after storage and shelf life. The controlled atmosphere treated fruits had no chilling injury symptoms even after storage at 4°C for 3 weeks while they were evident on non-controlled atmosphere treated fruit (Bautista and Silva, 1997). Sub-atmospheric pressures extended the storage life of guavas, mainly by inhibiting ripening (Salunkhe and Wu, 1974).

Horseradish, *Armoracia rusticana*

Controlled atmosphere storage was reported to have had a only a slight to no effect (SeaLand, 1991) and was not recommended for horseradish by Saltveit (1989). Weichmann (1981) studied increased levels of CO_2 of up to 7.5% during 6 months storage at 0–1°C and found no detrimental effects of CO_2 but no advantages over storage in air. He did, however, find that horseradish stored in 7.5% CO_2 had a higher respiration rate, and higher total sugar content than those stored in air.

Kiwifruit, Chinese gooseberry, Yang tao, *Actinidia chinensis*

Storage recommendations were 5% CO_2 with 2% O_2 by SeaLand (1991) and 0°C, 90% rh, 5% CO_2 and 2% O_2 by Lawton (1996). Storage in controlled atmosphere was shown to increase storage life by 30–40% with optimum conditions of 3–5% CO_2 and 2% O_2, but levels of CO_2 above 10% were found to be toxic to the fruit (Pratella *et al.*, 1985). Storage in 5% CO_2 with 2% O_2 at 0–5°C gave excellent results (Kader, 1985). Kader (1989, 1992) also recommended 0°C with 3–5% CO_2 and 1–2% O_2. Typical storage conditions were given as 0–5°C with 5–10% CO_2 and 1–2% O_2 by Bishop (1996).

Storage at 0°C and 90–95% rh in 3% CO_2 and 1–1.5% O_2 maintained fruit firmness during long-term storage, but 1% CO_2 with 0.5% O_2 gave the best storage conditions while maintaining an acceptably low level of incidence of rotting (Brigati *et al.*, 1989). Storage at 0.5–1°C with 92–95% rh and 4.5–5% CO_2 with 2–2.5% O_2 and ethylene at 0.03 ppm or less delayed flesh softening but strongly increased the incidence of *Botrytis* stem end rot (Tonini *et al.*, 1989). Controlled atmosphere storage in 2–5% O_2 with 0–4% CO_2 reduced ethylene production from the fruit (Wang *et al.*, 1994). Testoni *et al.* (1993) showed that storage in CO_2 concentrations higher than 5% resulted in irregular softening in the core. Brigati and Caccioni (1995) recommended 0.5 and 0.8°C less than 0.03 ppm ethylene, 95% rh with either 1% O_2 and 0.8% CO_2 or 2% O_2 and 4–5% CO_2 for 6 months. They also showed that the high level of CO_2 could lead to an increase in the incidence of *Botrytis cinerea*. *B. cinerea* can be controlled by the postharvest administration of certain fungicides, but the maximum permissible levels of pesticide residues legally permissible in kiwifruits varies widely between some European countries and does not allow a uniform control strategy to be set up. As an alternative or a help to fungicide treatments, the technique of curing can be used which involves putting the fruits in well-ventilated warehouses at 15–20°C for 48 h before refrigeration. Tonini and Tura (1997) showed that storage at –0.5°C in combination with 4.8% CO_2 and 1.8% O_2 reduced rots (*B. cinerea* and *Phialophora* spp.) and softening in kiwifruit compared to storage in air. The combination of either 1% CO_2 with 1% O_2 or 1.5% CO_2 with 1.5% O_2 was even more effective in controlling rots caused by *Phialophora* spp.

Intermittent storage at 0°C of 1 week in air and 1 week in air enriched with 10% or 30% CO_2 was shown to reduce fruit softening (Nicolas *et al.*, 1989).

Kohlrabi, *Brassica oleracea* variety *gongylodes*

Controlled atmosphere storage was reported to have had a slight to no effect (SeaLand, 1991).

Lanzones, Langsat, *Lansium domesticum*

Pantastico (1975) recommended 14.4°C in 0% CO_2 combined with 3% O_2 which gave a 16 day postharvest life compared to only 9 days at the same temperature in air. Pantastico also indicated that the skin of the fruit turns brown during retailing and if they are sealed in polyethylene film bags, where CO_2 can accumulate, the browning is aggravated.

Leek, *Allium ampeloprasum* variety *porrum*

Optimum storage conditions were reported to be 5–10% CO_2 with 1–3% O_2 at 0°C for up to 4–5 months (Kurki, 1979). CO_2 levels of 15–20% caused injury (Lutz and Hardenburg, 1968). Monzini and Gorini (1974) recommended 0°C with 15–25% CO_2 for 4.5 months. SeaLand (1991) recommended 3–5% CO_2 with 1–2% O_2. Kader (1985, 1992) recommended 0–5°C with 3–5% CO_2 and 1–2% O_2 which had a good effect but was claimed to be of no commercial use. Saltveit (1989) recommended 0–5°C with 5–10% CO_2 and 1–6% O_2 which had only a slight effect. Goffings and Herregods (1989) showed that at 0°C and 94–96% rh with 2% CO_2, 2% O_2 and 5% carbon monoxide leeks could be stored for up to 8 weeks compared to 4 weeks at the same temperature in air. Under those conditions the total losses were 19% while leeks stored in the same conditions but without carbon monoxide had 28% losses and those in air had 37% losses.

Lemon, *Citrus limon*

Storage recommendations were 5–10% CO_2 with 5% O_2 (SeaLand, 1991). The rate of colour change can be delayed with high CO_2 and low O_2 in the storage atmosphere but 10% CO_2 could impair their flavour (Pantastico, 1975). Lemons may be stored in good condition for 6 months under 10% O_2 with no CO_2 and the continuous removal of ethylene (Wild *et al.*, 1977). Storage at 10–15°C with 0–5% CO_2 and 5% O_2 was recommended by Kader (1985). Storage in 0–10% CO_2 and 5–10% O_2 at 10–15°C (Kader, 1989) had a good effect but was not used commercially (Kader, 1985, 1992). Controlled atmosphere storage was not considered beneficial but typical storage conditions were given as 10–15°C with 0–10% CO_2 and 5–10% O_2 by Bishop (1996).

Lettuce, *Lactuca sativa*

Storage in 1.5% CO_2 with 3% O_2 inhibited butt discoloration and pink rib at 0°C, but the effect did not persist during 5 days subsequent storage at 10°C

(Hardenburg *et al.*, 1990). CO_2 above 2.5% or O_2 levels below 1% can injure lettuce. For storage for a month 2% CO_2 with 3% O_2 is recommended (Hardenburg *et al.*, 1990). 0% CO_2 with 2–5% O_2 was also recommended by SeaLand (1991). Fellows (1988) recommended a maximum of 1% CO_2 and a minimum of 2% O_2. 0°C, 98% rh, 1.5% CO_2 and 3% O_2 was recommended by Lawton (1996). Storage in 0% CO_2 with 1–8% O_2 gave an extension in storage life and hypobaric storage increased their storage life from 14 days in conventional cold stores to 40–50 days (Haard and Salunkhe, 1975). Adamicki (1989) described successful storage of lettuce at 1°C with 3% CO_2 and 1% O_2 for 21 days with less loss in ascorbic acid than those stored in air. Brown stain on the mid ribs of leaves can be caused by levels of CO_2 of 2% or higher especially if combined with low O_2 (Haard and Salunkhe, 1975). Kader (1985) recommended 0–5°C with 0% CO_2 and 2–5% O_2 or 0–5°C with 0% CO_2 and 1–3% O_2 (Kader, 1992) which had a good effect and was of some commercial use when carbon monoxide was added at the 2–3% level. Saltveit (1989) recommended 0–5°C with 0% CO_2 and 1–3% O_2 for leaf, head and cut and shredded lettuce, the effect on the former was moderate and on the latter it had a high level of effect. Typical storage conditions were given as 0–5°C with 0% CO_2 and 1–3% O_2 by Bishop (1996). Tataru and Dobreanu (1978) showed that lettuce could be kept for 3 weeks in an atmosphere containing 3–5% CO_2 with 15% O_2.

Lime, *Citrus aurantifolia*

Controlled atmosphere storage recommendations were 0–10% CO_2 with 5% O_2 (SeaLand, 1991). Pantastico (1975) recommended storage in 7% O_2 which reduced the symptoms of chilling injury. However, Pantastico (1975) showed that controlled atmosphere storage of Tahiti limes increased decay rind scald and reduced juice content. Storage at 10–15°C with 0–10% CO_2 and 5% O_2 was recommended by Kader (1986) or 10–15°C with 0–10% CO_2 and 5% O_2 (Kader, 1989, 1992). These storage conditions were shown to increase the postharvest life of limes but were said to be not used commercially (Kader, 1985). Controlled atmosphere storage was not considered beneficial but typical storage conditions were given as 10–15°C with 0–10% CO_2 and 5–10% O_2 by Bishop (1996).

Packing limes in sealed polyethylene film bags inside cartons resulted in a weight loss of only 1.3% in 5 days, but all the fruits degreened more rapidly than those which were packed just in cartons where the weight loss was 13.8% (Thompson *et al.*, 1974b). This degreening effect could be countered by including an ethylene absorbent, made from potassium permanganate, in the bags.

Longan *Dimocarpus longana*

Zhang and Quantick (1997) stored the cultivar Shixia in 0.03 mm thick poly-ethylene film for 7 days at room temperature followed by 35 days at 4°C. Atmospheres of 1, 3, 10 or 21% O_2 were established in the bags and the former two were said to be effective in delaying peel browning, retaining total soluble solids and ascorbic acid content of the fruit, although taste panels detected a slight 'off-flavour' in fruit stored in 1% O_2.

Lychee, Litchi, *Litchi chinensis*

Kader (1993) in his review of storage conditions for tropical and subtropical fruit recommended 3–5% CO_2 combined with 5% O_2 at 7°C, with a range of 5–12°C, and that the benefits of reduced O_2 were good and those of increased CO_2 were moderate. Vilasachandran *et al.* (1997) stored the cultivar Mauritius at 5°C in air (control) or 5, 10 and 15% CO_2 combined with either 3 or 4% O_2. After 22 days all fruit were removed to air at 20°C for 1 day. Fruits stored in 15% CO_2 with 3% O_2 or 10% CO_2 with 3% O_2 were lighter in colour, retained total soluble solids better than the other treatments, but had the highest levels of 'off-flavours'. All the controlled atmosphere storage treatments had negligible levels of black spot and stem end rot compared to the controls. On the basis of the above the authors recommended 5% CO_2 with 3% O_2 or 5% CO_2 with 4% O_2.

Ragnoi (1989) working with the cultivar Hong Huai from Thailand showed that fruits packed in sealed 150 gauge polyethylene film bags containing 2 kg of fruit and sulphur dioxide pads could be kept in good condition at 2°C for up to 2 weeks while fruits that were stored without packaging were discoloured and unmarketable. Kader (1993) also claimed that modified atmosphere packaging was used to a limited extent.

Mango, *Mangifera indica*

General published controlled atmosphere storage conditions for mangoes vary with 5% CO_2 and 5% O_2 recommended by Pantastico (1975) and SeaLand (1991) and 10°C, 90% rh, 10% CO_2 and 5% O_2 by Lawton (1996). Typical storage conditions for mangoes were given as 10–15°C with 5–10% CO_2 and 3–5% O_2 by Bishop (1996). Storage in 10–15°C with 5% CO_2 and 5% O_2 (Kader, 1986) or with 5–10% CO_2 and 3–5% O_2 (Kader, 1989, 1992) was said to have a fair effect on storage but was not being used commercially (Kader, 1985). Kader (1993) in a review of modified atmosphere packaging and controlled atmosphere storage of tropical fruits recommended 13°C, with a range of 10–15°C, combined with 5–10% CO_2 and 3–5% O_2, or 5–7% O_2 for south-east Asian varieties. Kane and Marcellin (1979) recommended 10–12°C with 5% CO_2 and 5% O_2 for Julie and Amelie for 4 weeks which was said to reduce storage rots and give the best eating quality compared to those stored in air. Fuchs *et al.* (1978) described an experiment where storage at 14°C with 5% CO_2 and 2% O_2 kept the fruit green and firm for 3 weeks. Upon removal they attained full colour after 5 days at 25°C, but 9% of the fruit had fungal infections. An additional week in storage resulted in 40% of the fruit developing rots during ripening.

Controlled atmosphere storage conditions have been recommended for specific varieties. Hatton and Reeder (1966) recommended 13°C with 5% CO_2 and 5% O_2 for optimum storage of Keitt mango after which fruit ripened normally when exposed to air. In later work it was shown that Haden mangoes could be stored for 6 weeks under 2% O_2 and either 1 or 5% CO_2 at 10–11°C (Medlicott and Jeger, 1987). Kapur *et al.* (1962) recommended 7.5°C for Alphonso and Rajpuri. Optimum concentrations for Julie and Amelie mangoes have been reported as 5% O_2 and 5% CO_2 at 11°C for 4 weeks (Medlicott and Jeger, 1987). The optimum conditions for the Philippine variety, Carabao, was reported to be 5% CO_2 combined with 5% O_2 for 35–40 days (Mendoza, 1978). Storage at 8°C with 10% CO_2 and 6% O_2 was successful for 30 days for Haden and 35 days for Carlotta, Jasmin and San Quirino (Bleinroth *et al.*, 1977). Storage of mangoes was studied at combinations of 4, 6 and 8% CO_2 and 4, 6 and 8% O_2 at 13°C. Ripening of the cultivar Rad was delayed by controlled atmosphere storage for 2 weeks compared to samples stored in air. Fruits stored at 4% CO_2 and 6% O_2 were acceptable to panellists (Noomhorm and Tiasuwan, 1988). Storage of cultivars Tommy Atkins, Maya and Haden was extended to 6–8 weeks and of cultivar Keitt to 8–10 weeks under controlled atmosphere storage conditions. Even at the relatively high temperatures required (13–14°C) ripening was delayed. Lower storage temperatures caused chilling injury (Sive and Resnizky, 1985). Kensington fruits were stored under 36 different atmospheres (2, 4, 5, 6, 8 or 10% O_2 with 0, 2, 4, 6, 8 or 10% CO_2), with control fruits held in air, at 13°C for 33 days. Colour development was linearly retarded by decreasing O_2 (from 10 to 2%) and increasing CO_2 (from 0 to 4% only, no further effect from 4 to 10%) concentrations. Fruits from low O_2 atmospheres also had high titratable acidity contents after storage, but after 5 days at ambient conditions, these fruits continued to develop typical external colour and metabolize acid, thereby increasing palatability. Fruits from all atmospheres were soft after storage. Within the concentrations studied, the optimum atmosphere appeared to be around 4% CO_2 with 2–4% O_2 but the data does suggest that further research is required below 2% O_2 and above 10% CO_2, as well as at lower storage temperatures to control softening (McLauchlan *et al.*, 1994). Mango fruits (cultivar Kensington) were tested over two seasons for their response to low temperature (5, 7, 10 or 13°C) controlled atmosphere (0, 5 or 10% CO_2 and 5 or 21% O_2) storage. Fruits were removed from treatment after 0, 1, 3 and 5 weeks and ripened at 22°C for 7 days. After ripening, fruits were assessed for skin browning, pulp colour, titratable acidity, chlorophyll and ethanol concentration in pulp. Skin browning was observed only at 5°C, increasing in severity with storage time (from 1 week onwards). No skin browning was evident at higher temperatures. At 7°C, pulp colour was significantly depressed and titratable acidity increased after 3 weeks of storage. Over-ripeness was more common in fruits after 5 weeks at 13°C (and 10°C in the first season) and was associated with a decline in titratable acidity. Enhanced CO_2 atmospheres (5 and 10%) alleviated chilling injury symptoms. Increased pulp ethanol concentration was associated with CO_2 injury and over-ripening but not with chilling injury. Reduced O_2 concentration (5%) had no significant effect on chilling injury. The results indicate that Kensington fruits can be stored at 10°C without danger of serious chilling injury (O' Hare and Prasad, 1993).

Detrimental effects of controlled atmosphere storage have also been reported. Storage at O_2 levels of 1% resulted in the production of 'off-flavour' and skin discoloration but storage at 12°C with 5% CO_2 and 5% O_2 was possible for 20 days (Hatton and Reeder, 1966). Deol and Bhullar (1972) mentioned that there was increased decay in mangoes stored in either controlled atmosphere storage or modified atmosphere packaging compared to those stored unwrapped in air. This is in contrast with the work of Thompson (1971), Wardlaw (1937) and Kane and Marcellin (1979) who all showed that either controlled atmosphere storage or modified atmosphere packaging reduced postharvest decay of mangoes. The quality of Keitt mangoes stored for 6 days at 20°C under low O_2 atmosphere (0.3%) was evaluated before storage in modified atmospheres in three types of low density polyethylene films with different characteristics and after being stored for 30 days at 10 or 20°C. Both controlled atmosphere storage and modified atmosphere packaging delayed losses of weight and firmness. Fruits maintained good appearance with a significant delay in reaching maturity. However, a few fruits packaged in two of the three films developed a fermented flavour after 10 days at 20°C (Gonzalez-Aguilar et al., 1994). Mature, unripe mangoes of the cultivars Kent and Tommy Atkins were stored for 21 days at 12°C in air or 25, 45, 50 or 70% CO_2 plus either 3% or 20.8% O_2 in nitrogen or the same CO_2 concentrations mixed with air. The fruits were then transferred to air at 20°C for 5 days. CO_2 levels of 50 and 70% resulted in the highest ethanol production rates and symptoms of CO_2 injury, while the 3% O_2 concentration seemed to have little effect on ethanol production of either cultivar (Bender et al., 1994). Bleinroth et al. (1977) reported that fermentation of fruits occurred with alcohol production during storage at 8°C and > 10% CO_2 for 3 weeks.

Controlled atmosphere storage can be used to control pest infestation. Low O_2 concentrations (0.5%), alone or in conjunction with high CO_2 concentrations, can be used to control insects in fresh horticultural crops. However, most fruits and vegetables do not tolerate these extreme atmospheres for prolonged periods. In storage trials to control fruit fly in mango fruits, it was found that fruits exposed to 50% CO_2 with 2% O_2 for 5 days or 70–80% CO_2 with less than 0.1% O_2 for 4 days did not suffer adverse effects when they were subsequently ripened in air (Yahia *et al.*, 1989). In a later study, mangoes (cultivar Keitt) were stored for 0–5 days at 20°C in a continuous flow of an insecticidal low O_2 atmosphere (0.2–0.3%, balance N_2). Fruits were evaluated every day during storage in the low O_2 atmosphere and again after being held in air at 20°C for 5 days. There was no fruit injury or reduction in organoleptic fruit quality due to the low O_2 atmosphere, and fruits ripened normally. These results indicate that applying low O_2 atmospheres postharvest can be used to control insects in mangoes without adversely affecting fruit quality (Yahia and Tiznado Hernandez, 1993). Immersion in hot water (52°C for 5 min) plus benomyl (Benlate 50 WP 1 g l^{-1}) provided good control of stem end rot on mangoes following inoculation with either *Dothiorella dominicana* or *Lasiodiplodia* (*Botryodiplodia*) *theobromae* during storage for 14 days at 25–30°C. In the same storage conditions prochloraz (as Sportak 45EC), DPXH6573 (40EC), RH3866, (25EC) and calcium chloride did not control stem end rot (*D. dominicana*). During long-term storage in a controlled atmosphere (5% O_2, 2% CO_2, 13°C for 26 days followed by air for 11 days at 20°C) hot benomyl (as above) followed by prochloraz (45 EC, 0.55 ml l^{-1}, 25°C, 30 s) provided effective control of stem end rot and anthracnose. The addition of guar gum to hot benomyl improved control of stem end rot in the combination treatment. Hot benomyl alone was ineffective. Other diseases, notably Alternaria rot (*A. alternata*) and dendritic spot (*D. dominicana*) emerged as problems during controlled atmosphere storage. *A. alternata* and dendritic spot were also controlled by hot benomyl followed by prochloraz. *Penicillium expansum*, *Botrytis cinerea*, *Stemphylium vesicarium* and *Mucor circinelloides* are reported as postharvest pathogens of mango, cultivar Kensington Pride, for the first time (Johnson *et al.*, 1990a, 1990b).

Controlled atmosphere transport has also been described for mangoes. Green mature mango fruits (cultivar Irwin) were transported in a refrigerated truck (at 12°C) on a car ferry from Kume Island to Tokyo Port. The fruits were transported from Tokyo to Tsukuba University by land. The total journey took 4 days. On arrival the fruits were stored in controlled atmosphere storage (5% O_2 and 5% CO_2) at 8, 10 or 12°C. Fruit quality was retained at all temperatures for 1 month. No chilling injury was observed, even at the lowest temperature (Maekawa, 1990).

Modified atmosphere packaging can be used to extend the storage life of fruit and vegetables and it is used commercially to preserve minimally processed fruit such as mangoes (Fig. 31). Fruits stored in polyethylene film bags at 21°C had almost twice the storage life of fruits stored without wraps (Thompson, 1971). Mature fruits of the mango cultivar Keitt were stored at 20°C in air (controls) or in a modified atmosphere (modified atmosphere jars supplied with humidified CO_2 at 210 ml min^{-1} for 2 h before being sealed, resulting in 0.03–0.26% O_2 and 72–82% CO_2) for up to 4 days, or in

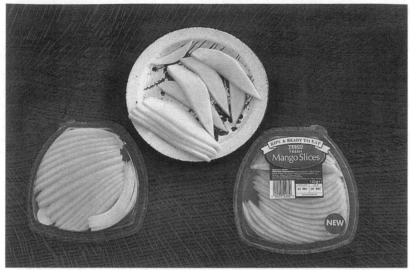

Fig. 31. Modified atmosphere packaging of minimally processed mangoes on sale in a UK supermarket in 1997. (Courtesy of Allan Hilton.)

controlled atmosphere (jars ventilated with 2% O_2 with 50% CO_2 with 48% nitrogen at a continuous rate of 210 ml min^{-1}) for up to 5 days. Flesh firmness losses (measured after 12 and 19 days of storage) were slightly less in controlled atmosphere storage and much less in modified atmospheres than in air. Both modified atmosphere and controlled atmosphere storage delayed fruit ripening. Fruits showed no signs of internal or external injuries or off-flavour, either immediately after removal from controlled atmosphere storage or modified atmosphere storage, or after transfer to air. There were no significant differences between air stored and controlled atmosphere storage stored fruits in all sensory attributes evaluated, but the overall acceptability of fruits stored in modified atmospheres for over 72 h was lower than that of fruits stored in air (Yahia and Vazquez Moreno, 1993).

Hypobaric storage has been studied for mangoes. Mature green mango fruits (cultivar O Krong) were hydrocooled from 32.2°C to 30, 20 or 15°C, and dipped in 0.7, 0.85 or 1.0% concentrations of a commercial wax formulation. The fruits were then stored at 20°C for 2–5 weeks or at 13°C for 4 weeks at atmospheric pressure (760 mmHg) or at hypobaric pressures (60, 100 or 150 mmHg). Weight loss, pulp firmness, sugar content, percentage of decay and the degree of skin colour change were measured at the end of the storage period. Quality after storage was best in fruits hydrocooled to 15°C, dipped in wax (any of the concentrations tested) and stored at 13°C at 60 or 100 mmHg (Ilangantileke and Salokhe, 1989).

Melon, *Cucumis melo*

Storage with 10–15% CO_2 and 3–5% O_2 was recommended for Cantaloupe and 5–10% CO_2 with 3–5% O_2 for Honeydew and Casaba (SeaLand, 1991). Martinez-Javega *et al.* (1983) stored the cultivar Tendral in 12% CO_2 with 10% O_2 at temperatures within the range of 2–17°C and found no effects on levels of decay or chilling injury compared to fruits stored at the same temperatures in air. Kader (1985, 1992) recommended 3–7°C with 10–15% CO_2 and 3–5% O_2 for Cantaloupes which had a good effect on storage but was said to have limited commercial use. For Honeydew, Kader (1985, 1992), recommended 10–12°C with 0% CO_2 and 3–5% O_2 which had a fair effect but was of no commercial use. Saltveit (1989) recommended 5–10°C with 10–20% CO_2 and 3–5% O_2 which had only a slight effect. Typical storage conditions were given as 2–7°C with 10–20% CO_2 and 3–5% O_2 by Bishop (1996) for Cantaloupes. Perez Zungia *et al.* (1983) compared storage of the cultivar Tendral in 0% CO_2 with 10% O_2 with storage in air and found that the controlled atmosphere stored fruit had a better overall quality.

Martinez-Javega *et al.* (1983) reported that sealing fruits individually in 0.017 mm thick polyethylene film bags reduced chilling injury during storage at 7–8°C.

Mushroom, *Agaricus bisporus*

Recommended optimum storage conditions were 5–10% CO_2 (Lutz and Hardenburg, 1968; Ryall and Lipton, 1979) or 10–15% CO_2 and 21% O_2 (SeaLand, 1991). Storage with CO_2 levels of 10–20% inhibited mould growth and retarded cap and stalk development, O_2 levels of less than 1% were injurious (Pantastico, 1975). Kader (1985, 1992) recommended 0–5°C with 10–15% CO_2 and 21% O_2 which had a fair effect but was of limited commercial use. Saltveit (1989) recommended 0–5°C with 10–15% CO_2 and 21% O_2 which had a moderate level of effect. Colour deterioration in harvested mushrooms was inhibited by storage at 0°C. Controlled atmosphere storage with low O_2 and high CO_2 inhibited cap opening and internal browning but caused a yellowing of the cap surface. The best atmosphere for maintaining quality was 8% O_2 and 10% CO_2 (Zheng and Xi, 1994). Tomkins (1966) levels of CO_2 of 10% can delay deterioration but levels of over 10% could cause a pinkish discoloration.

PVC overwraps on consumer sized punnets (about 400 g) greatly reduced weight loss, cap and stalk development, and discoloration especially when combined with refrigeration (Nichols, 1971). Modified atmosphere packaging in microporous film delayed mushroom development especially when combined with storage at 2°C (Burton and Twyning, 1989).

Natsudaidai, *Citrus natsudaidai*

Kajiura and Iwata (1972) showed that fruit was stored in nitrogen atmospheres containing O_2 concentrations of 0–21% for 2 months at 3 or 4°C or for 1 month at 20°C. At 3 and 4°C total soluble solids and titratable acidity were little affected by the O_2 concentration, but the fall in ascorbic acid was less marked at low O_2. O_2 levels below 3% resulted in a fermented flavour but the eating quality was similar in fruit stored at all O_2 concentrations above 5%. At 20°C there was more stem end rot but *Penicillium* development was reduced by low O_2. Titratable acidity, total soluble solids and ascorbic acid levels were similar at all O_2 levels but below 5% the fermented flavour developed. Two kinds of injury were encountered, at 3 and 4°C and below 1.5% O_2 the albedo became watery and yellow and the juice sacs turned from orange to yellow while CO_2 production fell. The other type of injury occurred at 20°C where the albedo became yellow at 0% O_2. Storage in 7% O_2 at 4°C was recommended.

Kajiura (1972) showed that *C. natsudaidai* fruit in 4°C had less button browning in 5% CO_2 and above. Total soluble solids were not affected by CO_2 concentration, but titratable acidity was reduced in 25% CO_2. At 20°C button browning was retarded above 3.5–5.0% CO_2 but there were no differences in total soluble solids content and titratable acidity between different CO_2 levels. An abnormal flavour and a sweet taste developed above 13 and 5% CO_2, respectively. Above 13% CO_2 the peel turned red.

In subsequent work, Kajiura (1973) stored at 4°C and either 98–100% rh or 85–95% rh in air mixed with 0, 5, 10 or 20% CO_2 for 50 days and found that high CO_2 retarded button browning at both humidities and the fruit developed granulation and loss of acidity at low humidity. High humidity combined with high CO_2 increased the water content of the peel and the ethanol content of the juice, and produced abnormal flavour and reduced the internal O_2 content of the fruit causing water breakdown. At the lower humidity, no injury occurred and CO_2 was beneficial, its optimum level being much higher. In another trial, fruit were stored in 0% CO_2 with 5% O_2 or 5% CO_2 with 7% O_2. In 0% CO_2 and 5% O_2 there was little effect on button browning or acidity but granulation was retarded and an abnormal flavour and ethanol accumulation occurred. In 5% CO_2 and 7% O_2 the decrease in acidity was retarded and there was abnormal flavour development at high humidity but at low humidity it provided the best quality stored fruit. Exposure to 60% CO_2 with 20% O_2 and 20% nitrogen produced little or no effect in natsudaidais (Kubo *et al.*, 1989a).

Nectarine, *Prunus persica* variety *nectarina*

Storage recommendations are similar to peach for temperature and controlled atmosphere storage (Hardenburg *et al.*, 1990), storage with 5%

CO_2 and 2.5% O_2 for 6 weeks (Hardenburg *et al.*, 1990). Storage at 0.5°C with 90–95% rh with 5% CO_2 with 2% O_2 was recommended for 14–28 days (SeaLand, 1991). Storage recommendations were also given as 0–5°C with 5% CO_2 and 1–2% O_2 (Kader, 1985) or 0–5°C with 3–5% CO_2 and 1–2% O_2 which had a good effect, but controlled atmosphere storage had limited commercial use. The storage life at –0.5°C in air was about 7 days, but fruit stored at the same temperature in atmospheres containing 1.5% CO_2 with 1.5% O_2 could be stored for 5–7 weeks (Van der Merwe, 1996). Conditioning the fruit prior to storage at 20°C in 5% CO_2 with 21% O_2 for 2 days until they reach a firmness of 5.5 kg was necessary. Folchi *et al.* (1994) stored the cultivar Independent in less than 0.1% CO_2 with 0.3% O_2 and showed an increase in aldehyde and ethanol content of fruit with increased storage time.

Anderson (1982) stored the cultivar Regal Grand at 0°C in 5% CO_2 with 1% O_2 for up to 20 weeks with intermittent warming to 18–20°C every 2 days which almost entirely eliminated the internal browning associated with storage in air at O_2.

Okra, Lady's finger, *Hibiscus esculentus*

Storage at 7.2°C with 5–10% CO_2 kept them in good condition (Pantastico, 1975). In other work 0% CO_2 with 3–5% O_2 was recommended by SeaLand (1991). Kader (1985) also recommended 8–10°C with 0% CO_2 and 3–5% O_2 which had a fair effect but was of no commercial use, but CO_2 at 5–10% was beneficial at 5–8°C. Saltveit (1989) recommended 7–12°C with 4–10% CO_2 and 21% O_2 but it had only a slight effect. Ogata *et al.* (1975) stored okra at 1°C in air or 3% O_2 combined with 3, 10 or 20% CO_2 or at 12°C in air or 3% O_2 combined with 3% CO_2. At 1°C there were no effects of any of the controlled atmosphere storage treatments, but at 12°C the controlled atmosphere storage treatment resulted in lower ascorbic acid retention but it improved the keeping quality. At levels of 10–12% CO_2 off-flavour may be produced (Anandaswamy *et al.*, 1963).

Storage work in modified atmosphere packs indicated that 5–10% CO_2 lengthens their shelf life by about a week (Hardenburg *et al.*, 1990).

Olive, *Olea europaea*

Storage at 8–10°C with 5–10% CO_2 and 2–5% O_2 had a fair effect, but was not being used commercially (Kader, 1985). Kader (1989, 1992) recommended 5–10°C with 0–1% CO_2 and 2–3% O_2. In other work Kader *et al.* (1989) found that green olives of the cultivar Manzanillo were damaged when exposed to CO_2 levels of 5% and above, but storage was extended in 2% O_2. They found that in air their storage life was 8 weeks at 5°C, 6 weeks at 7.5°C and 4 weeks at 10°C, but in 2% O_2 the storage life was extended to 12 weeks at 5°C and 9 weeks at 7.5°C. Garcia *et al.* (1993) stored the cultivar Picual at 5°C in air, 3% CO_2 with 20% O_2, 3% CO_2 with 5% O_2 or 1% CO_2 with 5% O_2. They found that 5% O_2 gave the best results in terms of retention of skin colour, firmness, and acidity but had a higher incidence of postharvest losses compared to those stored in air. They found no added advantage of increased CO_2 levels. Agar *et al.* (1997) compared storage of the cultivar Manzanillo in air or 2% O_2 with 98% nitrogen at 0, 2.2 and 5°C. They found that fruit firmness was not affected by the controlled atmosphere storage compared to air storage but decay was reduced in the low O_2 compared to air at all three temperatures. They concluded that they could be stored for a maximum of 4 weeks at 2.2–5°C in 2% O_2.

Onion, Bulb onion, *Allium cepa* variety *cepa*

Monzini and Gorini (1974) recommended 4°C with 5–10% CO_2 and 3–5% O_2 for 6 months. Fellows (1988) recommended a maximum of 10% CO_2 and a minimum of 1% O_2. Kader (1985) recommended a temperature range of 0–5°C, 1–2% O_2, 0% CO_2 and 75% rh. In later work Kader (1989) recommended 0–5°C with 0–5% CO_2 and 1–2% O_2. Saltveit (1989) recommended 0–5°C with 0–5% CO_2 and 0–1% O_2 which was claimed to have had only a slight effect. Controlled atmosphere storage of onions gave better results when stored early in the season just after curing. Smittle (1988) found that more than 99% of bulbs of the cultivar Granex were marketable after 7 months of storage at 1°C in 5% CO_2 and 3% O_2, although the weight loss was 9%. Onion stored at 1°C in 5% CO_2 and 3% O_2, kept well when removed from storage, while bulbs from the other storage treatments (10% CO_2 with 3 and 5% O_2) became unmarketable at a rate of about 15% per week due to internal breakdown during the first month out of store. Bulb quality, as measured by low pungency and high sugar, decreased slowly when onions were stored at 27°C or at 1 or 5°C with controlled atmosphere at 70–85% rh. Quality decreased rapidly when the cultivar Granex were stored in air at 1 or 5°C. Adamicki (1989) also described successful storage of onions at 1°C with 5% CO_2 and 3% O_2.

Controlled atmosphere storage conditions gave variable success and were not generally recommended, however, 5% CO_2 with 3% O_2 were shown to reduce sprouting and root growth (Hardenburg *et al.*, 1990). SeaLand (1991) recommended 0% CO_2 with 1–2% O_2. Typical storage conditions were given as 0°C and 65–75% rh with 5% CO_2 and 3% O_2 by Bishop (1996).

Adamicki and Kepka (1974) stated that Chroboczek (undated) found that onions stored in an atmosphere with 5–15% CO_2 at room temperature showed a decrease in the percentage of onions that sprouted. Adamicki and Kepka (1974) also found that there were no changes in the colour, flavour or chemical composition of onions after 2 months of storage even at the 15% CO_2 concentration. Chawan and Pflug (1968) examined storage of the onion cultivars Dowing Yellow Globe and Abundance in several combinations of CO_2 (5 and 10%) and O_2 (1, 3 and 5%) at temperature of 1.1, 4.4 and 10°C. The best combination was 10% CO_2 and 3% O_2 at 4.4°C and the next best was 5% CO_2 and 5% O_2 at 4.4°C. In 10% CO_2 and 3% O_2 at 4.4°C, bulbs showed no sprouting after 34 weeks, while bulbs in air storage had 10% sprouting for Dowing Yellow Globe and 15% for Abundance. Internal spoilage of onion bulbs was observed in 10% CO_2 and 3% O_2 at 1.1°C but none at 4.4°C. Adamicki and Kepka (1974) also observed a high level of internal decay in bulbs stored in the 10% CO_2 and 5% O_2 at 1°C, but none at 5°C in the same atmospheric composition. The number of internally decayed bulbs increased with length of storage. Their optimum results for long storage were in 5% CO_2 and 3% O_2 either at 1 or 5°C. Stoll (1974) stated that there were indications that the early trimmed lots did not store as well in controlled atmosphere conditions as in conventionally refrigerated rooms, and he obtained better storage life by storing the onions at 0°C than at 2°C and 4°C when the same 8% CO_2 and 1.5% O_2 concentration were used.

Sitton *et al.* (1997) also found that storage in high CO_2 caused injury but neck rot (*Botrytis* spp.) was reduced in CO_2 levels of greater than 8.9%. They found that storage in O_2 levels of 0.5–0.7% were firmer, of better quality and had less neck rot than those stored on O_2 levels of over 0.7%. However, these onions sprouted more quickly when removed to 20°C for 2 weeks compared to those stored in O_2 levels of over 0.7%.

There is evidence that controlled atmosphere storage can have residual effects on onions after they have been removed from store. Bulbs of cultivar Wolska stored for 163 or 226 days in the optimum controlled atmosphere conditions and then transferred to 20°C sprouted about 10 days later than those transferred from air storage (Adamicki and Kepka, 1974).

Waelti *et al.* (1992) converted a refrigerated highway trailer to a controlled atmosphere storage and they found that the cultivar Walla Sweet stored well for 84 days in 2% O_2 at 75–85% rh, but subsequent the shelf life was only 1 week.

CO_2 injury or internal spoilage of onion bulbs is a physiological disorder that can be seen internally, and can be induced by elevating the CO_2 concentration around the bulbs. This effect was aggravated by low temperature (less than 5°C). Chawan and Pflug (1968) observed internal spoilage of bulbs in storage in less than 10% CO_2 and 3% O_2 at 1.1°C, but there was no spoilage in the same controlled atmosphere at 4.4°C. It was stated that 'internal spoilage of the bulbs was due to an adverse combined effect of temperature and gas concentration'. Also Adamicki and Kepka (1974) reported that Böttcher had stated that for the cultivars Ogata and Inoue there was very strong internal spoilage at concentrations of CO_2 above 10%. Adamicki and Kepka (1974) found that there were 23–56% of internally spoiled bulbs when CO_2 concentrations were 10% or higher after storage at

1°C (but not at 5°C) for more than 220 days. Adamicki *et al.* (1977), mentioned that internal breakdown was observed in 68.6% of the bulbs when they were stored at 1°C in sealed polythene bags with a CO_2 concentration higher than 10%. Also they found similar disorder in onion bulbs stored at 5°C in sealed polythene bags, and this was probably due to the very high concentration of CO_2 amounting to 28.6%. Adamicki and Kepka (1974) indicate that the statement could be made that internal spoilage of onions was due to the combined effects of high CO_2 concentrations, low temperature and a relatively long period of storage. The effect of CO_2 on the internal spoilage of onion bulbs has been reported by Adamicki and Kepka (1974), Chawan and Pflug (1968) and Ogata and Inous (1957) but without giving any details or description of the internal spoilage. Smittle (1988) found that the cultivar Granex stored at 5°C with 10% CO_2 and 3% O_2 for 6 months had CO_2 injury, exhibited by internal breakdown of tissue. The level of injury was 6% after 2 weeks and 10% after 3 or 4 weeks of shelf life evaluation at room temperature. This is in general agreement with Adamicki (1974) and Chawan (1966) for the effect of CO_2 concentration and duration on internal breakdown.

Orange, *Citrus sinensis*

Storage at 5–10°C with 5% CO_2 and 10% O_2 had a fair effect on storage but was not being used commercially (Kader, 1985). Kader (1989, 1992) recommended 5–10°C with 0–5% CO_2 and 5–10% O_2. Storage of the cultivar Valencia, grown in Florida, for 12 weeks at 1°C in 0 or 5% CO_2 and 15% O_2 followed by 1 week in air at 21°C retained better flavour and had less skin pitting than those stored in air, but CO_2 levels of 2.5–5% especially when combined with 5 or 10% O_2 adversely affected flavour retention (Hardenburg *et al.*, 1990). SeaLand (1991) recommended 5% CO_2 with 10% O_2. Controlled atmosphere storage was not considered beneficial but typical storage conditions were given as 5–10°C with 0–5% CO_2 and 5–10% O_2 by Bishop (1996). Anon (1968) showed that controlled atmosphere storage could have deleterious effects on fruit quality, particularly through increased rind injury and decay or on fruit flavour. Chase (1969) also showed that controlled atmosphere storage can affect both fruit flavour, rind breakdown and rotting (Table 7.2).

Smoot (1969) showed that for the cultivar Valencia the controlled atmosphere combinations of 0% CO_2 with 15% O_2 and 0% CO_2 with 10% O_2 had less decay than those stored in air.

Table 7.2. The effects of controlled atmosphere storage on the quality of oranges.

% O_2	% CO_2	% Rotting[1]	Flavour score[2]	% Rind breakdown[3]
21 (air)	0	97	66	0
15	0	58	97	32
15	2.5	28	95	13
15	5	62	93	0
10	0	67	74	49
10	2.5	60	69	3
10	5	81	52	3
5	0	47	38	9
5	2.5	36	20	2
5	5	79	5	0

[1]After 20 weeks storage at 1°C plus 1 week at 21°C.
[2]After 12 weeks storage at 1°C plus 1 week at 21°C.
[3]After 20 weeks storage at 1°C plus 1 week at 21°C.

Oyster mushroom, *Pleurotus ostreatus*

Controlled atmosphere storage was shown to have little effect on increasing storage life at either 1°C or 3.5°C but at 8°C with a combination of 10% CO_2 and 2% O_2, 20% CO_2 with 21% O_2 or 30% CO_2 in 21% O_2 the oyster mushrooms had reduced respiration rate and kept them in good condition for longer than those stored in air (Bohling and Hansen, 1989). Henze (1989) recommended storage at 1°C and 94% rh with 30% CO_2 and 1% O_2 for about 10 days. In earlier work Pantastico (1975) recommended 5% CO_2 with 1% O_2 for 21 days. Modified atmosphere packaging is also used during marketing (Fig. 32).

Papaya, Pawpaw, *Carica papaya*

Typical storage conditions were given as 10–15°C with 5–8% CO_2 and 2–5% O_2 by Bishop (1996). At 18°C, storage at 10% CO_2 reduced decay (Hardenburg *et al.*, 1990) and 10% CO_2 with 5% O_2 was recommended by SeaLand (1991) for surface transport. Storage at 10–15°C with 10% CO_2 and 5% O_2 had a fair effect but was said to be 'not being used commercially' (Kader, 1985). Kader (1989, 1992) later recommended 10–15°C with 5–10% CO_2 and 3–5% O_2. Kader (1993) in a review of modified atmosphere packaging and controlled atmosphere storage of tropical fruits recommended 12°C, with a range of 10–15°C, combined with 5–8% CO_2 and 2–5% O_2.

For maximum storage life Nazeeb and Broughton (1978) showed that it was advantageous to remove ethylene produced in the store. Sankat and Maharaj (1989) found that fruit of the cultivar Known You Number 1 stored at 16°C were still in an acceptable condition after 17 days storage in 5% CO_2 and 1.5–2% O_2 compared to only 13 days in air. The cultivar Tainung

Fig. 32. Modified atmosphere packaging of oyster mushrooms on sale in a UK supermarket in 1997.

Number 1 stored in the same conditions as Known You Number 1 had a maximum life of 29 days in controlled atmosphere storage and were unacceptable after 17 days stored in air. In Hawaii storage in 2% O_2 and 98% nitrogen was shown to extend the storage life of fruit at 10°C compared to storage in air at the same temperature (Akamine and Goo, 1969). Akamine (1969) showed that storage of the cultivar Solo in 10% CO_2 for 6 days at 18°C reduced decay compared to fruits stored in air, but fruit stored in air after storage in 10% CO_2 decayed rapidly. Akamine and Goo (1969) showed that storage at 13°C in 1% O_2 for 6 days or 1.5% O_2 for 12 days ripened about 1 day slower than those stored in air. Chen and Paull (1986) also showed that reduced O_2 (1.5–5%) delayed ripening of the cultivar Kapaho Solo compared to those stored in air and that there was no further benefits on storage by increasing the CO_2 level to either 2 or 10%. Hatton and Reeder (1969) showed that after storage in 13°C for 3 weeks in 5% CO_2 with 1% O_2 followed by ripening at 21°C fruit were in a fair condition with little or no decay and were of good flavour. Cenci *et al.* (1997) showed that the cultivar Solo stored at 10°C in 8% CO_2 with 3% O_2 could be stored for 36 days and still have an adequate shelf life of 5 days at 25°C.

In Malaysia Broughton *et al.* (1977) and Nazeeb and Broughton (1978) showed that scrubbing ethylene from the store containing the cultivar Solo Sunrise had no effects on their storage life but they recommended that for optimum storage 20°C with 5% CO_2 with ethylene removal was the optimum condition for 7–14 days. They also found that chilling injury could occur at 15°C. In later work it was shown that ethylene application can accelerate ripening by 25–50% (Wills, 1990).

Storage at 20°C with 0% CO_2 and less than 0.4% O_2 as a method of fruit fly control in papaya fruits resulted in increased incidence of decay, abnormal ripening and the development of off-flavour after 5 days exposure (Yahia *et al.*, 1989).

Wills (1990) reported that modified atmosphere packaging in plastic film extended their storage life.

Parsley, *Petroselinum sativum*

Storage at 0–5°C with 8–10% CO_2 and 8–10% O_2 was recommended by Saltveit (1989) which had only a slight effect. Apeland (1971) recommended storage at 5°C and 11% CO_2 combined with 10% O_2. After 45 days in these conditions 95.6% was considered saleable compared to only 43.9% for those stored in air at 5°C.

Bangerth (1974) showed that hypobaric storage in 0.1 atmosphere (76 mmHg) at 2–3°C extended their life for up to 8 weeks and improved retention of ascorbic acid, chlorophyll and protein.

Passionfruit, *Passiflora edulis* forma *flavicarpa, P. edulis* forma *edulis*

Mohammed (1993) showed that both the yellow passionfruit and purple passionfruit stored best at 10°C in perforated low density polyethylene film bags, 0.025 mm thick.

Peach, *Prunus persica*

Storage in 5% CO_2 and 1% O_2 maintained quality and retarded internal breakdown for about twice as long as those stored in air (Hardenburg *et al.*, 1990). Controlled atmosphere storage plus intermittent warming to 18°C in air for 2 days every 3–4 weeks has shown promising results (Hardenburg *et al.*, 1990). SeaLand (1991) recommended 5% CO_2 with 2% O_2. Storage recommendations were given by Kader (1985) as 0–5°C with 5% CO_2 and 1–2% O_2 which had a good effect, but was claimed to be of limited commercial use. Kader (1989, 1992) recommended also 0–5°C and 3–5% CO_2 and 1–2% O_2 for both freestone and clingstone peaches. Typical storage conditions were given as 0–5°C with 3–5% CO_2 and 1–2% O_2 by Bishop (1996). The storage life of dessert peaches at –0.5°C in air was about 7 days, but fruit stored at the same temperature in atmospheres containing 1.5% CO_2 with 1.5% O_2 could be stored for 4 weeks (Van der Merwe, 1996). Conditioning the fruit prior to storage at 20°C in 5% CO_2 with 21% O_2 for 2 days until they reach a firmness of 5.5 kg was necessary. Anderson (1982) stored the cultivar Rio Oso Gem at 0°C in 5% CO_2 with 1% O_2 for up to 20 weeks with intermittent warming to 18–20°C every 2 days which almost entirely eliminated the internal browning associated with storage in air at O_2. Brecht *et al.* (1982) showed that there was a varietal controlled atmosphere storage interaction. They stored 5 cultivars at –1.1°C in 5% CO_2 with 2% O_2 and found that Loadel and Carolyn could be successfully stored for up to 4 weeks while Andross, Halford and Klamt could be stored for only comparatively short periods. Wade (1981) showed that storage of the cultivar J.H. Hale at 1°C resulted in chilling injury symptom (flesh discoloration and soft texture), but fruits stored at the same temperature in atmospheres containing 20% CO_2 resulted in only moderate levels even after 42 days. Bogdan *et al.* (1978) showed that the cultivars Elbarta and Flacara could be stored at 0°C for 3–4 weeks in air or about 6 weeks in 5% CO_2 with 3% O_2. Truter *et al.* (1994) showed that the cultivars Oom Sarel, Prof. Neethling and Kakamas could be successfully stored at –0.5°C either in 1.5% CO_2 with 1.5% O_2 or 5% CO_2 with 2% O_2 for 6 weeks with a mass loss of only 1%. This compared to mass losses of 20.7 and 12.6% during fruits stored in air at the same temperature for Oom Sarel or Prof. Neethling, respectively. Controlled atmosphere storage, however, did not affect decay incidence but fruits were of an acceptable quality even after 6 weeks storage and were also considered still highly suitable for canning.

Modified atmosphere packaging which gives a high concentration of CO_2 was tested on the cultivars Elegant Lady and O'Henry stored at 0°C. The atmospheres inside the packs varied between 10 and 25% for CO_2 and 1.5 and 10% for O_2. The rate of fruit softening and flesh browning were reduced in all the high CO_2 packages, but there was variation in the effects of modified atmosphere packaging on the development of mealiness and off-flavours (Zoffoli *et al.*, 1997).

For canning peaches Van der Merwe (1996) recommended –0.5°C with 1.5% CO_2 with 1.5% O_2 for up to 4 weeks.

Pear, *Pyrus communis*

Fellows (1988) recommended a maximum of 5% CO_2 and a minimum of 2% O_2 for the cultivar Bartlett. Koelet (1992) recommended 0.5°C with 0.5–1.0% CO_2 and 2–2.2% O_2. SeaLand (1991) recommended 0–1% CO_2 with 2–3% O_2. Lawton (1996) recommended 0°C, 93% rh, 0.5% CO_2 and 1.5% O_2. Storage at 0–5°C with 0–1% CO_2 and 2–3% O_2 was recommended (Kader, 1985) or 0–5°C with 0–3% CO_2 and 1–3% O_2 (Kader, 1992). After 6.5 months storage, at –0.5°C the average percentage of Beurre d'Anjou with symptoms of scald ranged from zero in fruits stored in 1.0% O_2 with 0% CO_2 or 1.5% O_2 with 0.5% CO_2 to 2% in fruits stored in 2.5% O_2 with 0.8% CO_2 and 100% in control fruits stored in air. Johnson (1994) recommended –1 to –0.5°C with less than 1% CO_2 and 2% O_2 for both Conference and Concorde. Sharples and Stow (1986) recommended −1 to −0.5°C with less than 1% CO_2, and 2% O_2 for Conference. Exposure of Anjou to 12% CO_2 for 2 weeks immediately after harvest had beneficial effects on the retention of ripening capacity (Hardenburg *et al.*, 1990). Other recommendations include 0.8–1% CO_2 and 2–2.5% O_2 or 0.1% CO_2 or less with 1–1.5% O_2 for Anjou have been recommended (Hardenburg *et al.*, 1990). Fidler and Mann (1972) recommended 0.5–1°C with 5% CO_2 and 5% O_2 for Conference and Doyenne du Comice and 0.5–1°C with 6% CO_2 and about 15% O_2 William's Bon Chretien for pears produced in England while Stoll (1972) recommended 0°C with 2% CO_2 and 2% O_2 for Conference, Doyenne du Comice and William's Bon Chretien for pears produced in Switzerland. Chen and Varga (1997) found that no controlled atmosphere storage regime containing 0.5–2% O_2 was feasible in controlling physiological disorders of D'Anjou pears during 6 or 8 months storage. They showed that storage in 0.5% O_2 with less than 0.1% CO_2 resulted in a high incidence of scald and black speck.

M. Herregods (1993, personal communication) and Meheriuk (1993) compared recommendations for controlled atmosphere storage for selected cultivars from different countries (Table 7.3).

Van der Merwe (1996) recommended specific conditions for storage of fruit produced in South Africa (Table 7.4).

Recommended storage conditions for various pear cultivars was given by Richardson and Meheriuk (1989) (Table 7.5).

Pea, Garden pea, Mange tout, Snow pea, Sugar pea, *Pisum sativum*

Monzini and Gorini (1974) recommended 0°C with 5% CO_2 and 21% O_2 for 20 days. Pantastico (1975) quoted that in storage at 0°C peas could be kept in good condition for 7–10 days while in 5–7% CO_2 with 5–10% O_2 they could be kept for 20 days. Saltveit (1989) recommended 0–10°C with 2–3% CO_2 and 2–3% O_2 for sugar peas which had only a slight effect. Storage in 5–7% CO_2 at 0°C maintained their eating quality for 20 days (Hardenburg *et al.*, 1990). Fellows (1988) recommended a maximum of 7% CO_2 and a minimum of 5% O_2.

Persimmon, Sharon fruit, Kaki, *Diospyros kaki*

Hulme (1971) recommended 1°C and 90–100% rh with 8% CO_2 and 3–5% O_2 for the cultivar Fuyu. Kitagawa and Glucina (1984) recommended 5–8% CO_2 and 2–3% O_2 at 0°C for 5–6 months for Fuyu. Storage at 0–5°C with 5–8% CO_2 and 3–5% O_2 was reported to have only a fair effect on storage and not commercially useful (Kader, 1985, 1992). Controlled atmosphere storage was recommended with 5–8% CO_2 with 3–5% O_2 by SeaLand

(1991). Typical storage conditions were given as 0–5°C with 5–8% CO_2 and 3–5% O_2 by Bishop (1996). The incidence of skin browning was reduced by storing fruits under controlled atmosphere conditions of 2% O_2 with 5% CO_2 and ethylene absorber compared to those stored in air (Lee *et al.*, 1993). In Korea the cultivar Fuyu could be stored for 5 months at 0–2°C in 9% CO_2 plus 2% O_2 (Chung *et al.*, 1994). Trees of the cultivar Triumph were sprayed with 50 mg l^{-1} of gibberellic acid 2 weeks prior to harvest were ten times less sensitive to ethylene than fruit from untreated trees when subsequently stored at 10°C in air or 1.5–2.0% CO_2 with 3.0–3.5% O_2 (Ben Arie *et al.*, 1989).

Table 7.3. Recommended controlled atmosphere storage conditions for selected pear cultivars by country and region (M. Herregods, 1993, personal communication).

	°C	CO_2	O_2
Comice			
Belgium	0 to –0.5	< 0.8	2–2.2
France	0	5	5
Germany (Westphalia)	0	3	2
Italy	–1 to 0.5	3–5	3–4
Spain	–0.5	3	3
Switzerland	0	5	3
USA (Oregon)	–1	0.1	0.5
Conference			
Belgium	–0.5	< 0.8	2–2.2
Denmark	–0.5 to 0	0.5	2–3
England	–1 to 5	< 1	2
Germany (Saxony)	1	1.1–1.3	1.3–1.5
Germany (Westphalia)	–1 to 0	1.5–2	1.5
Holland	–1 to –0.5	Much less than 1	3
Italy	–1 to –0.5	2–4	2–3
Italy	–1 to –1.5	1–1.5	6–7
Slovenia	0	3	3
Spain	–1	2	3.5
Switzerland	0	2	2
Packham's Triumph			
Australia (South)	–1	3	2
Australia (Victoria)	0	0.5	1
Australia (Victoria)	0	4.5	2.5
Germany (Westphalia)	–1 to 0	3	2
New Zealand	–0.5	2	2
Slovenia	0	3	3
Switzerland	0	2	2

In Korea fruits are stored for 6 months at 2–3°C with minimal rotting or colour change, and storage in the same conditions with ten fruits sealed in 60 μ thick polyethylene film bags was even better (A.K. Thompson, 1974, unpublished results). Astringency of persimmon fruits was removed by storing fruits in polyethylene bags which had been evacuated or with nitrogen or CO_2 atmospheres, but the fruits stored in the CO_2 atmosphere were susceptible to flesh browning. Fruits stored under vacuum or nitrogen atmosphere maintained high quality and firmness for 2 weeks at 20°C and 3

Table 7.4. Recommended controlled atmosphere storage conditions for pears (Van der Merwe, 1996).

Cultivar	Temperature (°C)	% O_2	% CO_2	Duration (months)
Bon Chretien	−0.5	1	0	4
Buerré Bosc	−0.5	1.5	1.5	4
Packham's Triumph	−0.5	1.5	1.5	9
Forelle	−0.5	1.5	0–1.5	7

Table 7.5. Recommended controlled atmosphere storage conditions for pears (Richardson and Meheriuk, 1989).

Cultivar	Temperature (°C)	% O_2	% CO_2	Remarks
Abate Fetel	−1 to 0	3–4	4–5	
Alexander Lucas	−1 to 0	3	1	Very sensitive to CO_2, store 6 months
Anjou	−1 to 0	1–2.5	0.03–2	
Bartlett	−1 to 0	1–3	0–3	
Buerré Bosc	−1 to 0	1–3	0.03–4	
Buerré Hardy	−1 to 0	2–3	0–5	
Clapps Favorite	0	2	0–1	
Conference	−1 to 0.5	1.5–3	0–3	
Decana Comizio	0	3	4–5	
Decana Inverno	0	3	5	
Diels Butterbirne	−1 to 0	2	2	Store 4–6 months
Doyenne du Comice	−1 to 1	2–3	0.8–5	
Gellerts Butterbirne	−1 to 0	2	1	Sensitive to CO_2, store 4–5 months
Kaiser Alexander	0–1	3	3–4	
Kosui	−0.5 to 5	1–2	–	
Nijiseiki	−1 to 1	3	≤1	
Packham's Triumph	−1 to 1	1–3	0–5	
Passe Crassane	−1 to 1	1.5–3	2–5	
Tsu Li	−0.5 to 1	1–2	–	
20th Century	−1 to 1	3	≤1	
Williams Bon Cheretien	−1 to 0	1–3	0–3	
Ya Li	−0.5 to 0.5	4–5	Up to 5	

months at 1°C. Flesh appearance and taste after 14 weeks storage at 1°C was best under nitrogen atmospheres (Pesis *et al.*, 1986).

Pineapple, *Ananas comosus*

Akamine and Goo (1971) found that the storage life of the cultivar Smooth Cayenne was significantly extended under 2% O_2 at 7.2°C. Kader *et al.* (1985) recommended 5% O_2 and 10% CO_2 at 10–15°C. Storage at 10–15°C with 10% CO_2 and 5% O_2 had a fair effect but was not being used commercially (Kader, 1989). Kader (1989, 1992) recommended 8–13°C with 5–10% CO_2 and 2–5% O_2. Kader (1993) recommended 10°C, with a range of 8–13°C, combined with 5–10% CO_2 and 2–5% O_2. SeaLand (1991) recommended 10% CO_2 with 5% O_2. If fruits are exposed to 3% O_2 in the first week of storage at 22°C followed by 8°C, the symptom will be effectively reduced. Paull and Rohrbach (1985) found that storage at 3% O_2 and 5% CO_2, or 3% O_2 and 0% CO_2 did not suppress the internal browning symptom (sometime referred to as black heart) in the cultivar Smooth Cayenne stored at 8°C. If fruits were exposed to 3% O_2 in the first week of storage at 22°C followed by 8°C, it was shown that the symptom could be effectively reduced. Dull *et al.* (1967) described an experiment where fruit were harvested at stage 4 maturity (less than 12% shell colour) and stored in atmospheres containing 21, 10, 5 or 2.5% O_2 with the balance nitrogen. As the O_2 concentration decreased, so did the respiration rate. Where 5 or 10% CO_2 was added to the atmosphere the only noticeable effect was a slight further reduction in respiration rate. Controlled atmosphere storage was not considered beneficial but typical storage conditions were given as 8–13°C with 5–10% CO_2 and 2–5% O_2 by Bishop (1996). Dull *et al.* (1967) also considered controlled atmosphere storage had no obvious effect on fruit quality or the maintenance of fruit appearance. For fruit harvested at full maturity and stored at 8°C in 2 with 0, 2 with 10 or 1 with 0% O_2 with % CO_2 or in air all the fruits were still in good condition after 3 weeks. However, in subsequent storage at 22°C to simulate marketing, some fruits in all the treatments developed internal translucency within 3 days and fruit of the control and 2% O_2 with 0% CO_2 developed internal browning or black heart disease (Haruenkit and Thompson, 1993). Controlled atmosphere storage can reduce but not eliminate internal browning (Table 7.6).

Storage under hypobaric conditions can extend the storage life up to 30–40 days (Staby, 1976).

Storage of the cultivar Mauritius under modified atmosphere, using polyethylene film bag, for 2 weeks at 10°C the black heart development was less than 10%. The final O_2 and CO_2 content in the bag were 10 and 7%, respectively (Hassan *et al.*, 1985). They also suggested pineapples should be kept in polyethylene film bags until consumption in order to avoid the developing of black heart disease.

Table 7.6. Effects of controlled atmosphere storage on the frequency and level of internal browning score (0 = none, 5 = maximum) on Smooth Cayenne pineapples stored at 8°C for 3 weeks and then 5 days at 20°C (Haruenkit and Thompson, 1996).

Gas composition (%)				
O_2	CO_2	N_2	Internal browning score	% Affected
1.3	0	98.7	0.8	75
2.2	0	97.8	0.8	75
5.4	0	94.6	3.8	100
1.4	11.2	87.4	0.3	50
2.3	11.2	86.5	0.3	25
20.8	0	79.2	4.1	100
LSD ($P = 0.05$)			1.4	–

Plantain, *Musa*

Storage recommendations were 2–5% CO_2 with 2–5% O_2 (SeaLand, 1991). Maintaining high humidity around the fruit can help to keep fruit in the pre-climacteric stage so that where fruit were stored in moist coir dust or individual fingers were sealed in polyethylene film they can remain green and pre-climacteric for over 20 days in Jamaican ambient conditions of about 28°C (Thompson *et al.*, 1972a). Such fruit ripened quickly and normally when removed from the plastic film. Since packing in moist coir proved as effective as modified atmosphere packaging it could be that the effects of modified atmospheres on plantains are due entirely or at least partially to humidity rather than changes in O_2 and CO_2 inside the bags (Thompson *et al.*, 1974d).

Plum, Prune, *Prunus domestica*

Storage at 0–5°C with 0–5% CO_2 and 1–2% O_2 was reported to have a good effect on fruit, but was not being used commercially (Kader, 1985). Van der Merwe (1996) recommended –0.5°C and 5% CO_2 with 3% O_2 for fruit ripened to a firmness of 5.5 kg before storage for up to 7–8 weeks depending on cultivar. In storage at 0.5°C physiological injury to the cultivar Monarch was some 25% less in 2.5% CO_2 with 5% O_2 than in air (Anon, 1968). The cultivars Santa Rosa and Songold, partially ripened to a firmness of approximately 4.5 kg prior to storage, then stored at 0.5°C in 4% O_2 with 5% CO_2 for 7 or 14 days was sufficient to keep plum fruits in an excellent condition for an additional 4 weeks in normal atmospheres at 7.5°C. Internal breakdown was almost eliminated by this treatment and the fruits were of excellent eating quality after ripening (Truter and Combrink, 1992). Tonini *et al.* (1993) showed that there was a reduction in the incidence of internal breakdown in the cultivar Stanley in atmospheres containing 20% CO_2 compared to those stored in air both at 0°C. Higher concentrations proved phytotoxic. For the cultivar Angelo, phytotoxicity was observed during storage at CO_2 concentrations higher than 2.5%. At 1°C with 12% CO_2 and 2% O_2 the cultivar Buhler Fruhzwetsche had a good appearance, taste and firmness after 4 weeks storage (Streif, 1989). No CO_2 injury was detected in Buhler Fruhzwetsche during storage at concentrations below 16% (Streif, 1989). SeaLand (1991) recommended 0–5% CO_2 with 2% O_2. Typical storage conditions were given as 0–5°C with 0–5% CO_2 and 1–2% O_2 by Bishop (1996). Fruits of the cultivar Angeleno were kept in air and in 0.25% or 0.02% O_2 at 0, 5, or 10°C. Exposures to the low O_2 atmospheres inhibited ripening, including reduction in ethylene production rate, retardation of skin colour changes and flesh softening, maintenance of titratable acidity and increased resistance to CO_2 diffusion. The most important detrimental effect of the low O_2 treatments was the development of an alcoholic off-flavour (Ke *et al.*, 1991a). In Korea the optimum conditions for the storage of the cultivar Soldam were found to be 0% CO_2 plus 1% O_2 for 12 weeks, without any quality changes (Chung *et al.*, 1994). Storage with 1% O_2 at 0.5°C with intermittent warming was shown to have beneficial effects (Hardenburg *et al.*, 1990). Naichenko and Romanshchak (1984) recommended storage at –2°C for the cultivar Vengerka Obyknovennaya. At that temperature fruits were kept in good condition for only 19 days in air but they could be stored for up to 125 days in 5% CO_2 with 3% O_2 and 92% nitrogen.

The cultivar Victoria fruits at different stages of ripeness were stored in 3 or 21% O_2 and 0, 4 or 7% CO_2 at 0.3 or 0.5°C for up to 4 weeks. Some fruits were harvested unripe, others when not yet fully ripe but of good colour. In 3% O_2 fruit colour development was delayed and fruits remained firm, but after 4 weeks they tasted better than those stored in 21% O_2. In high CO_2 fruit colour development was also delayed, but fruit flavour was not affected. Average wastage after 2, 3 and 4 weeks of storage was 9, 20 and 32%, respectively (Roelofs, 1993b).

Fruits of the cultivars Opal and Victoria, picked when not fully ripe, could be stored for up to 3 weeks at 1°C in 2% O_2 and 5% CO_2 (Roelofs and Breugem, 1994). With the cultivar Opal, the best storage conditions were 0°C with 1 or 2% O_2 with 5% CO_2, and with the cultivar Monsieur Hatif they were 0°C and 1 or 2% O_2 with 0% CO_2. Both cultivars could be stored for up to 3 weeks without excessive spoilage due to rots and fruit cracking (Wild and Roelofs, 1992). Folchi *et al.* (1994) stored the cultivar President in less than 0.1% CO_2 with 0.3% O_2 and showed an increase in aldehyde and ethanol content of fruit with increased storage time. Truter and Combrink (1997) stored the cultivars Laetitia, Casselman and Songold at 0°C for 8 weeks in various controlled atmosphere storage conditions followed by 7 days in air at 10°C. The best treatments of those investigated were 5% CO_2 with 3% O_2 for Laetitia, 5% CO_2 with 3% O_2 or partial controlled atmosphere storage for Casselman and 5% CO_2 with 3% O_2 for a maximum of 7 weeks for Songold.

Good control of plum fruit moth *Cydia funebrana* and the fungus *Monilia* spp. on fruit harvested in dry weather and at an optimum maturity stage could be achieved by storage for 1 month in normal cold storage at 0°C and for 6 weeks in controlled atmosphere (with 2% O_2 and 2.5% CO_2). Atmospheres with 1–1.5% O_2 could limit browning of the flesh, maintain a crisp and juicy texture and slow down rot development (Westercamp, 1995).

Pomegranate, *Punica granatum*

Little published work is available on controlled atmosphere storage, but it has been shown to affect the development of a postharvest superficial browning disorder called husk scald. The most effective control in the cultivar Wonderful was by storing late harvested fruits in 2% O_2 at 2°C. However, this treatment resulted in accumulation of ethanol, which caused off-flavour development but when the fruits were transferred to air at 20°C, ethanol and off-flavour dissipated (Ben Arie and Or, 1986). Fruits of the cultivar Hicaz were stored at 6, 8 or 10°C in controlled atmosphere regimes of 1% CO_2 with 3% O_2, 3% CO_2 with 3% O_2 or 6% CO_2 with 3% O_2 or in air (control), all at 85–90% rh. Minimal quality and weight losses were observed in fruits stored in the 6 to 3 controlled atmosphere storage regime for 6 months. Fruits stored in air at 6°C were acceptable after 5 months of storage. Fruits stored in air at 8 or 10°C had a storage life of 50 days, whereas fruits stored at these temperatures under controlled atmosphere storage conditions had a storage life of 130 days (Kupper *et al.*, 1994). Kader *et al.* (1984) showed that pomegranate fruit did not respond to pre-storage ethylene treatment.

Pomerac, Otaheite apple, *Syzygium malaccese*

Sankat and Basanta (1997) stored pomerac fruits at 5°C for up to 30 days in 5% CO_2 with 2% O_2, in 5% CO_2 with 4% O_2, in 5% CO_2 with 6% O_2, in 5% CO_2 with 8% O_2, 5% CO_2 with 1% O_2, 8% CO_2 with 1% O_2, 11% CO_2 with 1% O_2 and 14% CO_2 with 1% O_2. The optimum storage conditions were found to be 11% CO_2 with 1% O_2 and 14% CO_2 with 1% O_2 where fruits maintained their flavour and appearance for 25 days. Previous work is quoted where storage at 5°C in air resulted in a postharvest life of 20 days.

Potato, *Solanum tuberosum*

For optimum controlled atmosphere storage, Fellows (1988) recommended a maximum of 10% CO_2 and a minimum of 10% O_2. The amount of O_2 and CO_2 in the atmosphere of a potato store can affect the sprouting of the tubers, rotting, physiological disorders, respiration rate, sugar content and processing quality (Table 7.7).

If potatoes are to be processed into crisps or chips it is important that the levels of reducing sugars and amino acids are at a level to permit the Maillard reaction during frying. This gives them an attractive golden colour and their characteristic flavour. If levels of reducing sugars are too high the crisps or chips are too dark after frying and unacceptable to the processing industry (Schallenberger *et al.*, 1959; Roe *et al.*, 1990). Potatoes stored at low temperatures (around 4°C) have higher levels of sugars than those stored at higher temperatures of 7–10°C (Khanbari and Thompson, 1994). This can result in the production of dark coloured crisps from potatoes which have been stored at low temperatures. At low storage temperatures,

Table 7.7. Sugars (g per 100 g of dry weight) in tubers of three potato cultivars stored for 25 weeks under different controlled atmospheres at 5 and 10°C and reconditioned for 2 weeks at 20°C (Khanbari and Thompson, 1997).

	Gas combinations		5°C			10°C		
Cultivars	CO_2 (%)	O_2 (%)	Sucrose	RS	TS	Sucrose	RS	TS
Record	9.4	3.6	0.757	0.216	0.973	0.910	0.490	1.400
	6.4	3.6	0.761	0.348	1.109	1.385	1.138	2.523
	3.6	3.6	0.622	0.534	1.156	1.600	0.749	2.349
	0.4	3.6	0.789	0.510	1.299	0.652	0.523	1.175
	0.5	21.0	0.323	0.730	1.053	0.998	0.634	1.632
			0.650	**0.488**	**1.138**	**1.109**	**0.707**	**1.816**
Saturna	9.4	3.6	0.897	0.324	1.221	0.685	0.233	0.918
	6.4	3.6	0.327	0.612	0.993	0.643	0.240	0.883
	3.6	3.6	0.440	0.382	0.822	0.725	0.358	1.083
	0.4	3.6	0.291	0.220	0.511	0.803	0.117	0.920
	0.5	21.0	0.216	0.615	0.831	0.789	0.405	1.194
			0.434	**0.473**	**0.907**	**0.729**	**0.271**	**1.000**
Hermes	9.4	3.7	0.371	0.480	0.851	0.256	0.219	0.475
	6.4	3.7	0.215	0.332	0.547	1.364	0.472	1.836
	3.6	3.5	0.494	0.735	1.229	0.882	0.267	1.149
	0.4	3.6	0.287	0.428	0.715	0.585	0.303	0.888
	0.5	21.0	0.695	0.932	1.627	0.617	0.510	1.127
			0.412	**0.682**	**1.094**	**0.741**	**0.354**	**1.095**

RS, reducing sugars; TS, total sugars.
Numbers in bold represent mean values of gas combinations.

potato tubers accumulate reducing sugars, but the effects of CO_2 can influence their sugar levels (Table 7.8). With levels of 6% CO_2 and O_2 of 2–15% in store during 178 days at 8 and 10°C fry colour was very dark (Reust *et al.*, 1984). Mazza and Siemans (1990) studying 1–3.2% CO_2 found that darkening of crisps occurred after there was a rise in CO_2 levels in stores. At levels of CO_2 of 8–12% in stores the fry colour was darker than for potatoes stored in air (Schmitz, 1991). At O_2 levels of 2% reducing sugar levels were reduced or there was no accumulation at low temperature, but 5% O_2 was much less effective (Workman and Twomey, 1969). The cultivar Bintje were stored at 6°C in atmospheres containing 0, 3, 6 or 9% CO_2 and 21, 18, 15 or 12% O_2. During the early phase the unfavourable effect of high CO_2 concentration on chip colour increased to a maximum at 3 months. In the later phase chip colour deteriorated in tubers stored without CO_2 and CO_2 enrichment had less effect (Schouten, 1994). Khanbari and Thompson (1994) stored Record tubers at 4°C under low levels of CO_2 (0.7–1.8%) and low levels of O_2 (2.1–3.9%) and found that it resulted in a significantly lighter crisp colour, low sprout growth and fewer rotted tubers compared with simulated air storage of 0.9% CO_2 and 21.0% O_2.

Storing tubers in anaerobic conditions of total nitrogen prevented accumulation of sugars at low temperature but it had undesirable side-effects of the tubers (Harkett, 1971).

Schouten (1992) discussed the effects of controlled atmosphere storage on sprouting of stored potatoes. He found that sprout growth was stimulated in 3% CO_2 at 6°C, whereas some inhibition of growth occurred in 6% CO_2. Sprout growth was strongly inhibited in 1% O_2 at 6°C. However, pathological breakdown may develop at this O_2 content. Internal disorders were found at O_2 contents less than 1%. Stimulation of sprout growth occurred at slightly higher O_2 levels if controlled atmosphere storage started from the beginning of the season. Stimulation was less if low O_2 conditions were applied from January onwards. It is concluded that controlled atmosphere storage at 6°C was not an alternative to chemical control of sprout growth for ware potatoes (that is potatoes used for consumption). In the early phase of storage Schouten (1994) showed that sprouting was

Table 7.8. Effects of CO_2 levels in storage on reducing sugar content in potatoes.

CO_2 (%)	Effect	Reference
4	Higher reducing sugars than in air	Workman and Twomey, 1969
5	Prevents accumulation of reducing sugars but increased accumulation of sucrose	Denny and Thornton, 1941
6	Accumulation of sugars especially sucrose	Reust *et al.*, 1984
5–20	Initial reduction in reducing sugars but later increase to a higher level than air	Burton, 1987

Table 7.9. Sugar content of tubers from three potato cultivars after being stored for 25 weeks in high CO_2 (9.4% CO_2 and 3.7% O_2) and another 20 weeks in air at 5°C and then reconditioned for 2 weeks at 20°C (Khanbari and Thompson, 1997).

	Cultivars	Type of sugar g per 100 g of dry weight				
		Fructose	Glucose	Sucrose	RS	Grey level
Direct	Record	0.778	0.847	0.520	1.625	130.3
	Saturna	0.660	0.685	0.540	1.345	136.8
	Hermes	0.980	1.120	0.564	2.100	122.9
Recondition	Record	0.628	0.563	0.897	1.191	132.7
	Saturna	0.381	0.343	0.489	0.724	151.8
	Hermes	1.030	1.127	1.133	2.157	123.6

RS, reducing sugars.

stimulated by 3–6% CO_2 and inhibited by 9% CO_2. During the later phase all CO_2 concentrations inhibited sprouting. Khanbari and Thompson (1994) showed that high CO_2 with low O_2 combinations during storage completely inhibited sprout growth, but caused the darkest crisp colour, but after reconditioning tubers gave the same level of sprouting and crisps as light as the other controlled atmosphere storage combination.

Reconditioning of stored potato tubers can reduce their sugar content (Table 7.9) and improve their fry colour (Khanbari and Thompson, 1997). Schouten (1994) recommended reconditioning tubers at 15°C for 2–4 weeks and showed that the treatment improved chip fry colour, but reconditioning stimulated sprouting. Tubers stored in atmospheres of 0.7–1.8% CO_2 with 2.1–3.9% O_2 showed a significantly higher weight loss and shrinkage after reconditioning compared to tubers which had previously been stored in air (Khanbari and Thompson, 1994).

Tubers stored in high CO_2, especially at 10 or 15%, had earlier onset of rotting. At low concentrations of CO_2 (0.7–1.6%), and low O_2 (2–2.4%) there was also an increase in tuber rotting (Khanbari and Thompson, 1994).

Radish, *Raphanus sativus*

Low O_2 storage was shown to prolong their postharvest life (Haard and Salunkhe, 1975). Lipton (1972) showed that storage of topped radishes in 1% O_2 at 5 or 10°C reduced sprouting by about 50% compared to those stored in air, but although storage in 0.5% O_2 further reduced sprouting it resulted in injury to the radishes. Storage at 0.6°C with 5% O_2 or 5 or 10°C with 1–2% O_2 were recommended by Pantastico (1975). Saltveit (1989) recommended 0–5°C with 2–3% CO_2 and 1–2% O_2 for topped radishes but indicated that controlled atmosphere storage had only a slight effect.

Rambutan, *Nephellium lappaceum*

Kader (1993) in a review of modified atmosphere packaging and controlled atmosphere storage of tropical fruits recommended 10°C, with a range of 8–15°C, combined with 7–12% CO_2 and 3–5% O_2. Storage decay could be controlled by dipping the fruit 100 ppm benomyl fungicide.

Fruit have been stored successfully in polyethylene film bags (Mendoza *et al.*, 1972) with storage times of about 10 days in perforated bags and 12 days in non-perforated bags at 10°C. Muhammad (1972) recommended storage in sealed polyethylene film bags at 10°C for 12 days but at high temperatures (32–37°C) perforation of the bags was preferable. Under normal atmospheres, maximum shelf life for cultivars R162, Jit Lee and R156 was 15 days at 7.5°C, and 11–13 days at 10°C. Storage of cultivar R162 in 9–12% CO_2 retarded colour loss and extended shelf life by 4–5 days but storage in 3% O_2 or 5 ppm ethylene did not significantly affect the rate of colour loss (O'Hare *et al.*, 1994).

Raspberry, *Rubus idaeus*

Kader (1989) recommended 0–5°C with 15–20% CO_2 and 5–10% O_2. Storage with 20–25% CO_2 retarded the development of rots (Hardenburg *et al.*, 1990). Typical storage conditions were given as 0–5°C with 15–20% CO_2 and 5–10% O_2 by Bishop (1996). Black raspberries cultivar Bristol were stored at 18 or 0°C in air or 20% CO_2. Storage in 20% CO_2 greatly improved their postharvest quality at both storage temperatures by reducing grey mould development (Goulart *et al.*, 1992). Goulart *et al.* (1990) showed that when the cultivar Bristol stored at 5°C in 2.6, 5.4 or 8.3% O_2 and 10.5 and 19.6% CO_2, respectively, the percentage mass loss was greatest after 3 days in the control. When fruits were removed after 3 days and held for 4 days in air at 1°C, those stored in 15% CO_2 had less deterioration than any other treatment except for controls. Deterioration was greatest in the three O_2 and the 10% CO_2 treatments. Fruits removed after 7 days and held for up to 12 days at 1°C showed least deterioration with the 15% CO_2 treatment. In the second experiment, fruits of the red raspberry cultivar Heritage were harvested in mid-September, cooled to 2°C and transported to a store at around 4°C. On the next day they were transferred to controlled atmosphere storage of either 5% O_2 in nitrogen or 20% CO_2 in air plus an air control. The gas concentrations within the plastic bags, which were also included in the trial was 5% O_2 or 18.1% CO_2. They were held at 5°C for 3 or 5 days. The percentage of deterioration was greatest with fruits held in 5% O_2, followed by the control. No off-flavour were detected in either black or red raspberries held in a high CO_2 atmosphere.

The cultivars Glen Clova, Glen Moy, Glen Prosen and Willamette, stored in 2% O_2 and 10% CO_2 had reduced fruit rot (*Botrytis cinerea*), delayed ripening, slower colour development, slower breakdown of acids (including ascorbic acid), slower breakdown of total soluble solids and firmer fruits compared to storage in air (Callesen and Holm, 1989). Recommended storage temperature is -0.5–$0°C$ at 90–95% rh using high CO_2 and reduced O_2 is effective in reducing the incidence of rots and maintaining fruit quality (Robbins and Fellman, 1993).

Red currant, *Ribes sativum*

Typical storage conditions were given as 0–5°C with 12–20% CO_2 and 2–5% O_2 by Bishop (1996). Controlled atmosphere storage at 0.5°C, 90–95% rh with 2% O_2 and 18% CO_2 was recommended by Ellis (1995). The cultivars Rotet, Rondom, Rovada, Roodneus and Augustus were stored for 8–25 weeks at 1°C in 0, 10, 20 or 30% CO_2 in combination with 2% O_2 or high (not controlled, 15–21%) O_2. The fruits had been picked four times at weekly intervals during the normal harvest period for each cultivar. Weight loss and fruit rot were monitored during the storage period and shelf life. Drying through the fruit stalk and other weight losses were high during the first 8–11 weeks of storage, independent of the storage conditions. Within each cultivar the earliest harvested berries were more sensitive to rot. Increased CO_2 concentration reduced percentage fruit rot significantly and reduced O_2 concentration only reduced fruit rot at low CO_2 concentrations. Levels above 20% CO_2 resulted in internal breakdown and fruit discolouring in some cultivars after 13 weeks of storage. Low O_2 concentrations further increased these symptoms. Best storage results were obtained at 20% CO_2 with 2% O_2; after storage of more than 20 weeks in this atmosphere, fruit quality was still reasonable (Roelofs *et al.*, 1993b). Storage of soft fruits at 20% CO_2 slowed down respiration and decomposition by 50% and doubled fruit storability. It also suppressed rot fungi, both in low temperature storage and during marketing. Gaseous CO_2 in cylinders was used mainly during transport, while the cheaper dry ice was used in storage, particularly with storage temperatures of 0–10°C. CO_2 concentrations had to be reduced by half during prolonged storage. Storage at 0°C required only one-third of the concentration of CO_2 required at 7°C (Hansen, 1986). In ordinary cold storage of the cultivars Rovada (late ripening) and Stanza (mid-season ripening) at 0–1°C it was shown that the berries could be kept for 2–3 weeks while in controlled atmosphere storage in a 20% CO_2 concentration or more they could be kept for 8–10 weeks (Leeuwen and Van de Waart, 1991). The cultivar Rotet were stored at 1°C in either a controlled atmosphere of 10, 20 or 30% CO_2 all with 2% O_2 or a high CO_2 environment of 10, 20 or 30% CO_2 all with $> 15\%$ O_2. The optimum CO_2 concentration was found to be 20%. Ethanol accumulation was higher under controlled atmosphere storage than high CO_2 environment conditions. Fruit could be stored for 8–10 weeks under controlled atmosphere storage or high CO_2 environment conditions, compared with 1 month for fruits stored at 1°C in a normal atmosphere (Agar *et al.*, 1991). Six cultivars were stored at 1°C in 1, 2 or 21% O_2 and 0 or 10% CO_2, in all combinations, from 30 July to 13 December. The late-ripening cultivars Augustus and Roodneus had the least spoilage and fruits could have been stored satisfactorily until Christmas. Spoilage was least in 1% O_2 and 10% CO_2 (Roelofs, 1992). Seven cultivars (Rotet, Rondom, Rovada, Augustus, Roodneus, Cassa and Blanka) grown in tunnels were harvested four times at weekly intervals, starting when the berries were adequately coloured, and stored in combinations of CO_2 at 0, 20 or 25% combined with O_2 at 2 or 18–21%, at 1 or –0.5°C for up to 24 weeks. The effects on fruit quality were assessed immediately after storage and after holding at 20°C for 4 days. After 24 weeks the average incidence of fungal rots was lowest (about 5%) at –0.5°C and 20 or 25% CO_2. Rotting increased to about 35% at 1°C and 25% CO_2 and to about 50% at 1°C and 20% CO_2. With 0% CO_2 at either temperature the incidence of rots was about 95%. Fruit quality, however, was adversely affected by CO_2 at 20% and high O_2 which was in contrast to the findings in the previous year published by the author (Roelofs, 1994). The cultivar Rotet was stored at 1°C for 10 weeks in either 0 with 21 (control), 10 with 19, 20 with 17, 30 with 15, 10 with 2, 20 with 2 and 30 with 2 O_2 with CO_2 to determine the evolution of the chlorophyll content losses of the peduncles during storage. The chlorophyll content was most stable in controlled atmosphere storage of high CO_2 and low O_2. For fruits stored in air, the decrease in chlorophyll content coincided with a decrease in fruit quality (Agar *et al.*, 1994c).

Rhubarb, *Rheum rhaponticum*

Unsealed polyethylene film liners in crates reduced their weight loss (Lutz and Hardenburg, 1968) but there was no mention of modified atmosphere effects.

Runner bean, Green bean, French bean, Kidney bean, Snap bean, *Phaseolus vulgaris*

Storage at 0–5°C with 5–10% CO_2 and 2–3% O_2 had a fair effect on storage but was said to be of limited commercial use (Kader, 1985). Saltveit (1989) recommended 5–10°C with 4–7% CO_2 and 2–3% O_2, which had only a slight effect, but for those destined for processing 5–10°C with 20–30% CO_2 and 8–10% O_2 was recommended, but it had a moderate effect. Storage at 5–10% CO_2 with 2–3% O_2 retarded yellowing, also discoloration of the cut end of the beans can be prevented from discoloration by exposure to 20–30% CO_2 for 24 h (Hardenburg *et al.*, 1990). The same conditions were recommended by SeaLand (1991). High CO_2 levels could result in development of off-flavour, but storage at 45°F (7.2°C) combined with 5–10% CO_2 and 2–3% O_2 retarded yellowing (Anandaswamy and Iyengar, 1961). Snap beans may be held for up to 3 weeks in 8–18% CO_2, depending on the temperature. The cultivars Strike and Opus snap beans were stored for up to 21 days at 1, 4 or 8°C in controlled atmospheres with 2% O_2 in combination with up to 40% CO_2, then transferred to air at 20°C for up to 4 days. CO_2 concentrations of 20% or greater always caused severe injury, manifested as loss of tissue integrity followed by decay. At 1°C, 8% CO_2 was the maximum level tolerated, while 18% CO_2 caused injury at 4°C, but not at 8°C. (Costa *et al.*, 1994).

Salsify, *Tragopogen porrifolius*

Storage with 3% CO_2 and 3% O_2 at 0°C was recommended by Hardenburg *et al.* (1990).

Sapodilla, *Manilkara zapota*

Controlled atmosphere storage recommendations were 20°C with 5–10% CO_2 and scrubbing of ethylene (Broughton and Wang, 1979), although optimum temperature for storage in non-controlled atmosphere conditions varied from as low as 0°C (Thompson, 1996). In other work, fruits stored for 28 days at 10°C failed to ripen properly indicating chilling injury (Abdul-Karim *et al.*, 1993).

Satsuma mandarin, *Citrus unshiu,* Common mandarin, *C. reticulata*

Controlled atmospheres had only a slight or no effect on storage (SeaLand, 1991). At 25°C an atmosphere comprising 60% CO_2 with 20% O_2 with 20% nitrogen did not inhibit respiration in satsumas, but under 80% CO_2 with 20% air and 90% CO_2 with 10% air, respiration rates were significantly reduced (Kubo *et al.*, 1989b). Storage of *C. unshiu* in China in containers with a D45 M2 1 silicone window of 20–25 $cm^2\ kg^{-1}$ of fruit gave the optimum CO_2 concentration of less than 3% together with O_2 less than 10% (Hong *et al.*, 1983).

Ogaki *et al.* (1973) carried out experiments over a number of years and showed that the most suitable atmosphere for storing satsumas was 1% CO_2 combined with 6–9% O_2. 85–90% rh produced the best quality fruit and resulted in only 3% weight loss. Pre-storage treatment of 7–8°C and 80–85% rh was also recommended.

In two-season trials satsumas were stored for 3 months at 3°C and 92% rh in controlled atmosphere (2.8–6.5% O_2 plus 1% CO_2 and 93.5–97.2% nitrogen) or in ordinary storage (control). Weight losses were much lower in controlled atmosphere (1.2–1.5%) than in the control (6.5–6.7%) and fruit colour, flavour, aroma and consistency were also better in controlled atmosphere. Ascorbic acid content in the flesh and peel of controlled atmosphere stored fruits was 5.9 and 10.3% higher, respectively, than in the controls (Dubodel and Tikhomirova, 1985). Satsumas were stored at 2–3°C 90% rh in 3–6% O_2 and 1% CO_2 or in air (control). The total sugar content of fruit flesh decreased during storage in both experimental and control fruit by 10.7 and 13.0%, respectively, and losses in the peel were even greater. Reducing sugars decreased during storage, especially in the flesh of experimental fruits and in the peel of control fruits (Dubodel *et al.*, 1984).

Shiitake, *Lentinus edodes*

Storage in an unspecified temperature with 40% CO_2 and 1–2% O_2 extended their storage life four times longer than the control (Minamide, 1981 quoted by Bautista, 1990). Storage at 0°C with 5, 10, 15 or 20% CO_2 and 1, 5 and 10% O_2 plus an air control was studied by Pujantoro *et al.* (1993) who found that 1 or 5% O_2 levels resulted in poor storage.

Soursop, *Annona muricata*

It was shown that fruit in storage are damaged by exposure to low temperatures (Wardlaw, 1937) and Snowdon (1990) recommended 10–15°C. No specific data on controlled atmosphere storage could be found, but in a comparison between 16, 22 or 28°C the longest storage period (15 days) occurred in the 16°C with fruits packed in high density polyethylene bags without Ethysorb. This packaging treatment also resulted in the longest storage duration at the two higher temperatures, but the differences between treatments at those temperatures were much less marked (Guerra *et al.*, 1995).

Spinach, *Spinacia oleracea*

Wardlaw (1937) mentioned that storage in high concentrations of CO_2 could damage the tips of leaves. Kader (1985) recommended 0–5°C with 10–20% CO_2 and 21% O_2 which had a fair effect but was of no commercial use. Saltveit (1989) recommended 0–5°C with 5–10% CO_2 and 7–10% O_2 which had only a slight effect. Storage in 10–40% CO_2 and 10% O_2 was said to retard yellowing (Hardenburg *et al.*, 1990). 10–20% CO_2 and 21% O_2

was recommended by SeaLand (1991). Fellows (1988) recommended a maximum of 20% CO_2. Spinach could be stored for 30 days in 7% O_2 and 4.9% CO_2 (Niedzielski, 1984).

Spring onions, Escallion, Green onions, *Allium cepa*

Kader (1985) recommended 0–5°C with 10–20% CO_2 and 1–2% O_2 for green onions, but it had only a fair effect and was of limited commercial use. Saltveit (1989) recommended 0–5°C with 0–5% CO_2 and 2–3% O_2 which had only a slight effect. Optimum recommended storage conditions by Hardenburg *et al.* (1990) were 5% CO_2 with 1% O_2 at 0°C for 6–8 weeks. 10–20% CO_2 with 2–4% O_2 was recommended by SeaLand (1991).

Squash, *Cucurbita* spp.

Courgette, Summer squash, Zucchini, *Cucurbita pepo* variety *melopepo*, Marrow, Vegetable marrow, Winter squash, *Cucurbita maxima, C. mixta, C. moschata, C. pepo*, Pumpkin, *Cucurbita maxima, C. mixta, C. moschata, C. pepo*

Storage in low O_2 was claimed to be of little or no value at 5°C (Hardenburg *et al.*, 1990), but SeaLand (1991) recommended 5–10% CO_2 with 3–5% O_2. Storage in 1% O_2 reduced chilling injury symptoms at 2.5°C compared to storage in air (Wang and Kramer, 1989). Symptoms occurred after 9 days storage compared with 3 days storage in air. *C. pepo* cultivar Romanesco was stored for 19 days at 5°C in 21 ± 1% O_2 plus 0, 2.5, 5 or 10% CO_2 and then held for 4 more days at 13°C in air. High CO_2 levels inhibited the rate of CO_2 production and reduced the development of chilling injury symptoms at all three maturities (16, 20 and 22 cm long fruits). At the end of the 23-day storage period, 82% of the 22 cm fruits held in 10% CO_2 appeared saleable, but they had a slight off-flavour and were soft; 79% of the fruits held in 5% CO_2 were saleable, firm and free from off-flavour; samples from 2.5% CO_2 and air were unacceptable because of decay and pitting. The percentage of unsaleable fruits was greater for 20 cm than for 16 cm long fruit. It is suggested that CO_2 concentrations around 5% may be useful for storing zucchini squash at about 5°C, a temperature that normally causes chilling injury in zucchini (Mencarelli, 1987a). Storage had slight to no effect (SeaLand, 1991).

The extent of surface pitting of *C. pepo* cultivar Ambassador stored at 2.5°C was less under 1% O_2 than when kept in air (Wang and Ji, 1989). At 0°C squashes stored in 1% O_2 had reduced chilling injury compared to those stored in air (Wang and Kramer, 1989). Mencarelli *et al.* (1983) also showed that storage of zucchini squash at 2.5°C resulted in chilling injury, but the effect could be delayed by reducing the O_2 content in the store atmosphere to 1–4%. In later work *C. pepo* fruits were stored for 2 weeks in 1, 2 or 4% levels inhibited CO_2 and ethylene production rates, particularly at 5 and 10°C. Both rates increased during subsequent aeration for 2 days at 10°C, but much more in samples held previously at 2.5 or 5°C than in those from 10°C. About 75 and 55%, respectively, of the burst in CO_2 production during aeration in samples from 2.5 and 5°C was due to exposure to low temperature; the remainder was attributed to the effect of exposure to low O_2 levels. For ethylene production, the corresponding values were about 95 and 70%. All fruits stored at 5°C for 2 weeks were virtually free of chilling injury, surface mould, decay or off-flavour; almost all the fruits were rated good to excellent in appearance. About three-quarters of the fruits were still in this category after 2 additional days at 10°C. At 5°C the fruits did not benefit from storage in low O_2 atmospheres. Storage at 2.5°C induced severe chilling injury, which was ameliorated by holding the fruits in 1, 2 or 4% O_2 instead of 8 or 21% O_2. However, amelioration was no longer evident after 2 days of aeration at 10°C. Defects not related to chilling injury, especially off-flavour, developed most in fruits held at 10°C; the benefit from low O_2 levels was slight (Mencarelli *et al.*, 1989).

Strawberry, *Fragaria* × *ananassa*

Woodward and Topping (1972) showed that high levels of CO_2 in the storage atmosphere at 1.7°C reduced rotting caused by infection with *Botrytis cinerea*, but levels as high as 20% CO_2 were injurious to the fruit. However, they showed that there was a temperature interaction with CO_2 levels. In CO_2 concentrations up to 20% fruits stored at 3°C were in good condition after 10 days, but after 15–20 days there was a distinct loss of flavour which appeared to be the main limiting factor to storage in the UK. In later work storage at 0–5°C with 15–20% CO_2 and 10% O_2 (or 5–10% O_2; Kader, 1989) had an excellent effect on storage (Kader, 1985). Storage of 10–30% CO_2 or 0.5–2% O_2 are used to slow respiration rate and reduce disease levels, but 30% CO_2 or less than 2% O_2 was also reported that it may cause off-flavour to develop (Hardenburg *et al.*, 1990). SeaLand (1991) recommended 15–20% CO_2 with 10% O_2 and Lawton (1996) recommended 1°C, 90% rh, 20% CO_2 and 17% O_2. Typical storage conditions were given as 0–5°C with 15–20% CO_2 and 5–10% O_2 by Bishop (1996). Fellows (1988) recommended a maximum of 20% CO_2 and a minimum of 2% O_2. Van der Merwe (1996) recommended 1.5% CO_2 with 1.5% O_2 and showed that the storage life can be extended to 3 weeks compared to only 4 days in air.

Baab (1987) described experiments with the cultivar Elvira. Treatments with 20% CO_2 for 24 h before storage at 24°C, those cooled to 16°C before treating with 20% CO_2 for 12 h, and those exposed to 20% CO_2 for 24 h before returning to 24°C produced the least losses (3.1, 3.9 and 4.9%, respectively). Fruit stored continuously in air at 24°C, those cooled to 17°C before treating with 40% CO_2 for 4 h and those cooled to 17°C before treating with 20% CO_2 for 12 h had the greatest losses (18.9, 14.4 and 10.7%, respectively). In all treatments fruit colour and calyx colour were maintained, but CO_2 treated fruit had an off-flavour. However, 6 h after the CO_2 treatment the off-flavour disappeared, except for fruit exposed to 20% CO_2 for 114 h.

Storage at 20% CO_2 slowed down respiration and decomposition by 50% and doubled fruit storability compared to fruit stored in air. 20% CO_2 also suppressed rot fungi, both in low temperature storage and during marketing. Gaseous CO_2 in cylinders was used mainly during transport, while the cheaper dry ice was used in storage, particularly with storage temperatures of 0–10°C. CO_2 concentrations had to be reduced by half during prolonged storage. Storage at 0°C required only one-third of the concentration of CO_2 required at 7°C. There was a cultivar interaction with CO_2. High CO_2 concentrations improved the storability of the cultivar Elsanta more than that of Elvira (Hansen, 1986).

Fruits of the cultivar Redcoat were forced air cooled and stored in 0, 12, 15 or 18% CO_2 for 0, 2, 7 or 14 days. CO_2 increased firmness but fruits stored for more than 7 days softened rapidly under all CO_2 concentrations and fruits stored for 14 days were as soft as those not exposed to CO_2 at all. The amount of storage decay after 14 days was less under CO_2 than in the control. Organoleptic evaluation identified fruit quality differences between treatments but the results were not consistent (Smith et al., 1993).

Redcoat were stored at several temperatures and for various intervals in controlled atmospheres containing 0–18% CO_2 and 15–21% O_2. Controlled atmosphere storage of fresh fruits indicated that addition of CO_2 to the storage environment enhanced fruit firmness. Fruits kept under 15% CO_2 for 18 h were 48% firmer than untreated samples were initially. Response to increasing CO_2 concentration was linear. There was no response to changing O_2 concentration. Maximum enhancement of firmness was achieved at a temperature of 0°C; there was essentially no enhancement at 21°C. In some instances, there was a moderate enhancement of firmness as time in storage increased. CO_2 acted to reduce the quantity of fruit lost due to rot. Fruits that were soft and bruised after harvest became drier and firmer in a CO_2 enriched environment (Smith, 1992).

Fruits of various strawberry cultivars were stored at 0°C for 42 h in a controlled atmosphere of 15% CO_2 with 18% O_2 to study the effect on firmness. Compared with initial samples and control samples stored in air, firmness was increased in 21 of the 25 cultivars evaluated. The CO_2 treatment had no effect on colour, as measured by Hunter L, a and b values, or on total soluble solids or pH. There were significant differences among cultivars in firmness, colour (Hunter L, a and b values), total soluble solids and pH (Smith et al., 1993).

Fruits of 15 strawberry cultivars were picked at two different stages of maturity, precooled or not, and stored for 5 or 7 days at 0 or 20°C with 10 or 20% CO_2. Increased atmospheric CO_2 was associated with excellent control of storage decay caused by *Botrytis* and *Penicillium* species and it slowed down the fruit metabolism, thus preserving aroma and quality (Ertan *et al.*, 1990).

Fruits of the cultivar Chandler were kept in air, 0.25% O_2, 21% O_2 with 50% CO_2 or 0.25% O_2 with 50% CO_2 (balance nitrogen) at 5°C for 1–7 days to study the effects of controlled atmospheres on volatiles and fermentation enzymes. Acetaldehyde, ethanol, ethyl acetate and ethyl butyrate concentrations were greatly increased, while isopropyl acetate, propyl acetate and butyl acetate concentrations were reduced by the three controlled atmosphere storage compared with those of control fruits stored in air. Controlled atmosphere storage enhanced pyruvate decarboxylase and alcohol dehydrogenase activities but slightly decreased alcohol acetyl transferase activity. The results indicate that the enhanced pyruvate decarboxylase and alcohol dehydrogenase activities cause ethanol accumulation, which in turn drives the biosynthesis of ethyl esters; the increased ethanol concentration also competes with other alcohols for carboxyl groups for esterification reactions; the reduced alcohol acetyl transferase activity and limited availability of carboxyl groups due to ethanol competition decrease production of other acetate esters (Ke *et al.*, 1994b).

Sliced strawberries (cultivars Pajaro and G3) were dipped in solutions of citric acid, ascorbic acid and/or calcium chloride and were stored in air or in controlled atmosphere storage for 7 days at 2.5°C followed by 1 day in air at 20°C. Whole fruits were also stored under the same conditions. Fruit slices respired at a higher rate than whole fruits at both temperatures. Controlled atmosphere storage slowed down respiration and ethylene production rates of sliced fruits. Firmness of slices was maintained by storage in 12% CO_2 and in 0.5% O_2, or by dipping in 1% calcium chloride and storing in air or under controlled atmosphere storage conditions (Rosen and Kader, 1989).

Swede, Rutabaga, *Brassica napus* variety *napobrassica*

Controlled atmosphere storage was reported to have a slight to no effect on keeping quality (SeaLand, 1991). Coating with paraffin wax was reported to reduced weight loss but if this was too heavy the reduced O_2 supply could result in internal breakdown (Lutz and Hardenburg, 1968).

Sweetpotato, *Ipomoea batatas*

SeaLand (1991) indicated that controlled atmosphere storage had a slight or no effect compared to storage in air. Typical storage conditions were given as 0–5°C with 5–10% CO_2 and 2–4% O_2 by Bishop (1996). Storage in 2–3% CO_2 with 7% O_2 kept roots in better condition than those stored in air, but CO_2 levels above 10% or O_2 levels below 7% could give alcoholic or off-flavour to the roots (Hardenburg *et al.*, 1990). The cultivar Georgia Jet were stored at 15.5°C in air at 85% rh or in a 7% O_2 with 93% nitrogen atmosphere for 3 months. Controlled atmosphere storage had no effect on moisture content, β-carotene or total lipids (Charoenpong and Peng, 1990). Pantastico (1975) recommended 2–3% CO_2 with 7% O_2 because levels of CO_2 above 10% or levels of O_2 below 7% could result in off-flavour.

Sweetcorn, Babycorn, *Zea mays* variety *saccharata*

The respiration rates of sweetcorn was significantly reduced under 80% CO_2 with 20% air at both 15 and 25°C (Inaba *et al.*, 1989) Storage in high CO_2 retarded conversion of starch to sugar (Wardlaw, 1937). Kader (1985) recommended 0–5°C with 10–20% CO_2 and 2–4% O_2 which had a good effect but he claimed that controlled atmosphere storage was of no commercial use. Also Saltveit (1989) recommended 0–5°C with 5–10% CO_2 and 2–4% O_2 and again claimed that it had only a slight effect. Controlled atmosphere storage experiments have shown that cobs may be injured with more than 20% CO_2 or less than 2% O_2, but in an atmosphere of 2% O_2 the sucrose content remained higher (Hardenburg *et al.*, 1990). Other work recommended 10–20% CO_2 with 2–4% O_2 (SeaLand, 1991). Brash *et al.* (1992) subjected cobs of the cultivar Honey 'n' Pearl to controlled atmosphere storage at 0°C in 2.5% O_2 or levels of CO_2 in the range of 5–15% for 3–4 weeks followed by 3 days shelf life assessment at 15°C in simulated seafreight export trials in New Zealand. Low O_2 (2.5%) and moderate CO_2 (6–10%) ensured cobs retained sweetness and husk leaves stayed greener for longer during shelf life compared to air storage. Controlled atmosphere storage for 3 weeks also reduced insect numbers. Fellows (1988) recommended a maximum of 20% CO_2. Ryall and Lipton (1972) reported that storage in 5–10% CO_2 retarded sugar loss while 10% caused injury.

The use of controlled atmospheres to control insect pests on sweetcorn harvested in New Zealand was studied by Carpenter (1993) who concluded that cold storage (0–1°C) alone was as effective as treatment with controlled atmospheres.

Othieno *et al.* (1993) showed that the best modified atmosphere packaging proved to be unperforated film or film with the minimum perforation (one 0.4 mm hole per 6.5 cm²). However, cobs packed in polypropylene film with no perforations showed a reduced rate of toughening, weight loss and loss of sweetness, but they tended to develop 'off odours'.

Packing cobs with the sheath still attached but partly cut away to reveal the seeds below was the best treatment from both the storage and presentational aspects (Othieno *et al.*, 1993).

Sweetsop, Sugar apple, *Annona squamosa*

Storage of the cultivar Tsulin at 20°C with 5% CO_2 and 5% O_2 reduced ethylene production levels and delayed ripening (Tsay and Wu, 1989). In other work fruits of Tsulin were picked during the autumn at the green mature stage (with hard flesh). Storage in 5% O_2 with 5% CO_2 delayed ripening of the fruits and also inhibited ethylene evolution (Tsay and Wu, 1989). Kader (1993) in a review of modified atmosphere packaging and controlled atmosphere storage of tropical fruits recommended 15°C, with a range of 12–20°C, combined with 5–10% CO_2 and 3–5% O_2.

Taro, Dasheen, Eddoe, Cocoyam, *Xanthosoma* spp. *Colocasia* spp.

Modified atmosphere packaging in polyethylene film bags has been shown to reduce postharvest losses. Both *Xanthosoma* and *Colocasia* stored in Trinidad ambient conditions (27–32°C) showed reduced weight losses (Table 7.10), although optimum temperature for storage in non-controlled atmosphere conditions varied from 7 to 13°C (Thompson, 1996).

Table 7.10. Effects of packaging on the weight loss of stored taro (Passam, 1982).

Weeks in storage	Colocasia				Xanthosoma			
	Control	Dry coir	Moist coir	Polyethyl-ene film	Control	Dry coir	Moist coir	Polyethyl-ene film
2	11	7	0	1	8	7	0	1
4	27	21	3	5	15	19	4	3
6	35	30	7	9	28	28	21	6

Tomato, *Lycopersicon esculentum*

Wardlaw (1937) recommended 5% CO_2 with 5% O_2 at 12°C for fruits harvested at the yellow or tinted stage. Monzini and Gorini (1974) recommended 12°C with 2.5–5% CO_2 and 2.5% O_2 for 3 months. At 12.8°C fruit colour and flavour were maintained in an acceptable condition for 6 weeks in 0% CO_2 with 3% O_2, while in other work 1–2% CO_2 with 4–8% O_2 at 12.8°C for breaker or pink fruit was recommended (Pantastico, 1975). Kader (1985) recommended 12–20°C with 0% CO_2 and 3–5% O_2 for mature green fruits which had a good effect but was of limited commercial use. For partially ripe fruit Kader (1985) recommended 8–12°C with 0% CO_2 and 3–5% O_2 which had a good effect but was of limited commercial use. Saltveit (1989) recommended 12–20°C with 2–3% CO_2 and 3–5% O_2 but which had only a slight effect. Adamicki (1989) described successful storage of mature green tomatoes at 12–13°C either 0% CO_2 and 2% O_2, 5% CO_2 and 3% O_2 or 5% CO_2 and 5% O_2 for 6–10 weeks. Parsons et al. (1974) harvested tomatoes at the mature green stage and stored them at 55°F (12.8°C) for 6 weeks and found that 0% CO_2 in combination with 3% O_2 was better than those stored in air. Fruits in air were fully red in 6 weeks, those in 0% CO_2 in combination with 3% O_2 or those in 5% CO_2 in combination with 3% O_2 fruits were still pink with the latter combination having an additional delaying effect on ripening. Batu and Thompson (1995) recommended 13°C and 9.1% CO_2 with 5.5% O_2 for 60 days for fruit of the cultivar Criterium harvested at the pink stage of maturity. Dennis et al. (1979) stored fruits harvested at the mature green stage at 13°C and 93–95% rh in air, 5% CO_2 with 3% O_2 or 5% CO_2 with 5% O_2 for 6 to 10 weeks. Two greenhouse grown cultivars (Sonato and Sonatine) and three field grown (Fortuna, Hundrefold and Vico) were used. Controlled atmosphere stored fruit were found to ripen more evenly and to a better flavour when removed from storage than air stored fruit, with the 5% CO_2 with 3% O_2 treatment giving the best quality fruit.

Storage for mature green was 0% CO_2 with 3–5% O_2 or for breaker to light pink 0–3% CO_2 with 3–5% O_2 (SeaLand, 1991). Typical storage conditions were given as 10–15°C with 3–5% CO_2 and 3–5% O_2 by Bishop (1996). Fellows (1988) recommended a maximum of 2% CO_2 and a minimum of 3% O_2. Mature green fruits of the cultivars Saba and Saul were stored for 14 days at 12°C in a room equipped with an ethylene scrubber or 40 days at 12°C in a controlled atmosphere room (3% O_2 and 3% CO_2). At the end of storage the fruits had a marketable percentage of 90 and 85% for ethylene scrubber and controlled atmosphere storage, respectively. *Phytophthora* decay appeared on some fruits by the end of controlled atmosphere storage (Anelli et al., 1989). Tataru and Dobreanu (1978) showed that tomatoes kept best in 3% CO_2 with 3% O_2. Greenhouse grown mature green and pink tomatoes (cultivar Veegan) were stored in air (controls) or in 0.5, 3.0 or 5.0% O_2 in nitrogen or argon. Colour development was not affected by the background gases nitrogen or argon except at 3% O_2 where ripening was delayed by the argon mixture. All the levels of O_2 suppressed colour development in mature green fruits compared with air controls, with maximum suppression at 0.5% O_2. Most fruits held in 0.5% O_2 rotted after removal to air before colour development was complete. CO_2 and ethylene production were lower in mature green and pink fruits in all atmospheres than in air and were lower in argon than in nitrogen (Lougheed and Lee, 1989). Green mature tomato fruits (cultivar Ramy) were stored for 20 days at 4 or 8°C in 3% O_2 with 0% CO_2 with or without ethylene removal or in 1.5% O_2 with 0% CO_2 with ethylene removal (using potassium permanganate). Fruit ripening was delayed most in fruits stored in 1.5% O_2 with 0% CO_2 with ethylene removal (Francile, 1992). In controlled atmosphere storage the CO_2 levels affected the colour changes of the tomatoes. The colour of tomatoes harvested at the pink stage of maturity did not change when they were stored in 6.4% CO_2 with 5.5% O_2 and 9.1% CO_2 with 5.5% O_2 even after 50 days and in some cases after 70 days storage. The red colour development of the tomatoes exposed to less than 6.4% CO_2 increased, whereas red colour decreased with CO_2 levels above 9.1% during storage (Batu, 1995). Storage in atmospheres containing over 4% CO_2 or less than 4% O_2 gave uneven ripening (Pantastico, 1975). Li *et al.* (1973) showed that partial pressures of O_2 below 1% caused physiological injury during storage, owing presumably to anaerobic fermentation. The O_2 level for prolonged storage at 12–13°C was about 2–4% O_2; which delayed the respiratory peak. Salunkhe and Wu (1973) showed that exposing 'green-wrap' tomato fruits of the cultivar DX-54 to low O_2 atmospheres at 12.78°C inhibited ripening and increased storage life. Storage at 10% O_2 with 90% nitrogen resulted in a storage life of 62 days, and at 3% O_2 and 97% nitrogen it was 76 days. The maximum storage life was 87 days at 1% O_2 and 99% nitrogen. Low O_2 atmospheres inhibited fruit chlorophyll and starch degradation and also lycopene, β-carotene and soluble sugar synthesis. Parsons and Spalding (1972) inoculated fruit with soft rot bacteria and held them for 6 days at 55°F. Fruits kept better in controlled atmosphere storage with 3% O_2 with 5% CO_2 than in air. Decay lesions were smaller on fruits stored in controlled atmosphere storage than on those stored in air, but controlled atmosphere storage did not control decay. However, although at 55°F (12.8°C) inoculated tomatoes kept better in controlled atmosphere storage than in air, at 45°F (7.2°C) or 65°F (18.3°C) they kept equally well in controlled atmosphere storage and air.

Batu (1995) showed that there were interactions between modified atmosphere packaging and temperature in that modified atmosphere packaging was more effective in delaying ripening at 13°C than at 20°C. Modified atmosphere packaging also interacted with tomato harvest maturity where it was more effective in delaying ripening of fruit harvested at the mature green stage than for those harvested at a more advanced stage of maturity. There was an interaction between packaging films and ripening time of fruits. Tomatoes ripened later when they were sealed in films which were less permeable to O_2, CO_2 and water vapour than when they were sealed in higher permeable films. Packaging films also affected fruit firmness of tomatoes. All green tomatoes sealed in 25 μ thick polypropylene film were very firm even after 60 days of storage at 13 or 20°C, compared to other films. Tomatoes harvested at the mature green stage and sealed in 50 μ thick polyethylene or polypropylene films had delayed development of the red colour after 30 days of storage compared to fruit stored in air at the same temperature, and those tomatoes also had the lowest weight loss and the highest soluble solids after 60 days of storage. There were differences in ripening of tomatoes between modified atmosphere packaging and controlled atmosphere storage where mature green fruits in modified atmosphere packaging ripened earlier than the same fruits in controlled atmosphere storage. The differences were probably due to ethylene accumulation in modified atmosphere packaging which would not occur in controlled atmosphere storage because of the continuous gas flushing.

Vidigal *et al.* (1979) used the cultivars Angela and Kada harvested at the mature green stage and stored them at 10°C and 90% rh with atmospheres within the range of 3–20% CO_2 and 1–10% O_2. They found that the combination of 8% CO_2 and 5% O_2 gave the best results for both cultivars for up to 36 days.

Turnip rooted parsley, *Petroselinum crispum* subsp. *tuberosum*

Storage of turnip rooted parsley at 11% CO_2 and 10% O_2 helped to retain the green colour of the leaves during storage (Hardenburg *et al.*, 1990), although optimum temperature for storage in non-controlled atmosphere conditions was about 0°C (Thompson, 1996).

Turnip, *Brassica rapa* variety *rapa*

Controlled atmosphere storage of turnip was reported to have had only a slight to no effect (SeaLand, 1991).

Watermelon, *Citrullus lanatus*

Tamas (1992) showed that the storage life of watermelons (cultivar Crimson Sweet) was extended from 14 to 16 days to 42–49 days by cooling within 48 h of harvest and storing the fruits at 7–8°C and 85–90% rh in a controlled atmosphere of 2% CO_2 with 7% O_2. Changes in fruit appearance, flavour and aroma were minimal but changes in the colour intensity and consistency of the flesh were noticeable. In other work controlled atmosphere storage had only a slight or no effect (SeaLand, 1991).

Yam, *Dioscorea* spp.

Asiatic yam, Lisbon yam, White yam, *Dioscorea alata*, Asiatic yam, *Dioscorea esculentum*, Asiatic bitter yam, *Dioscorea hispida*, Bitter yam, *Dioscorea dumetorum*, Chinese yam, *Dioscorea opposita*, Potato yam, *Dioscorea bulbifera*, White yam, *Dioscorea rotundata*, Yampie, Cush cush, *Dioscorea trifida*, Yellow yam, *Dioscorea cayenensis*

Little information is available in the literature on controlled atmosphere storage but one report indicated that it had only a slight to no effect (SeaLand, 1991). Fully mature *D. opposita* tubers can be stored for long periods of time at high O_2 tensions at 5°C which was also shown to reduce browning of tubers (Imakawa, 1967).

Storage of tubers in sealed polyethylene film bags can have beneficial effects on storage of yam but small white spots were observed on the tuber surface which was shown to be parenchymatous proliferation through lenticels (Thompson *et al.*, 1973). *D. trifida* began to sprout within 3 weeks in storage at 20–29°C and 46–62% rh but sealing tubers in polyethylene film bags reduced losses (Table 7.11).

Table 7.11. Effects of packaging material on the quality of *D. trifida* after 64 days at 20–29°C and 46–62% rh. Fungal score was 0 = no surface fungal growth, 5 = tuber surface entirely covered with fungi. Necrotic tissue was estimated by cutting the tuber into two length ways and measuring the area of necrosis and expressing it as a per cent of the total cut surface (Thompson, 1972).

Packaging type	Weight loss (%)	Fungal score	Necrotic tissue (%)
Paper bags	23.6	0.2	5
Sealed polyethylene bags[1]	15.7	0.2	7
Ventilated polyethylene bags[2]	5.4	0.4	4

[1]Sealed 0.03 mm thick polyethylene bags.
[2]Polyethylene bags with 0.15% of the area as holes.

References

Abdel-Rahman, M. and Isenberg, F.M.R. (1974) Effects of growth regulators and controlled atmosphere on stored carrots. *Journal of Agricultural Science* 82, 245–249.

Abdul Rahman, A.S., Maning, N. and Dali, O. (1997) Respiratory metabolism and changes in chemical composition of banana fruit after storage in low oxygen atmosphere. *Seventh International Controlled Atmosphere Research Conference, 13–18 July 1997.* University of California, Davis, California 95616 USA (abstract) 51.

Abdul Raouf, U.M., Beuchat, L.R. and Ammar, M.S. (1993) Survival and growth of *Escherichia coli* 0157:H7 on salad vegetables. *Applied and Environmental Microbiology* 59, 1999–2006.

Abdul-Karim, M.N.B., Nor, L.M. and Hassan, A. (1993) The storage of sapadilla *Manikara achras* L. at 10, 15, and 20°C. *ACIAR Proceedings* 50, 443.

Abdullah, H. and Pantastico, E.B. (1990) *Bananas.* Association of Southeast Asian Nations-COFAF, Jakarta Indonesia.

Abdullah, H. and Tirtosoekotjo, S. (1989) *Association of Southeast Asian Nations Horticulture Produce Data Sheets.* Association of Southeast Asian Nations Food Handling Bureau, Kuala Lumpur.

Abe, Y. (1990) Active packaging – a Japanese perspective. In: Day, B.P.F. (ed.) *International Conference on Modified Atmosphere Packaging* Part 1. Campden Food and Drinks Research Association, Chipping Campden, Gloucestershire.

Adamicki, F. (1989) Przechowywanie warzyw w kontrolowanej atmosferze. Biuletyn Warzywniczy Supplement I, 107–113.

Adamicki, F., Dyki, B. and Malewski, W. (1977) Effect of CO_2 on the physiological disorders observed in onion bulbs during CA storage. *Quality of Plant Foods and Human Nutrition* 27, 239–248.

Adamicki, F. and Elkner, K. (1985) Effect of temperature and a controlled atmosphere on cauliflower storage and quality. *Biuletyn Warzywniczy* 28, 197–224.

Adamicki, F. and Kepka, A.K. (1974) Storage of onions in controlled atmospheres. *Acta Horticulturae* 38, 53–73.

Agar, I.T., Bangerth, F. and Streif, J. (1994a) Effect of high CO_2 and controlled atmosphere concentrations on the ascorbic acid, dehydroascorbic acid and total vitamin C content of berry fruits. Postharvest physiology of fruits, Kyoto, Japan, 21–27 August. *Acta-Horticulturae* 398, 93–100.

219

Agar, I.T., Cetiner, S., Garcia, J.M. and Streif, J. (1994b) A method of chlorophyll extraction from fruit peduncles: application to redcurrants *Turkish Journal of Agriculture and Forestry* 18, 209–212.

Agar, I.T., Streif, J. and Bangerth, F. (1994c) Effect of high CO_2 and controlled atmosphere (CA) storage on the keepability of blackberry cv, 'Thornfree'. *Commissions C2,D1,D2/3 of the International Institute of Refrigeration International Symposium.* 8–10 June. Istanbul, Turkey, pp. 271–280.

Agar, T., Hess-Pierce, B., Sourour, M.M. and Kader, A.A. (1997) Identification of optimum preprocessing storage conditions to maintain quality of black ripe 'Mananillo' olives. *Seventh International Controlled Atmosphere Research Conference, 13–18 July 1997.* University of California, Davis, California 95616 USA (abstract) 118.

Agar, T., Streif, J. and Bangerth, F. (1991) Changes in some quality characteristics of red and black currants stored under CA and high CO_2 conditions. *Gartenbauwissenschaft* 56, 141–148.

Aharoni, N., Yehoshua, S.B. and Ben-Yehoshua, S. (1973) Delaying deterioration of romaine lettuce by vacuum cooling and modified atmosphere produced in polyethylene packages. *Journal of the American Society for Horticultural Science* 98, 464–468.

Aharoni, Y. and Houck, L.G. (1980) Improvement of internal color of oranges stored in O_2-enriched atmospheres. *Scientia Horticulturae* 13, 331–338.

Aharoni, Y. and Houck, L.G. (1982) Change in rind, flesh, and juice color of blood oranges stored in air supplemented with ethylene or in O_2-enriched atmospheres. *Journal of Food Science* 47, 2091–2092.

Aharoni, Y., Nadel-Shiffman, M. and Zauberman, G. (1968) Effects of gradually decreasing temperatures and polyethylene wraps on the ripening and respiration of avocado fruits. *Israel Journal of Agricultural Research* 18, 77–82.

Ahrens, F.H. and Milne, D.L. (1993) Alternative packaging methods to replace SO_2 treatment for litchis exported by sea. Alternatiewe verpakkingsmetodes vir seeuitvoer van lietsjies om SO_2-behandeling te vervang. *Yearbook South African Litchi Growers' Association* 5, 29–30.

Akamine, E.K. (1969) Controlled atmosphere storage of papayas. *Proceedings 5th Annual Meeting, Hawaii Papaya Industry Association. 12–20 September, 1969.* Hawaii Cooperative Extension Service, Miscellaneous Publication, 764.

Akamine, E.K. and Goo, T. (1969) Effects of controlled atmosphere storage on fresh papayas *Carica papaya* L. variety Solo with reference to shelf life extension of fumigated fruits. *Hawaii Agricultural Experimental Station, Honolulu, Bulletin* 144.

Akamine, E.K. and Goo, T. (1973) Respiration and ethylene production during ontogeny of fruit. *Journal of the American Society for Horticultural Science* 98, 381–383.

Ali Niazee, M.T., Richardson, D.G., Kosittrakun, M. and Mohammad, A.B. (1989) Non-insecticidal quarantine treatments for apple maggot control in harvested fruits. *Proceedings of the Fifth International Controlled Atmosphere Research Conference.* Wenatchee, Washington, USA, June 14–16, 1989, vol. 1, pp. 193–205.

Alique, R. and Oliveira, G.S. (1994) Changes in sugars and organic acids in Cherimoya *Annona cherimola* Mill. fruit under controlled atmosphere storage. *Journal of Agricultural and Food Chemistry* 42, 799–803.

Allen, F.W. and McKinnon, L.R. (1935) Storage of Yellow Newtown apples in chambers supplied with artificial atmospheres. *Proceedings of the American Society for Horticultural Science* 32, 146.

Allen, F.W. and Smock, R.M. (1938) CO_2 storage of apples, pears, plums and peaches. *Proceedings of the American Society for Horticultural Science* 35, 193–199.

Allwood, M.E. and Cutting, J.G.M. (1994) Progress report: gas treatment of 'Fuerte' avocados to reduce cold damage and increase storage life. *Yearbook South African Avocado Growers' Association* 17, 22–26.

Alvarez, A.M. (1980) Improved marketability of fresh papaya by shipment in hypobaric containers. *HortScience* 15, 517–518.

Al-Zaemey, A.B.S., Falana, I.B. and Thompson, A.K. (1989) Effects of permeable fruit coatings on the storage life of plantains and bananas. *Aspects of Applied Biology* 20, 73–80.

Anandaswamy, B. *et al.* (1963) Pre-packaging of fresh produce. IV. Okra *Hibiscus esculentus*. *Food Science Mysore* 12, 332–335. (Horticultural Abstracts 1964.)

Anandaswamy, B. and Iyengar, N.V.R. (1961) Pre-packaging of fresh snap beans *Phaseolus vulgaris*. *Food Science* 10, 279.

Anderson, R.E. (1982) Long term storage of peaches and nectarine intermittently warmed during controlled atmosphere storage. *Journal of the American Society for Horticultural Science* 107, 214–216.

Anderson, R.E., Hardenburg, R.E. and Vaught, H.C. (1963) Controlled atmosphere storage studies with cranberries. *Proceedings of the American Society of Horticultural Science* 83, 416–422.

Andre, P., Blanc, R., Buret, M., *et al.* (1980a) Globe artichoke storage trials combining the use of vacuum pre-refrigeration, controlled atmospheres and cold. *Revue Horticole* 211, 33–40.

Andre, P., Buret, M., Chambroy, Y., Dauple, P., Flanzy, C. and Pelisse, C. (1980b) Conservation trials of asparagus spears by means of vacuum pre-refrigeration, associated with controlled atmospheres and cold storage. *Revue Horticole* 205, 19–25.

Andrich, G. and Fiorentini, R. (1986) Effects of controlled atmosphere on the storage of new apricot cultivars. *Journal of the Science of Food and Agriculture* 37, 1203–1208.

Andrich, G. Zinnai, A., Balzini, S., Silvestri, S. and Fiorentini, R. (1994) Anaerobic respiration rate of Golden Delicious apples controlled by environmental P_{CO_2}. *Commissions C2, D1, D2/3 of the International Institute of Refrigeration International Symposium.* 8–10 June Istanbul, Turkey, pp. 233–242.

Anelli, G., Mencarelli, F. and Massantini, R. (1989) One time harvest and storage of tomato fruit: technical and economical evaluation. *Acta Horticulturae* 287, 411–415.

Anon. (1919) Food Investigation Board. *Department of Scientific and Industrial Research Report for the Year 1918.*

Anon. (1920) Food Investigation Board. *Department of Scientific and Industrial Research Report for the Year 1920*, pp. 16–25.

Anon. (1941) Annual Report for 1940. *Massachusetts Agricultural Experiment Station*, Bulletin 378.

Anon. (1958) Food Investigation Board. *Department of Scientific and Industrial Research Report for the Year 1957*, pp. 35–36.

Anon. (1964a) *Hygrometric Tables*, Part III: *Aspirated Psychrometer Readings Degrees Celsius*. London, Her Majesty's Stationery Office.

Anon. (1964b) State Storage Regulations for Certified Controlled Atmosphere of Apples. *US Department of Agriculture, Federal Extension Service*, December.

Anon. (1968) Fruit and vegetables. *American Society of Heating, Refrigeration and Air-conditioned Engineering, Guide and Data Book*, no. 361.

Anon. (1974) *Atmosphere Control in Fruit Stores*. Ministry of Agriculture, Fisheries and Food, Agricultural Development and Advisory Service, Short Term Leaflet 35.

Anon. (1978) Banana CA storage. *Bulletin of the International Institue of Refrigeration* 18, 312.

Anon. (1993) Onion in store without MH. Horticultural Development Council Progress Report. *Grower* April 12–13.

Anon. (1997) Squeezing the death out of food. *New Scientist* 2077, 28–32.

Apeland, J. (1971) Factors affecting respiration and colour during storage of parsley. *Acta Horticulturae* 20, 43–53.

Apeland, J. (1985) Storage of Chinese cabbage *Brassica campestris* L. *pekinensis* Lour Olsson in controlled atmospheres. *Acta Horticulturae* 157, 185–191.

Apelbaum, A., Aharoni, Y. and Temkin-Gorodeiski, N. (1977) Effects of sub-atmospheric pressure on the ripening processes of banana fruits. *Tropical Agriculture Trinidad* 54, 39–46.

Aracena, J.J., Sargent, S.A., Brecht, J.K., Campbell, C.A. and Saltveit, M.E. (1993) Environmental factors affecting vascular streaking, a postharvest physiological disorder of cassava root *Manihot esculenta* Crantz. *Acta Horticulturae* 343, 297–299.

Arpaia, M.L., Mitchell, F.G., Kader A.A. and Mayer, G. (1985) Effects of 2% O_2 and various concentrations of CO_2 with or without ethylene on the storage performance of kiwifruit. *Journal of the American Society for Horticultural Science* 110, 200–203.

Artes Calero, F., Escriche, A., Guzman, G. and Marin, J.G. (1981) Violeta globe artichoke storage trials in a controlled atmosphere. Essais de conservation d'artichaut 'Violeta' en atmosphere controlee. *30 Congresso Internazionale di Studi sul Carciofo*, 1073–1085.

Awad, M., Oliveira, A.I. de and Correa, D. de L. (1975) The effect of Ethephon, GA and partial vacuum on respiration in bananas (*Musa acuminata*). *Revista de Agricultura, Piracicaba, Brazil* 50, 109–113.

Awad, M.A.G., Jager, A., Roelofs, F.P., Scholtens, A. and De-Jager, A. (1993) Superficial scald in Jonagold as affected by harvest date and storage conditions. *Acta Horticulturae* 326, 245–249.

Baab, G. (1987) Improvement of strawberry keeping quality. *Obstbau* 12, 265–268.

Badran, A.M. (1969) Controlled atmosphere storage of green bananas. *US Patent 17 3 June*, 450, 542.

Ballantyne, A., Stark, R. and Selman, J.D. (1988) Modified atmosphere packaging of broccoli florets. *International Journal of Food Science and Technology* 23, 353–360.

Bandyopadhyay, C. and Tewari, G.M.J. (1976) Lachrymatory factor in spring onion. *Journal of the Science of Food and Agriculture* 27, 733–735.

Bangerth, F. (1974) Hypobaric storage of vegetables. *Acta Horticulturae* 38, 23–32.

Bangerth, F. (1984) Changes in sensitivity for ethylene during storage of apple and banana fruits under hypobaric conditions. *Scientia Horticulturae* 24, 151.

Barmore, C.R. (1987) Packaging technology for fresh and minimally processed fruits and vegetables. *Journal of Food Quality* 10, 207–217.

Bastrash, S., Makhlouf, J., Castaigne, F. and Willemot, C. (1993) Optimal controlled atmosphere conditions for storage of broccoli florets. *Journal of Food Science* 58, 338–341, 360.

Batu, A. (1995) Controlled atmosphere storage of tomatoes. PhD thesis, Silsoe College, Cranfield University, UK.

Batu, A. and Thompson, A.K. (1994) The effects of harvest maturity, temperature and thickness of modified atmosphere packaging films on the shelf life of tomatoes. *Commissions C2, D1, D2/3 of the International Institute of Refrigeration International Symposium.* 8–10 June Istanbul, Turkey, pp. 305–316.

Batu, A. and Thompson, A.K. (1995) *Effects of controlled atmosphere storage on the extension of postharvest qualities and storage life of tomatoes.* Workshop of the Belgium Institute for Automatic Control, Ostend. June 1995, pp. 263–268.

Baumann, H. (1989) Adsorption of ethylene and CO_2 by activated carbon scrubbers. *Acta Horticulturae* 258, 125–129.

Bautista, O.K. (1990) *Postharvest Technology for Southeast Asian Perishable Crops.* Technology and Livelihood Resource Center.

Bautista, P.B. and Silva, M.E. (1997) Effects of CA treatments on guava fruit quality. *Seventh International Controlled Atmosphere Research Conference, 13–18 July 1997.* University of California, Davis, California 95616, USA (abstract) 113.

Beaudry, R., Schwallier, P. and Lennington, M. (1993) Apple maturity prediction: an extension tool to aid fruit storage decisions. *HortTechnology* 3, 233–239.

Beaudry, R.M. and Gran, C.D. (1993) Using a modified-atmosphere packaging approach to answer some post-harvest questions: factors influencing the lower O_2 limit. *Acta Horticulturae* 326, 203–212.

Ben Arie, R. (1996) Fresher via boat than airplane. *Peri News* 11, 1996.

Ben Arie, R. and Guelfat-Reich, S. (1976) Softening effects of CO_2 treatment for the removal of astringency from stored persimmon fruits. *Journal of the American Society for Horticultural Science* 101, 179–181.

Ben Arie, R., Levine, A., Sonego, L. and Zutkhi,Y. (1993) Differential effects of CO_2 at low and high O_2 on the storage quality of two apple cultivars. *Acta Horticulturae* 326, 165–174.

Ben Arie, R. and Or, E. (1986) The development and control of husk scald on 'Wonderful' pomegranate fruit during storage. *Journal of the American Society for Horticultural Science* 111, 395–399.

Ben Arie, R., Roisman, Y., Zuthi, Y. and Blumenfeld, A. (1989) Gibberellic acid reduces sensitivity of persimmon fruits to ethylene. *Advances in Agricultural Biotechnology* 26, 165–171.

Ben Arie, R. and Sonego, L. (1985) Modified-atmosphere storage of kiwifruit *Actinidia chinensis* Planch with ethylene removal. *Scientia Horticulturae* 27, 263–273.

Bender, R.J., Brecht, J.K. and Campbell, C.A. (1994) Responses of 'Kent' and 'Tommy Atkins' mangoes to reduced O_2 and elevated CO_2. 107th Annual meeting of the Florida State Horticultural Society, Orlando, Florida, USA, 30 October to 1 November 1994. *Proceedings of the Florida State Horticultural Society* 107, 274–277.

Ben-Yehoshua, S., Fang, D.Q., Rodov, V., Fishman, S. and Fang, D.Q. (1995) New developments in modified atmosphere packaging part II. *Plasticulture*, 107, 33–40.

Berard, J.E. (1821) Memoire sur la maturation des fruits. *Annales de Chimie et de Physique* 16, 152–183.

Berard, L.S. (1985) Effects of CA on several storage disorders of winter cabbage. Controlled atmospheres for storage and transport of perishable agricultural commodities. *4th National Controlled Atmosphere Research Conference, July 1985*, pp. 150–159.

Berard, L.S., Vigier, B., Crete, R. and Chiang, M. (1985) Cultivar susceptibility and storage control of grey speck disease and vein streaking, two disorders of winter cabbage. *Canadian Journal of Plant Pathology* 7, 67–73.

Berard, S.L., Vigier, B. and Dubuc Lebreux, M.A. (1986) Effects of cultivar and controlled atmosphere storage on the incidence of black midrib and necrotic spot in winter cabbage. *Phytoprotection* 67, 63–73.

Berger, H., Galletti, Y.L., Marin, J., Fichet, T. and Lizana, L.A. (1993) Efecto de atmosfera controlada y el encerado en la vida postcosecha de Cherimoya *Annona cherimola* Mill. cv. Bronceada. *Proceedings of the Interamerican Society for Tropical Horticulture* 37, 121–130.

Berrang, M.E., Brackett, R.E. and Beuchat, L.R. (1989) Growth of *Aeromonas hydrophila* on fresh vegetables stored under a controlled atmosphere. *Applied and Environmental Microbiology* 55, 2167–2171.

Berrang, M.E., Brackett, R.E. and Beuchat, L.R. (1990) Microbial, color and textural qualities of fresh asparagus, broccoli, and cauliflower stored under controlled atmosphere. *Journal of Food Protection* 53, 391–395.

Berry, G. and Aked, J. (1997) Controlled atmosphere alternatives to the post-harvest use of sulphur dioxide to inhibit the development of *Botrytis cinerea* in table grapes. *Seventh International Controlled Atmosphere Research Conference, 13–18 July 1997*, University of California, Davis, California 95616 USA (abstract) 100.

Bertolini, P. (1972) Preliminary studies on cold storage of cherries. *Universita di Bologna, Notiziatio CRIOF* 3.11.

Bertolini, P., Pratella, G.C., Tonini, G. and Gallerani, G. (1991) Physiological disorders of 'Abbe Fetel' pears as affected by low-O_2 and regular controlled atmosphere storage. *Technical innovations in freezing and refrigeration of fruits and vegetables*. Paper presented at a Conference held in Davis, California, USA, 9–12 July 1989, pp. 61–66.

Betts, G.D. (ed.) (1996) *A Code of Practice for the Manufacture of Vacuum and Modified Atmosphere Packaged Chilled Foods*. Campden and Chorleywood Food Research Association Guideline 11.

Biale, J.B. (1950) Postharvest physiology and biochemistry of fruits. *Annual Review of Plant Physiology* 1, 183–206.

Bishop, D. (1996) Controlled atmosphere storage. In: Dellino, C.J.V. (ed.) *Cold and Chilled Storage Technology*. Blackie, London.

Bishop, D.J. (1994) Application of new techniques to CA storage. *Commissions C2, D1, D2/3 of the International Institute of Refrigeration International Symposium*. 8–10 June, Istanbul, Turkey, pp. 323–329.

Blednykh, A.A., Akimov, Yu. A., Zhebentyaeva, T.N. and Untilova, A.E. (1989) Market and flavour qualities of the fruit in sweet cherry following storage in a controlled atmosphere. *Byulleten' Gosudarstvennogo Nikitskogo Botanicheskogo Sada* 69, 98–102.

Bleinroth, E.W., Garcia, J.L.M. and Yokomizo (1977) Conseracao de quatro variedades de manga pelo frio e em atmosfera controlada. *Coletanea de Instituto de Tecnologia de Alimentos* 8, 217–243.

Blythman, J. (1996) *The Food We Eat*. London, Michael Joseph.

Bogdan, M., Ionescu, L., Panait, E. and Niculescu, F. (1978) Research on the technology of keeping peaches in cold storage and in modified atmosphere. *Lucrari Stiintifice, Institutul de Cerctari Pentru Valorifocarea Legumelor si Fructelor* 9, 53–60.

Bohling, H. and Hansen, H. (1977) Storage of white cabbage *Brassica oleracea* var. *capitata* in controlled atmospheres. *Acta Horticulturae* 62, 49.

Bohling, H. and Hansen, H. (1989) Studies on the metabolic activity of oyster mushrooms *Pleurotus ostreatus* Jacq. *Acta Horticulturae* 258, 573–577.

Boon-Long, P., Achariyaviriya, S. and Johnson, G.I. (1994) Mathematical modelling of modified atmosphere conditions. Development of postharvest handling technology for tropical tree fruits: a workshop held in Bangkok, Thailand, 16–18 July 1992. *ACIAR Proceedings*, 58, 63–67.

Bouman, H. (1987) CA storage of head cabbage has a future with long storage. *Groenten en Fruit* 42, 156–157.

Bower, J.P., Cutting, J.G.M. and Truter, A.B. (1990) Container atmosphere, as influencing some physiological browning mechanisms in stored Fuerte avocados. *Acta Horticulturae* 269, 315–321.

Brackmann, A. (1989) Effect of different CA conditions and ethylene levels on the aroma production of apples. *Acta Horticulturae* 258, 207–214.

Brackmann, A., Streif, J. and Bangerth, F. (1993) Relationship between a reduced aroma production and lipid metabolism of apples after long-term controlled-atmosphere storage. *Journal of the American Society for Horticultural Science* 118, 243–247.

Brackmann, A., Streif, J. and Bangerth, F. (1995) Influence of CA and ULO storage conditions on quality parameters and ripening of pre-climacteric and climacteric harvested apple fruits. II. Effect on ethylene, CO_2, aroma and fatty acid production. *Gartenbauwissenschaft* 60, 1–6, 23.

Bramlage, W.J., Bareford, P.H., Blanpied, G.D., *et al.* (1977) CO_2 treatments for 'McIntosh' apples before CA storage. *Journal of the American Society for Horticultural Science* 102, 658–662.

Brash, D.W., Corrigan, V.K. and Hurst, P.L. (1992) Controlled atmosphere storage of 'Honey 'n' Pearl' sweet corn. *Proceedings Annual Conference Agronomy Society of New Zealand* 22, 35–40.

Brecht, J.K., Kader, A.A., Heintz, C.M. and Norona, R.C. (1982) Controlled atmosphere effects on quality of California canning apricots and clingstone peaches. *Journal of Food Science*, 47, 432–436.

Brewster, J.L. (1987) The effect of temperature on the rate of sprout growth and development within stored onion bulbs. *Annals of Applied Biology* 111, 463–464.

Brigati, S. and Caccioni, D. (1995) Influence of harvest period, pre- and post-harvest treatments and storage techniques on the quality of kiwifruits. *Rivista di Frutticoltura e di Ortofloricoltura* 57, 41–43.

Brigati, S., Pratella, G.C. and Bassi, R. (1989) CA and low O_2 storage of kiwifruit: effects on ripening and disease. *International Controlled Atmosphere Conference Fifth Proceedings 14 to 16 June 1989, Wenatchee, Washington USA. Vol. 2. Other Commodities and Storage Recommendations*, pp. 41–48.

Brooks, C. (1932) Effect of solid and gaseous CO_2 upon transit disease of certain fruits and vegetables. *US Department of Agriculture, Technical Bulletin* 318, September, 6.

Brooks, C. (1940) Modified atmospheres for fruits and vegetables in storage and in transit. *Refrigerating Engineering* 40, 233–236.

Brooks, C. and Cooley, J.S. (1917) Effect of temperature, aeration and humidity on Jonathan-spot and scald of apples in storage. *Journal of Agricultural Research* 12, 306–307.

Brooks, C. and McColloch, L.P. (1938) Stickiness and spotting of shelled green lima beans. *USDA Technical Bulletin* 625.

Broughton, W.J., Chan, B.E. and Kho, H.L. (1978) Maturation of Malaysian fruits. II. Storage conditions and ripening of banana *Musa sapientum* L. variety 'Pisang Emas'. *Malaysian Agricultural Research and Development Institute, Research Bulletin* 7, 28–37.

Broughton, W.J., Hashim, A.W., Shen, T.C. and Tan, I.K.P. (1977) Maturation of Malaysian fruits. I. Storage conditions and ripening of papaya *Carica papaya* L. cv. Sunrise Solo. *MARDI Research Bulletin* 5, 59–72.

Burg, S.P. (1975) Hypobaric storage and transportation of fresh fruits and vegetables. In: Haard, N. F. and Salunkhe, D. K. (eds) *Postharvest Biology and Handling of Fruits and Vegetables*. AVI Publishing, Westpoint, Connecticut, pp. 172–188.

Burg, S.P. (1993) Current status of hypobaric storage. *Acta Horticulturae* 326, 259–266.

Burg, S.P. and Burg, E.A. (1967) Molecular requirements for the biological activity of ethylene. *Plant Physiology* 42, 144–152.

Burton, K.S. and Twyning, R.V. (1989) Extending mushroom storage life by combining modified atmosphere packaging and cooling. *Acta Horticulturae* 258, 565–571.

Burton, W.G. (1958) The effect of the concentration of CO_2 and O_2 in the storage atmosphere upon the sprouting of potatoes at 10°C. *European Potato Journal* 1, 47–57.

Burton, W.G. (1968) The effect of O_2 concentration upon sprout growth on the potato tuber. *European Potato Journal* 11, 249–65.

Burton, W.G. (1974a) The O_2 uptake, in air and in 5% O_2, and the CO_2 out-put, of stored potato tubers. *Potato Research* 17, 113–37.

Burton, W.G. (1974b) Some biophysical principles underlying the controlled atmosphere storage of plant material. *Annals of Applied Biology* 78, 149–168.

Burton, W.G. (1982) *Postharvest Physiology of Food Crops*. Longmans, London.

Butchbaker, A.F., Nelson, D.C. and Shaw, R. (1967) Controlled atmosphere storage of potatoes. *Transactions of the American Society of Agricultural Engineers* 10, 534.

Calara, E.S. (1969) The effects of varying CO_2 levels in the storage of 'Bungulan' Bananas. BS thesis, University of the Philippines, Los Baños, Laguna.

Callesen, O. and Holm, B.M. (1989) Storage results with red raspberry. *Acta Horticulturae* 262, 247–254.

Cameron, A.C., Boylan-Pett, W. and Lee, J. (1989) Design of modified atmosphere packaging systems: modelling O_2 concentrations within sealed packages of tomato fruits. *Journal of Food Science* 54, 1413–1421.

Cantwell, M. (1995) Post-harvest management of fruits and vegetable stems. In: Ingelese, B.G. and Pimienta-Barrios, E. (eds) *Agro-Ecology, Cultivation and Uses of Cactus Pear. FAO Plant Production and Protection Paper* 132, 120–136.

Cantwell, M.I., Reid, M.S., Carpenter, A., Nie X. and Kushwaha, L. (1995) Short-term and long-term high CO_2 treatments for insect disinfestation of flowers and leafy vegetables. Harvest and postharvest technologies for fresh fruits and vegetables. *Proceedings of the International Conference*, Guanajuato, Mexico, 20–24 February, pp. 287–292.

Carpenter, A. (1993) Controlled atmosphere disinfestation of fresh Supersweet sweet corn for export. *Proceedings of the Forty Sixth New Zealand Plant Protection Conference*, 10–12 August, pp. 57–58.

Cenci, S.A., Soares, A.G., Bilbino, J.M.S. and Souza, M.L.M. (1997) Study of the storage of Sunrise 'Solo' papaya fruits under controlled atmosphere. *Seventh International Controlled Atmosphere Research Conference, 13–18 July 1997*, University of California, Davis, California 95616, USA (abstract) 112.

Ceponis, M.J. and Cappellini, R.A. (1983) Control of postharvest decays of blueberries by carbon dioxide-enriched atmospheres. *Plant Disease* 67, 169–171.

Chachin, K. and Ogata, K. (1971) Effect of delay between harvest and irradiation and of storage temperature on the sprouting inhibition of onion by gamma irradiation. *Journal of Food Science and Technology Japan* 18, 378.

Chambroy, V., Souty, M., Reich, M., Breuils, L., Jacquemin, G. and Audergon, J.M. (1991) Effects of different CO_2 treatments on post harvest changes of apricot fruit. *Acta Horticulturae* 293, 675–684.

Champion, V. (1986) Atmosphere control – an air of the future. *Cargo System* November issue, 28–33.

Chapon, J.F. and Trillot, M. (1992) Apples. Long-term storage in northern Italy. *Infos Paris* 78, 42–46.

Charoenpong, C. and Peng, A.C. (1990) Changes in beta-carotene and lipid composition of sweetpotatoes during storage. *OARDC Special Circular, Ohio Agricultural Research and Development Center* 121, 15–20.

Chase, W.G. (1969) Controlled atmosphere storage of Florida citrus. *Proceedings of the First International Citrus Symposium* 3, 1365–1373.

Chawan, T. and Pflug, I.J. (1968) Controlled atmosphere storage of onion. *Michigan Agricultural Station Quarterly Bulletin* 50, 449–475.

Cheah, L.H., Irving, D.E., Hunt, A.W. and Popay, A.J. (1994) Effect of high CO_2 and temperature on *Botrytis* storage rot and quality of kiwifruit. *Proceedings of the Forty Seventh New Zealand Plant Protection Conference*. Waitangi Hotel, New Zealand, 9–11 August 1994, pp. 299–303.

Chen, N. and Paull, R.E. (1986) Development and prevention of chilling injury in papaya fruits. *Journal of the American Society of Horticultural Science* 111, 639.

Chen, P.M., Mellenthin, W.M., Kelly, S.B. and Facteau, T.J. (1981) Effects of low oxygen and temperature on quality retention of 'Bing' cherries during prolonged storage. *Journal of the American Society for Horticultural Science* 105, 533–535.

Chen, P.M. and Varga, D.M. (1997) Determination of optimum controlled atmosphere regimes for the control of physiological disorders of 'D'Anjou' pears after short-term, mid-term and long-term storage. *Seventh International Controlled Atmosphere Research Conference, 13–18 July 1997.* University of California, Davis, California 95616, USA (abstract) 9.

Chu, C.L. (1992) Postharvest control of San Jose scale on apples by controlled atmosphere storage. *Postharvest Biology and Technology* 1, 361–369.

Chung, D.S., Son, Y.K., Chung, D.S. and Son, Y.K. (1994) Studies on CA storage of persimmon *Diospyros kaki* T. and plum *Prunus salicina* L.. *RDA Journal of Agricultural Science, Farm Management, Agricultural Engineering, Sericulture, and Farm Products Utilization* 36, 692–698.

Church, I.J. and Parsons, A.L. (1995) Modified atmosphere packaging technology: a review. *Journal of the Science of Food and Agriculture* 67, 143–152.

Church, N. (1994) Developments in modified-atmosphere packaging and related technologies. *Trends in Food Science and Technology* 5, 345–352.

Claypool, L.L. and Allen, F.W. (1951) The influence of temperature and O_2 level on the respiration and ripening of Wickson plums. *Hilgardia* 21, 129–160.

Claypool, L.L. and Ozbek, S. (1952) Some influences of temperature and CO_2 on the respiration and storage life of the Mission fig. *Proceedings of the American Society of Horticultural Science* 60, 226–230.

Colelli, G. and Martelli, S. (1995) Beneficial effects on the application of CO_2-enriched atmospheres on fresh strawberries *Fragaria* × *ananassa* Duch. *Advances in Horticultural Science* 9, 55–60.

Colelli, G., Mitchell, F.G. and Kader, A.A. (1991) Extension of postharvest life of 'Mission' figs by CO_2 enriched atmospheres. *HortScience* 26, 1193–1195.

Cooper, T., Retamales, J. and Streif, J. (1992) Occurrence of physiological disorders in nectarine and possibilities for their control. *Erwerbsobstbau* 34, 225–228.

Coquinot, J.P. and Richard, L. (1991) Methods of controlling scald in the apple Granny Smith without chemicals. *Neuvieme Colloque sur les Recherches Fruitieres, 'La Maitrise de la Qualite des Fruits Frais', Avignon, 4 to 6 Decembre 1990,* pp. 373–380.

Corrales-Garcia, J. (1997) Physiological and biochemical responses of 'Hass' avocado fruits to cold-storage in controlled atmospheres. *Seventh International Controlled Atmosphere Research Conference, 13–18 July 1997.* University of California, Davis, California 95616, USA (abstract) 50.

Costa, M.A.C., Brecht, J.K., Sargent, S.A. and Huber, D.J. (1994) Tolerance of snap beans to elevated CO_2 levels. *107th Annual Meeting of the Florida State Horticultural Society, Orlando, Florida, USA, 30 October to 1 November 1994. Proceedings of the Florida State Horticultural Society* 107, pp. 271–273.

Couey, H.M. and Wells, J.M. (1970) Low O_2 and high CO_2 atmospheres to control postharvest decay of strawberries. *Phytopathology* 60, 47–49.

Crank, J. (1975) *The Mathematics of Diffusion*, 2nd edn. Oxford, Clarendon Press.

Currah, L. and Proctor, F.J. (1990) Onions in tropical regions. *Natural Resources Institute Bulletin* 35.

Dalrymple, D.G. (1954) World's most modern apple storage. *American Fruit Grower* November, 20, 6–7.

Dalrymple, D.G. (1967) *The Development of Controlled Atmosphere Storage of Fruit.* Division of Marketing and Utilization Sciences, Federal Extension Service, US Department of Agriculture.

Damen, P. (1985) Verlengen afzetperiode vollegrondsgroenten. *Groenten en Fruit* 40, 82–83.

Daniels, J.A., Krishnamurthi, R. and Rizvi, S.S. (1985) A review of effects of CO_2 on microbial growth and food quality. *Journal of Food Protection* 48, 532–537.

Day, B.P.F. (1994) Modified atmosphere packaging and active packaging of fruits and vegetables. *Minimal processing of foods, 14–15 April 1994, Kirkkonummi, Finland, VTT-Symposium*, 142, pp. 173–207.

Day, B.P.F. (1996) High O_2 modified atmosphere packaging for fresh prepared produce. *Postharvest News and Information* 7, 31N–34N.

De Buckle, T.S., Castelblanco, H., Zapata, L.E., Bocanegra, M.F., Rodriguez, L.E. and Rocha, D. (1973) Preservation of fresh cassava by the method of waxing. *Review de Instituto Investigaciones de Tecnologia* 15, 33–47.

De la Plaza, J.L. (1980) Controlled atmosphere storage of Cherimoya. *Proceedings of the International Congress on Refrigeration* 3, 701.

De la Plaza, J.L., Muñoz Delgado, L. and Inglesias, C. (1979) Controlled atmosphere storage of Cherimoya. *Bulletin Instiute International de Friod* 59, 1154.

DeEll, J.R. and Prange, R.K. (1993) Postharvest physiological disorders, diseases and mineral concentrations of organically and conventionally grown McIntosh and Cortland apples. *Canadian Journal of Plant Science* 73, 223–230.

DeEll, J.R., Prange, R.K. and Murr, D.P. (1995) Chlorophyll fluorescence as a potential indicator of controlled-atmosphere disorders in 'Marshall' McIntosh apples. *HortScience* 30, 1084–1085.

Delate, K.M. and Brecht, J.K. (1989) Quality of tropical sweetpotatoes exposed to controlled-atmosphere treatments for postharvest insect control. *Journal of the American Society for Horticultural Science* 114, 963–968.

Delate, K.M., Brecht, J.K. and Coffelt, J.A. (1990) Controlled atmosphere treatments for control of sweetpotato weevil Coleoptera: Curculionidae in stored tropical sweetpotatoes. *Journal of Economic Entomology* 82, 461–465.

Dennis, C., Browne, K.M. and Adamicki, F. (1979) Controlled atmosphere storage of tomatoes. *Acta Horticulturae* 93, 75–83.

Denny, F.E. and Thornton, N.C. (1941) CO_2 prevents the rapid increase in the reducing sugar content of potato tubers stored at low temperatures. *Contributions of the Boyce Thompson Institute Plant Research* 12, 79–84.

Deol, I.S. and Bhullar, S.S. (1972) Effects of wrappers and growth regulators on the storage life of mango fruits. *Punjab Horticultural Journal* 12, 114.

Deschene, A., Paliyath, G., Lougheed, E.C., Dumbroff, E.B. and Thompson, J.E. (1991) Membrane deterioration during postharvest senescence of broccoli florets: modulation by temperature and controlled atmosphere storage. *Postharvest Biology and Technology* 1, 19–31.

Dewey, D.H., Ballinger, W.E. and Pflug, I.J. (1957) Progress Report on the Controlled Atmosphere Storage of Jonathan Apples. *Quarterly Bulletin Michigan Agricultural Experiment Station* 39, 692.

Dick, E. and Marcellin, P. (1985) Effect of high temperatures on banana development after harvest. Prophylactic tests. Effects des temperatures elevees sur l'evolution des bananes apres recolte. Tests prophylactiques. *Fruits* 40, 781–784.

Dilley, D.R. (1990) Historical aspects and perspectives of controlled atmosphere storage. In: Calderon, M. and Barkai-Golan, R. (eds) *Food Preservation by Modified Atmospheres*. CRC Press, Boca Raton, pp. 187–196.

Dillon, M., Hodgson, F.J.A., Quantick, P.C. and Taylor, D.J. (1989) Use of the APIZYM testing system to assess the state of ripeness of banana fruit. *Journal of Food Science* 54, 1379–1380.

Dimick, P.S. and Hoskins, J.C. (1983) Review of apple flavour. *State of the Art. CRC Critical Review, Food Science and Nutrition* 18, 387–409.

Dohring, S. (1997) Over sea and over land putting CA research and technology to work for international shipments of fresh produce. *Seventh International Controlled Atmosphere Research Conference, 13–18 July 1997*. University of California, Davis, California 95616 USA (abstract) 23.

Dori, S., Burdon, J.N., Lomaniec, E. and Pesis, E. (1995) Effect of anaerobiosis on aspects of avocado fruit ripening. *Acta Horticulturae* 379, 129–136.

Dourtoglou, V.G., Yannovits, N.G., Tychopoulos, V.G. and Vamvakias, M.M. (1994) Effect of storage under CO_2 atmosphere on the volatile, amino acid, and pigment constituents in red grape *Vitis vinifera* L. var. Agiogitiko. *Journal of Agricultural and Food Chemistry* 42, 338–344.

Dowker, B.D., Fennell, J.F.M., Horobin, J.F., *et al.* (1980) Onions. *Report of the National Vegetable Reseach Station for 1979*, 55.

Drake, S.R. (1993) Short-term controlled atmosphere storage improved quality of several apple cultivars. *Journal of the American Society for Horticultural Science* 118, 486–489.

Drake, S.R. and Eisele, T.A. (1994) Influence of harvest date and controlled atmosphere storage delay on the color and quality of 'Delicious' apples stored in a purge-type controlled-atmosphere environment. *HortTechnology* 4, 260–263.

Dubodel, N.P., Panyushkin, Y.A., Burchuladze, A.S. and Buklyakova, N.N. (1984) Changes in sugars of mandarin fruits in controlled atmosphere storage. *Subtropicheskie Kul'tury* 1, 83–86.

Dubodel, N.P. and Tikhomirova, N.T. (1985) Controlled atmosphere storage of mandarins. *Sadovodstvo* 6, 18.

Dull, G.G., Young, R.R. and Biale, J.B. (1967) Respiratory patterns in fruit of pineapple, *Ananas comosus*, detached at different stages of development. *Plant Physiology* 20, 1059.

Eaks, I.L. (1956) Effects of modified atmospheres on cucumbers at chilling and non-chilling temperatures. *Proceedings of the American Society for Horticultural Science* 67, 473.

Eaves, C.A. (1935) The present status of gas storage research with particular reference to studies conducted in Great Britain and preliminary trials undertaken at the Central Experiment Farm, Canada. *Scientific Agriculture* 15, 548–554.

Eaves, C.A. and Forsyth, F.R. (1968) The influence of light modified atmospheres and benzimidazole on Brussels sprouts. *Journal of Horticultural Science*, 43, 317.

Eeden, S.J., van, Combrink, J.C., Vries, P.J. and Calitz, F.J. (1992) Effect of maturity, diphenylamine concentration and method of cold storage on the incidence of superficial scald in apples. *Deciduous Fruit Grower* 42, 25–28.

Eksteen, G.J., Bodegom, P. van and Van Bodegom, P. (1989) Current state of CA storage in Southern Africa. *Proceedings of the Fifth International Controlled Atmosphere Research Conference*. Wenatchee, Washington, USA, 14–16 June, 1989, 1, pp. 487–494.

Eksteen, G.J. and Truter, A.B. (1989) Transport simulation test with avocados and bananas in controlled atmosphere containers. *Yearbook of the South African Avocado Growers' Association* 12, 26–32.

Ellis, G. (1995) Potential for all-year-round berries. *The Fruit Grower* December 17–18.

Eris, A., Turkben, C., Ozer, M.H., Henze, J. and Sass, P. (1994) A research on controlled atmosphere CA storage of peach cv. Hale Haven. *Acta Horticulturae* 368, 767–776.

Ertan, U., Ozelkok, S., Celikel, F. and Kepenek, K. (1990) The effects of pre-cooling and increased atmospheric concentrations of CO_2 on fruit quality and post-harvest life of strawberries. *Bahce* 19, 59–76.

Ertan, U., Ozelkok, S., Kaynas, K. and Oz, F. (1992) Comparative studies on the effect of normal and controlled atmosphere storage conditions on some important apple cultivars. I. Flow system. *Bahce* 21, 77–90.

Escribano, M.I., Del Cura, B., Muñoz, M.T. and Merodio, C. (1997) High CO_2-low temperature interaction on ribulose 1,5-biphosphate carboxylase and polygalacturonase protein levels in cherimoya fruit. *Seventh International Controlled Atmosphere Research Conference, 13–18 July 1997*, University of California, Davis, California 95616 USA (abstract) 115.

Eun, J.B., Kim, J.D. and Park, C.Y. (1997) Storage effects of LDPE film embedded with silver-coated ceramic in enoki mushroom. *Seventh International Controlled Atmosphere Research Conference, 13–18 July 1997*. University of California, Davis, California 95616 USA (abstract) 141a.

Evelo, R.G. (1995) Modelling modified atmosphere systems. COST 94. The postharvest treatment of fruit and vegetables: systems and operations for postharvest quality. *Proceedings of a Workshop, Milan, Italy, 14–15 September 1993*, pp. 147–153.

Ezeike, G.O.I., Ghaly, A.E., Ngadi, M.O. and Okafor, O.C. (1989) Engineering features of a controlled underground yam storage structure. *Paper American Society of Agricultural Engineers* 89, 5020.

Farber, J.M. (1991) Microbiological aspects of modified-atmosphere packaging technology – a review. *Journal of Food Protection* 54, 58–70.

Fellers, P.J. and Pflug, I.J. (1967) Storage of pickling cucumbers. *Food Technology* 21, 74.

Fellows, P.J. (1988) *Food Processing Technology*. Ellis Horwood, London.

Feng, S.Q., Chen, Y.X., Wu, H.Z. and Zhou, S.T. (1991) The methods for delaying ripening and controlling postharvest diseases of mango. *Acta Agriculturae Universitatis Pekinensis* 17, 61–65.

Ferrar, P. (ed.) (1988) Transport of fresh fruit and vegetables. *Proceedings of a Workshop held at CSIRO Food Research Laboratory, North Ryde, Sydney, Australia, 5–6 February 1987. ACIAR Proceedings* 23.

Fidler, J.C. (1963) Refrigerated storage of fruits and vegetables in the UK, the British Commonwealth, the USA and South Africa. *Ditton Laboratory Memoir* 93.

Fidler, J.C. (1970) Recommended conditions for the storage of apples. *Report of the East Malling Research Station for 1969*, 189–190.

Fidler, J.C. and Mann, G. (1972) *Refrigerated Storage of Apples and Pears – a Practical Guide*. UK, Commonwealth Agricultural Bureau.

Fidler, J.C., Wilkinson, B.G., Edney, K.L. and Sharples R.O. (1973) The biology of apple and pear storage. *Commonwealth Agricultural Bureaux Research Review* 3.

Firth, J. (1958) Controlled atmosphere regulations. *Proceedings of the One Hundred and Third Meeting of the New York State Horticultural Society*, pp. 192–193.

Fisher, D.V. (1939) Storage of Delicious apples in artificial atmospheres. *Proceedings of the American Society for Horticultural Science* 37, 459–462

Flagge, H.H. (1942) Controlled atmosphere storage for Jonathan apples as affected by restricted ventilation. *Refrigerating Engineering* 43, 215–220.

Folchi, A., Pratella, G.C., Bertolini, P., Cazzola, P.P. and Eccher Zerbini, P. (1994) Effects of oxygen stress on stone fruits. COST 94. *The Post-harvest Treatment of Fruit and Vegetables: Controlled Atmosphere Storage of Fruit and Vegetables. Proceedings of a Workshop, Milan, Italy, April 22–23 1993*, pp. 107–119.

Folchi, A., Pratella, G.C., Tian, S.P. and Bertolini, P. (1995) Effect of low O_2 stress in apricot at different temperatures. *Italian Journal of Food Science* 7, 245–253.

Forney, C.F., Rij, R.E. and Ross, S.R. (1989) Measurement of broccoli respiration rate in film wrapped packages. *HortScience* 24, 111–113.

Francile, A.S. (1992) Controlled atmosphere storage of tomato. *Rivista di Agricoltura Subtropicale e Tropicale* 86, 411–416.

Francile, A.S. and Battaglia, M. (1992) Control of superficial scald in Beurre d'Anjou pears and Granny Smith apples. *Rivista di Agricoltura Subtropicale e Tropicale* 86, 397–410.

Frenkel, C. (1975) Oxidative turnover of auxins in relation to the onset of ripening in Bartlett pear. *Plant Physiology* 55, 480–484.

Frenkel, C. and Patterson, M.E. (1974) Effect of CO_2 on ultrastructure of 'Bartlett pears'. *HortScience* 9, 338–340.

Fuchs, Y., Zauberman, G. and Yanko, U. (1978) Controlled atmosphere storage of mango. *Ministry of Agriculture, Institute for Technology and Storage of Agricultural Products: Scientific Activities 1974–1977*, 184.

Fulton, S.H. (1907) The cold storage of small fruits. *US Department of Agriculture, Bureau of Plant Industry, Bulletin*, 108, 17 September.

Gadalla, S.O. (1997) Inhibition of sprouting of onions during storage and marketing. PhD thesis, Cranfield University, UK.

Gallerani, G., Pratella, G.C., Cazzola, P.P. and Eccher-Zerbini, P. (1994) Superficial scald control via low-O_2 treatment timed to peroxide threshold value. COST 94. *The Post-harvest Treatment of Fruit and Vegetables: Controlled Atmosphere Storage of Fruit and Vegetables. Proceedings of a workshop, Milan, Italy, April 22–23 1993*, pp. 51–60.

Gane, R. (1934) Production of ethylene by some ripening fruits. *Nature* 134, 1008.

Garcia, J.M., Castellano, J.M., Morilla, A., Perdiguero, S. and Albi, M.A. (1993) CA-storage of Mill olives. *Controlled Atmosphere Storage of Fruit and Vegetables, Proceedings of a Workshop 22–23 April 1993, Milan, Italy COST 94*, pp. 83–87.

Gariepy, Y., Raghavan, G.S.V., Plasse, R., Theriault, R. and Phan, C.T. (1985) Long term storage of cabbage, celery, and leeks under controlled atmosphere. *Acta Horticulturae* 157, 193–201.

Gariepy, Y., Raghavan, G.S.V. and Theriault, R. (1984) Use of the membrane system for long-term CA storage of cabbage. *Canadian Agricultural Engineering* 26, 105–109.

Gariepy, Y., Raghavan, G.S.V. and Theriault, R. (1987) Cooling characteristics of cabbage. *Canadian Agricultural Engineering* 29, 45–50.

Gariepy, Y., Raghavan, G.S.V., Theriault, R. and Munroe, J.A. (1988) Design procedure for the silicone membrane system used for controlled atmosphere storage of leeks and celery. *Canadian Agricultural Engineering* 30, 231–236.

Garrett, M. (1992) Applications of controlled atmosphere containers. *BEHR'S Seminare Hamburg 16–17 November 1992, Munich, Germany.*

Geeson, J.D. (1984) *Improved long term storage of winter white cabbage and carrots.* London, Agricultural and Food Research Council Fruit, Vegetable and Science, 19, 21.

Geeson, J.D. (1989) Modified atmosphere packaging of fruits and vegetables. *Acta Horticulturae* 258, 143–150.

Gill, C.O. and DeLacy, K.M. (1991) Growth of *Escherichia coli* and *Salmonella typhimurium. International Journal of Food Microbiology* 13, 21–30.

Gill, C.O. and Reichel, M.P. (1989) Growth of the cold-tolerant pathogens *Yersinia enterocolitica, Aeromonas hydrophila* and *Listeria monocytogenes* on high-pH beef packaged under vacuum or carbon dioxide. *Food Microbiology* 6, 223–230.

Girard, B. and Lau, O.L. (1995) Effect of maturity and storage on quality and volatile production of 'Jonagold' apples. *Food Research International* 28, 465–471.

Goffings, G. and Herregods, M. (1989) Storage of leeks under controlled atmospheres. *Acta Horticulturae* 258, 481–484.

Goffings, G., Herregods, M. and Sass, P. (1994) The influence of the storage conditions on some quality parameters of Jonagold apples. *Acta Horticulturae* 368, 37–42.

Golias, J. (1987) Methods of ethylene removal from vegetable storage chambers. *Bulletin Vyzkumny a Slechtitelsky Ustav Zelinarsky Olomouc* 31, 51–60.

Gonzalez Aguilar, G., Vasquez, C., Felix, L., Baez, R., Siller, J. and Ait, O. (1994) Low O_2 treatment before storage in normal or modified atmosphere packaging of mango. Postharvest physiology, pathology and technologies for horticultural commodities: recent advances. *Proceedings of an International Symposium held at Agadir, Morocco, 16–21 January 1994*, pp. 185–189.

Goodenough, P.W. and Thomas, T.H. (1980) Comparative physiology of field grown tomatoes during ripening on the plant or retarded ripening in controlled atmosphere. *Annals of Applied Biology* 94, 445.

Goodenough, P.W. and Thomas, T.H. (1981) Biochemical changes in tomatoes stored in modified gas atmospheres. I. Sugars and acids. *Annals of Applied Biology* 98, 507.

Gorini, F. (1988) Storage and postharvest treatment of brassicas. I. Broccoli. *Annali dell'Istituto Sperimentale per la Valorizzazione Tecnologica dei Prodotti Agricoli, Milano*, 19, pp. 279–294.

Goulart, B.L., Evensen, K.B., Hammer, P. and Braun, H.L. (1990) Maintaining raspberry shelf life: Part 1. The influence of controlled atmospheric gases on raspberry postharvest longevity. *Pennsylvania Fruit News* 70, 12–15.

Goulart, B.L., Hammer, P.E., Evensen, K.B., Janisiewicz, B. and Takeda, F. (1992) Pyrrolnitrin, captan with benomyl, and high CO_2 enhance raspberry shelf life at 0 or 18°C. *Journal of the American Society for Horticultural Science* 117, 265–270.

Graell, J. and Recasens, I. (1992) Effects of ethylene removal on 'Starking Delicious' apple quality in controlled atmosphere storage. *Postharvest Biology and Technology* 2, 101–108.

Graell, J., Recasens, I., Salas, J. and Vendrell, M. (1993) Variability of internal ethylene concentration as a parameter of maturity in apples. *Acta Horticulturae* 326, 277–284.

Graham, D. (1990) Chilling injury in plants and fruits: some possible causes with means of amelioration by manipulation of postharvest storage conditions. *Proceedings of the International Congress of Plant Physiology, New Delhi, India.* 15–20 February 1988, 2, pp. 1373–1384.

Grierson, W. (1971) Chilling injury in tropical and subtropical fruits: IV. The role of packaging and waxing in minimizing chilling injury of grapefruit. *Proceedings of the Tropical Region, American Society for Horticultural Science* 15, 76–88.

Guerra, N.B., Livera, A.V.S., da Rocha, J.A.M.R. and Oliveira, S.L. (1995) Storage of soursop *Annona muricata*, L. in polyethylene bags with ethylene absorbent. *Proceedings of a Conference held in Guanajuato Mexico 20-24 February 1995.* Kushwaha, L., Serwatowski, R. and Brook, R. (eds) *Harvest and Postharvest Technologies for Fresh Fruits and Vegetables.* American Society of Agricultural Engineers St Joseph, USA, pp. 617–622.

Haard, N.F. and Salunkhe, D.K. (1975) *Symposium: Postharvest Biology and Handling of Fruits and Vegetables.* AVI Publishing, Westpoint.

Haffner, K.E. (1993) Storage trials of 'Aroma' apples at the Agricultural University of Norway. *Acta Horticulturae* 326, 305–313.

Hansen, H. (1975) Storage of Jonagold apples – preliminary results of storage trials. *Erwerbsobstbau* 17, 122–123.

Hansen, H. (1977) Storage trials with less common apple varieties. *Obstbau Weinbau* 14, 223–226.

Hansen, H. (1986) Use of high CO_2 concentrations in the transport and storage of soft fruit. *Obstbau* 11, 268–271.

Hansen, K., Poll, L., Olsen, C.E. and Lewis, M.J. (1992) The influence of O_2 concentration in storage atmospheres on the post-storage volatile ester production of 'Jonagold' apples. *Lebensmittel Wissenschaft and Technologie* 25, 457–461.

Hansen, M., Olsen, C.E., Poll, L. and Cantwell, M.I. (1993) Volatile constituents and sensory quality of cooked broccoli florets after aerobic and anaerobic storage. *Acta Horticulturae* 343, 105–111.

Harb, J., Streif, J., Bangerth, F. and Sass, P. (1994) Synthesis of aroma compounds by controlled atmosphere stored apples supplied with aroma precursors: alcohols, acids and esters. *Acta Horticulturae* 368, 142–149.

Hardenburg, R.E. (1955) Ventilation of packaged produce. Onions are typical of items requiring effective perforation of film bags. *Modern Packaging* 28, 140 199–200.

Hardenburg, R.E. and Anderson, R.E. (1962) Chemical control of scald on apples grown in eastern United States. *United States Department of Agriculture, Agricultural Research Service* 51–54.

Hardenburg, R.E., Anderson, R.E. and Finney, E.E. Jr (1977) Quality and condition of 'Delicious' apples after storage at 0°C and display at warmer temperatures. *Journal of the America Society for Horticultural Science* 102, 210–214.

Hardenburg, R.E., Watada, A.E. and Wang, C.Y. (1990) The commercial storage of fruits, vegetables and florist and nursery stocks. *United States Department of Agriculture, Agricultural Research Service, Agriculture Handbook*, 66.

Harjadi, S.S. and Tahitoe, D. (1992) The effects of plastic film bags at low temperature storage on prolonging the shelf life of rambutan *Nephelium lappacem* cv Lebak Bulus. *Acta Horticulturae* 321, 778–785.

Harkett, P.J. (1971) The effect of O_2 concentration on the sugar content of potato tubers stored at low temperature. *Potato Research* 14, 305–311.

Harman, J.E. (1988) Quality maintenance after harvest. *New Zealand Agricultural Science* 22, 46–48.

Harman, J.E. and McDonald, B. (1983) Controlled atmosphere storage of kiwifruit: effect on storage life and fruit quality. *Acta Horticulturae* 138 195–201.

Harris, C.M. and Harvey, J.M. (1973) Quality and decay of California strawberries stored in CO_2 enriched atmospheres. *Plant Disease Reporter* 57, 44–46.

Hartmans, K.J., Es, A. van and Schouten, S. (1990) Influence of controlled atmosphere CA storage on respiration, sprout growth and sugar content of cv. Bintje during extended storage at 4°C. *11th Triennial Conference of the European Association for Potato Research, Edinburgh, UK, July 8–13 1990*, pp. 159–160.

Haruenkit, R. and Thompson, A.K. (1993) Storage of fresh pineapples. *International Conference on Postharvest Handling of Tropical Fruit, Chiang Mai Thailand*, July.

Haruenkit, R. and Thompson, A.K. (1996) Effect of O_2 and CO_2 levels on internal browning and composition of pineapples Smooth Cayenne. *Proceedings of the International Conference on Tropical Fruits, Kuala Lumpur, Malaysia*. 23–26 July 1996, pp. 343–350.

Hassan, A., Atan, R.M. and Zain, Z.M. (1985) Effect of modified atmosphere on black heart development and ascorbic acid content in 'Mauritius' pineapple *Ananas comosus* cv. Mauritius during storage at low temperature. *Association of Southeast Asian Nations Food Journal* 1, 15–18.

Hatfield, S.G.S. (1975) Influence of post-storage temperature on the aroma production by apples after controlled-atmosphere storage. *Journal of Science Food and Agriculture* 26, 1611–1612.

Hatfield, S.G.S. and Patterson, B.D. (1974) Abnormal volatile production by apples during ripening after controlled atmosphere storage. In: *Facteurs and Regulation de la Maturation des Fruits*. Colleques Internationaux, CNRS, Paris, pp. 57–64.

Hatton, T.T. and Cubbedge, R.H. (1977) Effects of pre-storage CO_2 treatment and delayed storage on stem end rind breakdown of 'Marsh' grapefruit. *HortScience* 12, 120–121.

Hatton, T.T. and Cubbedge, R.H. (1982) Conditioning Florida grapefruit to reduce chilling injury during low temperature storage. *Journal of the American Society for Horticultural Science* 107, 57

Hatton, T.T., Cubbedge, R.H. and Grierson, W. (1975) Effects of pre-storage carbon dioxide treatments and delayed storage on chilling injury of 'Marsh' grapefruit. *Proceedings of the Florida State Society for Horticultural Science* 88, 335.

Hatton, T.T. and Reeder, W.F. (1966) Controlled atmosphere storage of 'Keitt' mangoes. *Proceedings of the Caribbean region of the American Society for Hoticultural Science* 10, 114–119.

Hatton, T.T. and Reeder, W.F. (1968) Controlled atmosphere storage of papayas. *Proceedings of the Tropical Region of the American Society for Horticultural Science* 13, 251–256.

Hatton, T.T. and Reeder, W.F. (1969) Responses of Florida avocados, mangoes and limes to storage in several controlled atmospheres. *Proceedings of the Controlled Atmosphere Research Conference, Michigan State University, East Lansing,* 72.

Hatton, T.T., Reeder, W.F. and Kaufman, J. (1966) Maintaining market quality of fresh lychees during storage and transit. *USDA Marketing Research Report* 770.

Hatton, T.T. and Spalding, D.H. (1990) Controlled atmosphere storage of some tropical fruits. In: Calderon, M. and Barkai-Golan, R. (eds) *Food Preservation by Modified Atmospheres* CRC Press, Boca Raton, pp. 301–313.

Health and Safety Executive (HSE) (1991) *Confined Spaces*. HSE, UK, Information Sheet 15.

Hellickson, M.L., Adre, N., Staples, J. and Butte, J. (1995) Computer controlled evaporator operation during fruit cool-down. Technologias de cosecha y post-cosecha de frutas y hortalizas. *Proceedings of a Conference held in Guanajuato Mexico 20–24 February. Harvest and Postharvest Technologies for Fresh Fruits and Vegetables,* pp. 546–553.

Henderson, J.R. and Buescher, R.W. (1977) Effects of sulfur dioxide and controlled atmospheres on broken-end discoloration and processed quality attributes in snap beans. *Journal of the American Society for Horticultural Science* 102, 768–770.

Henze, J. (1989) Storage and transport of *Pleurotus* mushrooms in atmospheres with high CO_2 concentrations. *Acta Horticulturae* 258, 579–584.

Hesselman, C.W. and Freebairn, H.T. (1969) Rate of ripening of initiated bananas as influenced by oxygen and ethylene. *Journal of the American Society for Horticultural Science* 94, 635.

Hill Jr, G.R. (1913) Respiration of fruits and growing plant tissue in certain cases, with reference to ventilation and fruit storage. *Cornell University, Agricultural Experiment Station, Bulletin* 330.

Hobson, G.E. and Grierson, D. (1993) Tomato. In: Seymour, G.B., Taylor, J.E. and Tucker, G.A. *Biochemistry of Fruit Ripening*. Chapman & Hall, London, pp. 405–442.

Hofman, P.J. and Smith, L.G. (1993) Preharvest effect on postharvest quality of sub-tropical and tropical fruits. *Proceedings of an International Conference, Chiang Mai, Thailand. Postharvest Handling of Tropical Fruits 19–23 July 1993. ACIAR Proceedings* 50, 261–268.

Hong, Q.Z., Sheng, H.Y., Chen, Y.F. and Yang, S.J. (1983) Effects of CA storage with a silicone window on Satsuma oranges. *Journal of Fujian Agricultural College* 12, 53–60.

Hotchkiss, J.H. and Banco, M.J. (1992) Influence of new packaging technologies on the growth of microorganisms in produce. *Journal of Food Protection* 55, 815–820.

Houck, L.G., Aharoni, Y. and Fouse, D.C. (1978) Colour changes in orange fruit stored in high concentrations of O_2 and in ethylene. *Proceedings of the Florida State Horticultural Society* 91, 136–139.

Hribar, J., Plestenjak, A., Vidrih, R., Simcic, M. and Sass, P. (1994) Influence of CO_2 shock treatment and ULO storage on apple quality. International symposium on postharvest treatment of horticultural crops. *Acta Horticulturae* 368, 634–640.

Hughes, P.A., Thompson, A.K., Plumbley, R.A. and Seymour, G.B. (1981) Storage of capsicums *Capsicum annum* L. Sendt under controlled atmosphere, modified atmosphere and hypobaric conditions. *Journal of Horticultural Science* 56, 261–265.

Hulme, A.C. (1956) CO_2 injury and the presence of succinic acid in apples. *Nature* 178: 218.

Hulme, A.C. (1970) *The Biochemistry of Fruits and their Products*, Vol. 1. Academic Press, London.

Hulme, A.C. (1971) *The Biochemistry of Fruits and their Products*, Vol. 2. Academic Press, London.

Ilangantileke, S. and Salokhe, V. (1989) Low pressure atmosphere storage of Thai mango. *Proceedings of the Fifth International Controlled Atmosphere Research Conference, Wenatchee, Washington, USA*, 14–16 June, 2, pp. 103–117.

Imakawa, S. (1967) Studies on the browning of Chinese yam. *Memoir of the Faculty of Agriculture, Hokkaido University, Japan* 6, 181

Inaba, A., Kiyasu, P. and Nakamura, R. (1989) Effects of high CO_2 plus low O_2 on respiration in several fruits and vegetables. *Scientific Reports of the Faculty of Agriculture, Okayama University* 73, 27–33.

Ionescu, L., Millim, K., Batovici, R., Panait, E. and Maraineanu, L. (1978) Resrach on the storage of sweet and sour cherries in cold stores with normal and controlled atmospheres. *Lucrari Stiintifice, Institutul de Cerctari Pentru Valorifocarea Legumelor si Fructelor* 9, 43–51.

Isenburg, F.M. and Sayle, R.M. (1969) Modified atmosphere storage of Danish cabbage. *Proceedings of the American Society of Horticultural Science* 94, 447–449.

Isenberg, F.M.R. (1979) Controlled atmosphere storage of vegetables. *Horticultural Review* 1, 337–394.

Isenberg, F.M.R., Thomas, T.H., Abed-Rahaman, M., Pendergrass, A., Carroll, J.C. and Howell, L. (1974) The role of natural growth regulators in rest, dormancy and regrowth of vegetables during winter storage. *Acta Horticulturae* 38, 95–125.

Isenberg, F.M.R., Thomas, T.H. and Pendergrass, M. (1977) Hormone and histo-logical differences between normal and malic hydrazide treated onions stored over winter. *Acta Horticulturae* 62, 95–122.

Ito, S., Kakiuchi, N., Izumi, Y. and Iba, Y. (1974) Studies on the controlled atmos-phere storage of satsuma mandarin. *Bulletin of the Fruit Tree Research Station B Okitsu* 1, 39–58.

Izumi, H., Watada, A.E. and Douglas, W. (1996a) Optimum O_2 or CO_2 atmosphere for storing broccoli florets at various temperatures. *Journal of the American Society for Horticultural Science* 121, 127–131.

Izumi, H., Watada, A.E., Nathanee, P.K. and Douglas, W. (1996b) Controlled atmos-phere storage of carrot slices, sticks and shreds. *Postharvest Biology and Tech-nology* 9, 165–172.

Jankovic, M. and Drobnjak, S. (1994) The influence of cold room atmosphere com-position on apple quality changes. Part 2. Changes in firmness, mass loss and physiological injuries. *Review of Research Work at the Faculty of Agriculture, Belgrade* 39, 73–78.

Jeffery, D., Smith, C., Goodenough, P.W., Prosser, T. and Grierson, D. (1984) Ethylene independent and ethylene dependent biochemical changes in ripening tomatoes. *Plant Physiology* 74, 32.

Jobling, J., McGlasson, W.B., Miller, P. and Hourigan, J. (1993) Harvest maturity and quality of new apple cultivars. *Acta Horticulturae* 343, 53–55.

Johnson, D.S. (1994) Storage conditions for apples and pears. *East Malling Research Association Review 1994 to 1995.*

Johnson, D.S., Dover, C.J. and Pearson, K. (1993) Very low oxygen storage in relation to ethanol production and control of superficial scald in Bramley's Seedling apples. *Acta Horticulturae* 326, 175–182.

Johnson, D.S. and Ertan, U. (1983) Interaction of temperature and O_2 level on the respiration rate and storage quality of Idared apples. *Journal of Horticultural Science* 58, 527–533.

Johnson, G., Boag, T.S., Cooke, A.W., Izard, M., Panitz, M. and Sangchote, S. (1990a) Interaction of post harvest disease control treatments and gamma irradiation on mangoes. *Annals of Applied Biology* 116, 245–257.

Johnson, G.I., Sangchote, S. and Cooke, A.W. (1990b) Control of stem end rot *Dothiorella dominicana* and other postharvest diseases of mangoes cultivar Kensington Pride during short and long term storage. *Tropical Agriculture* 67, 183–187.

Kader, A.A. (1985) Modified atmosphere and low-pressure systems during transport and storage. In: Kader, A.A., Kasmire, R.F., Mitchell, F.G., Reid, M.S., Sommer, N.F. and Thompson, J.F. (eds) *Postharvest Technology of Horticultural Crops.* Cooperative Extension, University of California, Division of Agriculture and Natural Resources, pp. 59–60.

Kader, A.A. (1986) Biochemical and physiological basis for effects on controlled and modified atmospheres on fruits and vegetables. *Food Technology* 40, 99–104.

Kader, A.A. (1989) A summary of CA requirements and recommendations for fruit other than pome fruits. *International Controlled Atmosphere Conference Fifth Proceedings, 14–16 June 1989.* Wenatchee, Washington USA. Vol. 2. *Other Commodities and Storage Recommendations*, pp. 303–328.

Kader, A.A. (1992) *Postharvest technology of horticultural crops.* Division of Agriculture and Natural Resources, University of California, 2nd edn, Oakland, California, USA. ANR Publications 3311.

Kader, A.A. (1993) Modified and controlled atmosphere storage of tropical fruits. *Postharvest handling of tropical fruits. Proceedings of an International Conference, Chiang Mai, Thailand. 19–23 July 1993, ACIAR Proceedings* 50, 239–249.

Kader, A.A. (1997) A summary of CA requirements and recommendations for fruits other than pome fruits. *Seventh International Controlled Atmosphere Research Conference, 13–18 July 1997.* University of California, Davis, California 95616, USA (abstract) 49.

Kader, A.A., Brecht, P.E., Woodruff, R. and Morris, L.L. (1973) Influence of carbon monoxide, CO_2 and O_2 levels on brown stain, respiration rate and visual quality of lettuce. *Journal of the American Society for Horticultural Science* 98, 485–488.

Kader, A.A., Chastagner, G.A., Morris, L.L. and Ogawa, J.M. (1978) Effects of carbon monoxide on decay, physiological responses, ripening, and composition of tomato fruits. *Journal of the American Society for Horticultural Science* 103, 665–670.

Kader, A.A., Chordas, A. and Elyatem, S. (1984) Responses of pomegranates to ethylene treatment and storage temperature. *California Agriculture* 38, 14–15.

Kader, A.A., Nanos, G.D. and Kerbel, E.L. (1989a) Responses of 'Manzanillo' olives to controlled atmosphere storage. *International Controlled Atmosphere Conference Fifth Proceedings, 14–16 June 1989.* Wenatchee, Washington, USA. Vol. 2. *Other Commodities and Storage Recommendations*, pp. 119–125.

Kader, A.A., Zagory, D. and Kerbel, E.L. (1989b) Modified atmosphere packaging of fruits and vegetables. *Critical Review in Food Science and Nutrition* 28, 1–30.

Kaji, H., Ikebe, T. and Osajima, Y. (1991) Effects of environmental gases on the shelf life of Japanese apricot. *Journal of the Japanese Society for Food Science and Technology* 38, 797–803.

Kajiura, I. (1972) Effects of gas concentrations on fruit. V. Effects of CO_2 concentrations on natsudaidai fruit. *Journal of the Japanese Society for Horticultural Science* 41, 215–222.

Kajiura, I. (1973) The effects of gas concentrations on fruits. VII. A comparison of the effects of CO_2 at different relative humidities, and of low O_2 with and without CO_2 in the CA storage of natsudaidai. *Journal of the Japanese Society for Horticultural Science* 42, 49–55.

Kajiura, I. and Iwata, M. (1972) Effects of gas concentrations on fruit. IV. Effects of O_2 concentration on natsudaidai fruits. *Journal of the Japanese Society for Horticultural Science* 41, 98–106.

Kale, P.N., Warade, S.D. and Jagtap, K. (1991) Effect of different cultural practices on storage life of onion bulbs. *Onion News Letter for the Tropics* 3, 25–26.

Kamath, O.C., Kushad, M.M. and Barden, J.A. (1992) Postharvest quality of 'Virginia Gold' apple fruit. *Fruit Varieties Journal* 46, 87–89.

Kapur, K.I., Verman, R.A. and Tripathi, M.P. (1985) Effect of Maturity and processing on quality of pulp slices and juice of mango c.v. Dashehari. *Indian Food Packer* 39, 60–67.

Kapur, N.S., Rao, K.S. and Srivastava, H.C. (1962) Refrigerated gas storage of mangoes. *Food Science India* 11, 228–231.

Karaoulanis, G. (1968) The effect of storage under controlled atmosphere conditions on the aldehyde and alcohol contents of oranges and grape. *Annual Report of the Ditton Laboratory* 1967–1968.

Karmarker, D.V. and Joshi, B.M. (1941) Investigations on the storage of onions. *Indian Journal of Agricultural Science* 11, 82–94.

Kawagoe, Y., Morishima, H., Seo, Y. and Imou, K. (1991) Development of a controlled atmosphere-storage system with a gas separation membrane. Part 1. Apparatus and its performance. *Journal of the Japanese Society of Agricultural Machinery* 53, 87–94.

Kaynas, K., Ozelkok, S. and Surmeli, N. (1994) Controlled atmosphere storage and modified atmosphere packaging of cauliflower. *Commissions C2,D1,D2/3 of the International Institute of Refrigeration International Symposium.* 8–10 June Istanbul, Turkey, pp. 281–288.

Kays, S.J. (1991) *Postharvest Physiology of Perishable Plant Products.* AVI Publishing, New York.

Ke, D.Y., Goldstein, L., O'Mahony, M. and Kader, A.A. (1991a) Effects of short-term exposure to low O_2 and high CO_2 atmospheres on quality attributes of strawberries. *Journal of Food Science* 56, 50–54.

Ke, D.Y., El Wazir, F., Cole, B., Mateos, M. and Kader, A.A. (1994a) Tolerance of peach and nectarine fruits to insecticidal controlled atmospheres as influenced by cultivar, maturity, and size. *Postharvest Biology and Technology* 4, 135–146.

Ke, D.Y. and Kader, A.A. (1989) Tolerance and responses of fresh fruits to O_2 levels at or below 1%. *Proceedings of the Fifth International Controlled Atmosphere Research Conference*. Wenatchee, Washington, USA, 14–16 June 1989, 2, pp. 209–216.

Ke, D.Y. and Kader, A.A. (1992a) External and internal factors influence fruit tolerance to low O_2 atmospheres. *Journal of the American Society for Horticultural Science* 117, 913–918.

Ke, D.Y. and Kader, A.A. (1992b) Potential of controlled atmospheres for postharvest insect disinfestation of fruits and vegetables. *Postharvest News and Information* 3, 31N–37N.

Ke, D.Y., Rodriguez-Sinobas, L. and Kader, A.A. (1991b) Physiological responses and quality attributes of peaches kept in low O_2 atmospheres. *Scientia-Horticulturae* 47, 295–303.

Ke, D.Y, Rodriguez Sinobas, L. and Kader, A.A. (1991c) Physiology and prediction of fruit tolerance to low O_2 atmospheres. *Journal of the American Society for Horticultural Science* 116, 253–260.

Ke, D., Zhou, L. and Kader, A.A. (1994) Mode of O_2 and CO_2 action on strawberry ester biosynthesis. *Journal of the American Society for Horticultural Science* 119, 971–975.

Kelly, M.O. and Saltveit Jr, M.E. (1988) Effect of endogenously synthesized and exogenously applied ethanol on tomato fruit ripening. *Plant Physiology* 88, 143–147.

Kerbel, E., Ke, D. and Kader A.A. (1989) Tolerance of 'Fantasia' nectarine to low O_2 and high CO_2 atmospheres. *Technical Innovations in Freezing and Refrigeration of Fruits and Vegetables*. Paper presented at a conference held in Davis, California, USA, 9–12 July, pp. 325–331.

Kester, J.J. and Fennema, O.R. (1986) Edible films and coatings – a review. *Food Technology* 40, 46–57.

Ketsa, S. and Raksritong, T. (1992) Effects of PVC film wrapping and temperature on storage life and quality of 'Nam Dok Mai' mango fruits on ripening. *Acta Horticulturae* 321, 756–763.

Khanbari, O.S. and Thompson, A.K. (1994) The effect of controlled atmosphere storage at 4°C on crisp colour and on sprout growth, rotting and weight loss of potato tubers. *Potato Research* 37, 291–300.

Khanbari, O.S. and Thompson, A.K. (1997) Effect of controlled atmosphere, temperature and cultivar on sprouting and processing quality of stored potatoes. *Potato Research* 38, 523–531.

Khitron Ya, I. and Lyublinskaya, N.A. (1991) Increasing the effectiveness of storing table grape. *Sadovodstvo i Vinogradarstvo* 7, 19–21.

Kidd, F. (1916) The controlling influence of CO_2: Part III. The retarding effect of CO_2 on respiration. *Proceedings of the Royal Society, London* 89B, 136–156.

Kidd, F. (1919) Laboratory experiments on the sprouting of potatoes in various gas mixtures nitrogen, O_2 and CO_2. *New Phytologist* 18, 248–252.

Kidd, F. and West, C. (1917a) The controlling influence of CO_2. IV. On the production of secondary dormancy in seeds of *Brassica alba* following a treatment with CO_2, and the relation of this phenomenon to the question of stimuli in growth processes. *Annals of Botany* 34, 439–446.

Kidd, F. and West, C. (1917b) The controlling influence of CO_2. IV. On the production of secondary dormancy in seeds of *Brassica alba* following treatment with CO_2, and the relation of this phenomenon to the question of stimuli in growth processes. *Annals of Botany* 31, 457–487.

Kidd, F. and West, C. (1920) The role of the seed-coat in relation to the germination of immature seed. *Annals of Botany* 34, 439–446.

Kidd, F. and West, C. (1923) Brown Heart – a functional disease of apples and pears. *Food Investigation Board Special Report* 12, 3–4.

Kidd, F. and West, C. (1925) The course of respiratory activity throughout the life of an apple. *Report of the Food Investigation Board London for 1924*, pp. 27–34.

Kidd, F. and West, C. (1927a) A relation between the concentration of O_2 and CO_2 in the atmosphere, rate of respiration, and the length of storage of apples. *Report of the Food Investigation Board London for 1925, 1926*, pp.41–42.

Kidd, F. and West, C. (1927b) A relation between the respiratory activity and the keeping quality of apples. *Report of the Food Investigation Board London for 1925, 1926*, pp. 37–41.

Kidd, F. and West, C. (1927c) Gas storage of fruit. *Special Report of the Food Investigation Board, Department of Scientific and Industrial Research* 30.

Kidd, F. and West, C. (1928) Forecasting the life of an apple. *Report of the Food Investigation Board London for 1927*, pp. 23–27.

Kidd, F. and West, C. (1930) The gas storage of fruit. II. Optimum temperatures and atmospheres. *Journal of Pomology and Horticultural Science* 8, 67–77.

Kidd, F. and West, C. (1934) Injurious effects of pure O_2 upon apples and pears at low temperatures. *Report of the Food Investigation Board London for 1933*, pp. 74–77.

Kidd, F. and West, C. (1935a) Gas storage of apples. *Report of the Food Investigation Board London for 1934*, pp. 103–109.

Kidd, F. and West, C. (1935b) The refrigerated gas storage of apples. *Department of Scientific and Industrial Research, Food Investigation Leaflet* 6.

Kidd, F. and West, C. (1936) The cold storage of English-grown Conference and Doyenne de Comice pears. *Report of the Food Investigation Board London for 1935*, pp. 85–96.

Kidd, F. and West, C. (1937) Recent advances in the work on refrigerated gas storage. *Journal of Pomology and Horticultural Science* 14, 304–305.

Kidd, F. and West, C. (1938) The action of CO_2 on the respiratory activity of apples. *Report of the Food Investigation Board London for 1937*, pp. 101–102.

Kidd, F. and West, C. (1939) The gas storage of Cox's Orange Pippin apples on a commercial scale. *Report of the Food Investigation Board London for 1938*, pp. 153–156.

Kidd, F. and West, C. (1945) Respiratory activity and duration of life of apples gathered at different stages of development and subsequently maintained at a constant temperature. *Plant Physiology Lancaster* 20, 467–504.

Kidd, F. and West, C. (1949) Resistance of the skin of the apple fruit to gaseous exchange. *Report of the Food Investigation Board London for 1939*, pp. 64–68.

Kitagawa, H. and Glucina, P.G. (1984) Persimmon culture in New Zealand. *New Zealand Department of Scientific and Industrial Research, Information Series* 159.

Klein, J.D. and Lurie, S. (1992) Pre-storage heating of apple fruit for enhanced post-harvest quality: interaction of time and temperature. *HortScience* 27, 326–328.

Klieber, A. and Wills, R.B.H. (1991) Optimisation of storage conditions for 'Shogun' broccoli. *Scientia Horticulturae* 47, 201–208.

Kluge, K. and Meier, G. (1979) Flavour development of some apple cultivars during storage. *Gartenbau* 26, 278–279.

Knee, M. (1973) Effects of controlled atmosphere storage on respiratory metabolism of apple fruit tissue. *Journal of the Science of Food and Agriculture* 24, 289–298.

Knee, M. (1990) Ethylene effects on controlled atmosphere storage of horticultural crops. In: Calderon, M. and Barkai-Golan, R. (eds) *Food Preservation by Modified Atmospheres*. CRC Press, Boca Raton, pp. 225–235.

Knee, M. and Bubb, M. (1975) Storage of Bramley Seedling apples. II. Effects of source of fruit, picking date and storage conditions on the incidence of storage disorders. *Journal of Horticultura Science* 50, 121–128.

Knee, M. and Looney, N.E. (1990) Effect of orchard and postharvest application of daminozide on ethylene synthesis by apple fruit. *Journal of Plant Growth Regulation* 9, 175–179.

Koelet, P.C. (1992) *Industrial Refrigeration*. London, MacMillan.

Kollas, D.A. (1964) Preliminary investigation of the influence of controlled atmosphere storage on the organic acids of apples. *Nature* 204, 758–759.

Kosiyachinda, P. (1968) Postharvest technology of mangosteen, durian and rambutan, part 2. *Keha Karnkaset* 10, 37–41.

Koyakumaru, T., Adachi, K., Sakoda, K., Sakota, N. and Oda, Y. (1994) Physiology and quality changes of mature green mume *Prunus mume* Sieb. et Zucc. fruits stored under several controlled atmosphere conditions at ambient temperature. *Journal of the Japanese Society for Horticultural Science* 62, 877–887.

Koyakumaru, T., Sakoda, K., Ono, Y. and Sakota, N. (1995) Respiratory physiology of mature-green mume *Prunus mume* Sieb. et Zucc. fruits of four cultivars held under various controlled atmospheres at ambient temperature. *Journal of the Japanese Society for Horticultural Science* 64, 639–648.

Kubo, Y., Inaba, A., Kiyasu, H. and Nakamura, R. (1989a) Effects of high CO_2 plus low O_2 on respiration in several fruits and vegetables. *Scientific Reports of the Faculty of Agriculture, Okayama University* 73, 27–33.

Kubo, Y., Inaba, A. and Nakamura, R. (1989b) Effects of high CO_2 on respiration in various horticultural crops. *Journal of the Japanese Society for Horticultural Science* 58, 731–736.

Kupper, W., Pekmezci, M. and Henze, J. (1994) Studies on CA-storage of pomegranate *Punica granatum* L., cv. Hicaz. Postharvest physiology of fruits, Kyoto, Japan, 21–27 August. *Acta-Horticulturae* 398, 101–108.

Kurki, L. (1979) Leek quality changes during CA storage. *Acta Horticulturae* 93, 85–90.

Lallu, N., Billing, D. and McDonald, B. (1997) Shipment of kiwifruit under CA conditions from New Zealand to Europe. *Seventh International Controlled Atmosphere Research Conference, 13–18 July 1997*. University of California, Davis, California 95616 USA (abstract) 28.

Lange, E. (1985) Recent advances in low O_2 and low ethylene storage of apples in Poland. *Controlled Atmosphere for Storage of Perishable Agricultural Commodities. Proceedings of the Fourth National Controlled Atmosphere Conference North Carolina State University*, 295.

Lange, E., Nowacki, J. and Saniewski, M. (1993) The effect of methyl jasmonate on the ethylene producing system in pre-climacteric apples stored in low oxygen and high carbon dioxide atmospheres. *Journal of Fruit and Ornamental Plant Research* 1, 9–14.

Lannelongue, M. and Finne, G. (1986) Effect of carbon dioxide on growth-rates of selected microorganisms isolated from black drum (*Pogonias cromis*). *Journal of Food Protection* 49, 806–810.

Laszlo, J.C. (1985) The effect of controlled atmosphere on the quality of stored table grape. *Deciduous Fruit Grower* 35, 436–438.

Lau, O.L. (1983) Effects of storage proceedures and low O_2 and CO_2 atmospheres on storage quality of 'Spartan' apples. *Journal of the American Society for Horticultural Science* 108, 953–957.

Lau, O.L. and Yastremski, R. (1993) The use of 0.7% storage O_2 to attenuate scald symptoms in 'Delicious' apples: effect of apple strain and harvest maturity. *Acta Horticulturae* 326, 183–189.

Lawton, A.R. (1996) *Cargo Care*. Cambridge Refrigeration Technology, 140 Newmarket Road, Cambridge, UK.

Leberman, K.W., Nelson, A.L. and Steinberg, M.P. (1968) Postharvest changes of broccoli stored in modified atmospheres. I. Respiration of shoots and colour of flower heads. *Food Technology* 22, 143.

Lee, B.Y., Kim, Y.B. and Han, P.J. (1983) Studies on controlled atmosphere storage of Korean chestnut, *Castanea crenata* var. *dulcis* Nakai. *Research-Reports, Office of Rural Development, S. Korea, Soil Fertilizer, Crop Protection, Mycology and Farm Products Utilization*, 25, pp. 170–181.

Lee, D.S., Hagger, P.E., Lee, J. and Yam, K.L. (1991) Model for fresh produce repiration in modified atmospheres based on the principles of enzyme kinetics. *Journal of Food Science* 56, 1580–1585.

Lee, K.S. and Lee, D.S. (1996) Modified atmosphere packaging of a mixed prepared vegetable salad dish. *International Journal of Food Science and Technology* 31, 7–13.

Lee, S.K., Shin, I.S. and Park, Y.M. (1993) Factors involved in skin browning of non astringent 'Fuju' persimmon. *Acta Horticulturae* 343, 300–303.

Levin, A., Sonego, L., Zutkhi, Y., Ben-Arie, R., and Pech, J.C. (1992) Effects of CO_2 on ethylene production by apples at low and high O_2 concentrations. Cellular and molecular aspects of the plant hormone ethylene. *Proceedings of the International symposium on cellular and molecular aspects of biosynthesis and action of the plant hormone ethylene, Agen, France*. August 31–September 4, 1992. *Current Plant Science and Biotechnology in Agriculture* 16, pp. 150–151.

Li, Y., Wang, V.H., Mao, C.Y. and Duan, C.H. (1973) Effects of oxygen and carbon dioxide on after ripening of tomatoes. *Acta Botanica Sinica* 15, 93–102.

Lidster, P.D., Lawrence, R.A., Blanpied, G.D. and McRae, K.B. (1985) Laboratory evaluation of potassium permanganate for ethylene removal from controlled atmosphere apple storages. *Transactions of the American Society of Agricultural Engineers* 28, 331–334.

Lill, R.E. and Corrigan, V.K. (1996) Asparagus responds to controlled atmospheres in warm conditions. *International Journal of Food Science and Technology* 31, 117–121.

Lindsey, R.T. and Neale, M.A. (1977) *Proceedings of the Vegetable Cooling and Storage Conference*. National Agricultural Centre Stoneleigh.

Lips, J., Cappelle, W. and Moermans, R. (1989) MA storage techniques for long-term storage of endive roots (*Cichorium intybus* L. var. *foliosum*). *International Controlled Atmosphere Conference Fifth Proceedings 14–16 June 1989*. Wenatchee, Washington, USA. Vol. 2. *Other Commodities and Storage Recommendations*, pp. 135–140.

Lipton, W.J. (1967) Some effects of low O_2 atmospheres on potato tubers. *American Potato Journal* 44, 292.

Lipton, W.J. (1968) Market quality of asparagus – effects of maturity at harvest and of high CO_2 atmospheres during simulated transit. *USDA Marketing Research Report* 817.

Lipton, W.J. (1972) Market quality of radishes stored in low O_2 atmospheres. *Journal of the American Society for Horticultural Science* 97, 164.

Lipton, W.J. and Harris, C.M. (1974) Controlled atmosphere effects on the market quality of stored broccoli *Brassica oleracea* L Italica group. *Journal of the American Society for Horticultural Science* 99, 200–205.

Lipton, W.J. and Mackey, B.E. (1987) Physiological and quality responses of Brussels sprouts to storage in controlled atmospheres. *Journal of the American Society for Horticultural Science* 112, 491–496.

Liu, F.W. (1976a) Storing ethylene pretreated bananas in controlled atmosphere and hypobaric air. *Journal of the American Society for Horticultural Science* 101, 198–201.

Liu, F.W. (1976b) Banana response to low concentrations of ethylene. *Journal of the American Society for Horticultural Science* 101, 222–224.

Lizana, A. and Figuero, J. (1997) Effect of different CA on post harvest of Hass avocado. *Seventh International Controlled Atmosphere Research Conference, July 13–18 1997*. University of California, Davis, California 95616 USA (abstract) 114.

Lizana, L.A., Fichet, T. Videla, G., Berger, H. and Galletti, Y.L. (1993) Almacenamiento de aguacates pultas cv. Gwen en atmosfera controlada. *Proceedings of the Interamerican Society for Tropical Horticulture* 37, 79–84.

Lopez-Briones, G., Varoquaux, P., Bareau, G. and Pascat, B. (1993) Modified atmosphere packaging of common mushroom. *International Journal of Food Science and Technology* 28, 57–68.

Lopez-Briones, G., Varoquaux, P., Chambroy, Y., Bouqant, J., Bareau, G. and Pascat, B. (1992) Storage of common mushrooms under controlled atmospheres. *International Journal of Food Science and Technology* 27, 493–505.

Lougheed, E.C. (1987) Interactions of O_2, CO_2, temperature, and ethylene that may induce injuries in vegetables. *HortScience* 22, 791–794.

Lougheed, E.C. and Lee, R. (1989) Ripening, CO_2 and C_2H_4 production, and quality of tomato fruits held in atmospheres containing nitrogen and argon. *Proceedings of the Fifth International Controlled Atmosphere Research Conference*. Wenatchee, Washington, USA, 14–16 June, 2, pp. 141–150.

Luo, Y. and Mikitzel, L.J. (1996) Extension of postharvest life of bell peppers with low O_2. *Journal of the Science of Food and Agriculture* 70, 115–119.

Lurie, S., Pesis, E. and Ben-Arie, R. (1991) Darkening of sunscald on apples in storage is a non-enzymatic and non-oxidative process. *Postharvest Biology and Technology* 1, 119–125.

Lurie, S., Zeidman, M., Zuthi, Y. and Ben Arie, R. (1992) Controlled atmosphere storage to decrease physiological disorders in peaches and nectarine. *Hassadeh* 72, 1118–1122.

Lutz, J.M. and Hardenburg, R.E. (1968) The commercial storage of fruits, vegetables and florist and nursery stocks. *United States Department of Agriculture, Agriculture Handbook* 66.

Lyons, J.M. and Rappaport, L. (1962) Effect of controlled atmospheres on storage quality of Brussel sprouts. *Proceedings of the American Society for Horticultural Science* 81, 324.

Maekawa, T. (1990) On the mango CA storage and transportation from subtropical to temperate regions in Japan. *Acta Horticulturae* 269, 367–374.

Magness, J.R. and Diehl, H.C. (1924) Physiological studies on apples in storage. *Journal of Agricultural Research* 27, 33–34.

Magomedov, M.G. (1987) Technology of grape storage in regulated gas atmosphere. *Vinodelie i Vinogradarstvo SSSR* 2, 17–19.

Maharaj, R. and Sankat, C.K. (1990a) The shelf life of breadfruit stored under ambient and refrigerated conditions. *Acta Horticulturae* 269, 411–424.

Maharaj, R. and Sankat, C.K. (1990b) Storability of papayas under refrigerated and controlled atmosphere. *Acta Horticulturae* 269, 375–385.

Makhlouf, J., Castaigne, F., Arul, J., Willemot, C. and Gosselin, A. (1989a) Long term storage of broccoli under controlled atmosphere. *HortScience* 24, 637–639.

Makhlouf, J., Willemot, C., Arul, J., Castaigne, F. and Emond, J.P. (1989b) Regulation of ethylene biosynthesis in broccoli flower buds in controlled atmospheres. *Journal of the American Society for Horticultural Science* 114, 955–958.

Malcolm, G.L. and Gerdts, D.R. (1995) Review and prospects for use of controlled atmosphere technology in Mexican agribusiness. *Proceedings of a Conference held in Guanajuato Mexico, 20–24 February 1995*. Kushwaha, L., Serwatowski, R. and Brook, R. (eds) *Harvest and Postharvest Technologies for Fresh Fruits and Vegetables*, pp. 530–537.

Mannheim, C. (1980) Development and trends in the preservation of perishable produce. *Eighth Ami Shachori Memorial Lecture, 23 September, London*.

Marcellin, P. (1973) Controlled atmosphere storage of vegetables in polyethylene bags with silicone rubber windows. *Acta Horticulturae* 38, 33–45.

Marcellin, P. and Chaves, A. (1983) Effects of intermittent high CO_2 treatment on storage life of avocado fruits in relation to respiration and ethylene production. *Acta Horticulturae* 138, 155–163.

Marcellin, P. and LeTeinturier, J. (1966) Etude d'une installation de conservation de pommes en atmosphère controleé. *International Instiution of Refrigeration Bulletin, Annex 1966–1*, 141–152.

Marcellin, P., Pouliquen, J. and Guclu, S. (1979) Refrigerated storage of Passe Crassane and Comice pears in an atmosphere periodically enriched in CO_2 preliminary tests. *Bulletin de l'Institut International du Froid* 59, 1152.

Marchal, J. and Nolin, J. (1990) Fruit quality. Pre- and post-harvest physiology. *Fruits*, Special Issue, 119–122.

Martinez-Cayuela, M., Plata, M.C., Sanchez-de-Medina, L., Gil, A. and Faus, M.J. (1986) Changes in various enzyme activities during ripening of cherimoya in controlled atmosphere. *ARS Pharmaceutica* 27, 371–380.

Martinez-Javega, J.M., Jimenez Cuesta, M. and Cuquerella, J. (1983) Conservacion frigoric del melon 'Tendral'. *Anales del Instituto Nacional de Investigaciones Agrarias Agricola* 23, 111–124.

Mateos, M., Ke, D., Cantwell, M. and Kader, A.A. (1993) Phenolic metabolism and ethanolic fermentation of intact and cut lettuce exposed to CO_2-enriched atmospheres. *Postharvest Biology and Technology* 3, 225–233.

Mattheis, J.P., Buchanan, D.A. and Fellman, J.K. (1991) Change in apple fruit volatiles after storage in atmospheres inducing anaerobic metabolism. *Journal of Agricultural and Food Chemistry* 39, 1602–1605.

Mattus, G.E. (1963) Regular and automatic CA storage. *Virginia Fruit* June 1963, 41.

Mazza, G. and Siemens, A.J. (1990) CO_2 concentration in commercial potato storages and its effect on quality of tubers for processing. *American Potato Journal* 67, 121–32.

McDonald, J.E. (1985) Storage of broccoli. *Annual Report, Research Station, Kentville, Nova Scotia*, p. 114.

McGill. J.N., Nelson, A.I. and Steinberg, M.P. (1966) Effect of modified storage atmosphere on ascorbic acid and other quality characteristics of spinach. *Journal of Food Science* 31, 510.

McGlasson, W.B. and Wills, R.B.H. (1972) Effects of O_2 and CO_2 on respiration, storage life and organic acids of green bananas. *Australian Journal of Biological Sciences* 25, 35–42.

McLauchlan, R.L., Barker, L.R. and Johnson, G.I. (1994) Controlled atmospheres for Kensington mango storage: classical atmospheres. *Development of postharvest handling technology for tropical tree fruits: a workshop held in Bangkok, Thailand. 16–18 July 1992. ACIAR Proceedings*, 58, pp. 41–44.

Medlicott, A.P. and Jeger, M.J. (1987) The development and application of postharvest treatments to manipulate ripening of mangoes. In: Prinsley, R.T. and Tucker, G. (eds) *Mangoes – a Review*. Commonwealth Science Council, London.

Meheriuk, M. (1989a) CA storage of apples. *International Controlled Atmosphere Conference Fifth Proceedings, 14–16 June 1989*, Wenatchee, Washington USA. Vol. 2. *Other Commodities and Storage Recommendations*, pp. 257–284.

Meheriuk, M. (1989b) Storage chacteristics of Spartlett pear. *Acta Horticulturae* 258, 215–219.

Meheriuk, M. (1993) CA storage conditions for apples, pears and nashi. *Proceedings of the Sixth International Controlled Atmosphere Conference, 15–17 June 1983, Cornell University, USA*, pp. 819–839.

Meheriuk, M. and Herregods, M. (1993) Apple storage conditions. *Sixth Annual Controlled Atmosphere Research Conference, Cornell University, Ithaca, New York, USA*.

Meinl, G., Nuske, D. and Bleiss, W. (1988) Influence of ethylene on cabbage quality under long term storage conditions. *Gartenbau* 35, 265.

Meir, S., Akerman, M., Fuchs, Y. and Zauberman, G. (1993) Prolonged storage of Hass avocado fruits using controlled atmosphere. *Alon Hanotea* 47, 274–281.

Mencarelli, F. (1987a) Effect of high CO_2 atmospheres on stored zucchini squash. *Journal of the American Society for Horticultural Science* 112, 985–988.

Mencarelli, F. (1987b) The storage of globe artichokes and possible industrial uses. *Informatore Agrario* 43, 79–81.

Mencarelli, F., Fontana, F. and Massantini, R. (1989) Postharvest practices to reduce chilling injury CI on eggplants. *Proceedings of the Fifth International Controlled Atmosphere Research Conference*. Wenatchee, Washington, USA, 14–16 June, 1989, Vol. 2. Pullman, Washington, USA; Washington State University, pp. 49–55.

Mencarelli, F., Lipton, W.J. and Peterson, S.J. (1983) Responses of 'zucchini' squash to storage in low O_2 atmospheres at chilling and non-chilling temperatures. *Journal of the American Society for Horticultural Science* 108, 884–890.

Mendoza, D.B. (1978) Postharvest handling of major fruits in the Philippines. *Aspects of Postharvest Horticulture in ASEAN*, pp. 23–30.

Mendoza, D.B., Pantastico, E.B. and Javier, F.B. (1972) Storage and handling of rambutan (*Nephelium lappaceum* L.). *Philippines Agriculturist* 55, 322–332.

Mercer, M.D. and Smittle, D.A. (1992) Storage atmospheres influence chilling injury and chilling injury induced changes in cell wall polysaccharides of cucumber. *Journal of the American Society for Horticultural Science* 117, 930–933.

Merodio, C. and De la Plaza, J.L. (1989) Interaction between ethylene and CO_2 on controlled atmosphere storage of 'Blanca de Aranjuez pears'. *Acta Horticulturae* 258, 81–88.

Mertens, H. (1985) Storage conditions important for Chinese cabbage. *Groenten en Fruit* 41, 17, 62–63.

Mertens, H. (1988) Ethylene and respiratory metabolism of cauliflower *Brassica oleracea* L. convar. *botrytis* in controlled atmosphere storage. *Acta Horticulturae* 258, 493–501.

Mertens, H. and Tranggono (1989) Ethylene and respiratory metabolism of cauliflower *Brassica olereacea* L. convar. *Botrytis* in controlled atmosphere storage. *Acta Horticulturae* 258, 493–501.

Miccolis, V. and Saltveit, M.E. Jr (1988) Influence of temperature and controlled atmosphere on storage of 'Green Globe' artichoke buds. *HortScience* 23, 736–741.

Miller, E.V. and Brooks, C. (1932) Effect of CO_2 content of storage atmosphere on carbohydrate transformation in certain fruits and vegetables. *Journal of Agricultural Research*, 45, 449–459.

Miller, E.V. and Dowd, O.J. (1936) Effect of CO_2 on the carbohydrates and acidity of fruits and vegetables in storage. *Journal of Agricultural Research* 53, 1–7.

Mitcham, E., Zhou, S. and Bikoba, V. (1997) Development of carbon dioxide treatment for Californian table grapes. *Seventh International Controlled Atmosphere Research Conference, 13–18 July 1997*. University of California, Davis, California 95616 USA (abstract) 65.

Mohammed, M. (1993) Storage of passionfruit in polymeric films. *Proceedings of the Interamerican Society for Tropical Horticulture* 37, 85–88.

Monzini, A. and Gorini, F.L. (1974) Controlled atmosphere in the storage of vegetables and flowers. *Annali dell'Istituto Sperimentale per la Valorizzazione Tecnologica dei Prodotti Agricoli* 5, pp. 277–291.

Moreno, J. and De la Plaza, J.L. (1983) The respiratory intensity of cherimoya during refrigerated storage: a special case of climacteric fruit. *Acta Horticulturae* 138, 179.

Morris, L.L. and Kader, A.A. (1977) Physiological disorders of certain vegetables in relation to modified atmosphere. *Second National Controlled Atmosphere Research Conference Proceedings, Michigan State University Horticultural Report* 28, pp. 266–267.

Naichenko, V.M. and Romanshchak, S.P. (1984) Growth regulators in fruit of the plum cultivar Vengerka Obyknovennaya during ripening and long term storage. *Fizioliya I Biokhimiya Kul'turnykhRastenii* 16, 143–148.

Nair, H. and Tung, H.F. (1988) Postharvest physiology and Storage of Pisang Mas. *Proceedings of the UKM simposium Biologi Kebangsaan ketiga, Kuala Lumpur, November 1988*, 22–24.

Neale, M.A., Lindsay, R.T. and Messer, H.J.M. (1981) An experimental cold store for vegetables. *Journal of Agricultural Engineering Research* 26, 529–540.

Nerd, A. and Mizrahi, Y. (1993) Productivity and postharvest behaviour of black sapote in the Israeli Negeve desert. *Proceedings of an International Conference, Chiang Mai, Thailand.* 19–23 July 1993. *ACIAR Proceedings* 50, 441 (abstract only).

Neuwirth, G.R. (1988) Respiration and formation of volatile flavour substances in controlled atmosphere-stored apples after periods of ventilation at different times. *Archiv fur Gartenbau* 36, 417–422.

Nichols, R. (1971) A review of the factors affecting the deterioration of harvested mushrooms. *Glasshouse Crops Research Institute, Littlehampton United Kingdom Report*, 174.

Nicolas, J., Rothan, C. and Duprat, F. (1989) Softening of kiwifruit in storage. Effects of intermittent high CO_2 treatments. *Acta Horticulturae* 258, 185–192.

Niedzielski, Z. (1984) Selection of the optimum gas mixture for prolonging the storage of green vegetables. Brussels sprouts and spinach. *Industries Alimentaires et Agricoles* 101, 115–118.

Nielsen, L.W. (1968) Accumulation of respiratory CO_2 around potato tubers in relation to bacterial soft rot. *American Potato Journal* 45, 174.

Noomhorm, A. and Tiasuwan (1988) Effect of controlled atmosphere storage for mango. *Paper, American Society of Agricultural Engineers* 88 6589.

Norwood, C. Thornton (1930) CO_2 storage of fruits, vegetables and flowers. *Industrial and Engineering Chemistry* 22, 1186–1189.

Nuske, D. and Muller, H. (1984) Preliminary results on the industrial type storage of headed cabbage under CA storage conditions. *Nachrichtenblatt fur den Pflanzenschutz in der DDR* 38, 185–187.

O' Hare, T.J. and Johnson, G.I. (1992) Postharvest physiology and storage of rambutan: a review. In: Highley, E. (ed.). *Development of Postharvest Handling Technology for Tropical Tree Fruits: a Workshop held in Bangkok, Thailand.* 16–18 July 1992. *ACIAR Proceedings*, 58, pp. 15–20.

O' Hare, T.J. and Prasad, A. (1993) The effect of temperature and CO_2 on chilling symptoms in mango. *Acta Horticulturae* 343, 244–250.

O' Hare, T.J., Prasad, A. and Cooke, A.W. (1994) Low temperature and controlled atmosphere storage of rambutan. *Postharvest Biology and Technology* 4, 147–157.

Ogaki, C., Manago, M., Ushiyama, K. and Tanaka, K. (1973) Studies on controlled atmosphere storage of satsumas. I. Gas concentration, relative humidity, wind velocity and pre-storage treatment. *Bulletin of the Kanagawa Horticultural Experiment Station* 21, 1–23.

Ogata, K. and Inous, T. (1957) Studies on the storage of onions. *Proceedings of the XIX International Horticultual Congress, Warsaw*, September 1974.

Ogata, K., Yamauchi, N. and Minamide, T. (1975) Physiological and chemical studies on ascorbic acid in fruits and vegetables. 1. Changes in ascorbic acid content during maturation and storage of okra. *Journal of the Japanese Society for Horticultural Science* 44, 192–195.

Othieno, J.K., Thompson, A.K. and Stroop, I.F. (1993) Modified atmosphere packaging of vegetables. *Post-harvest Treatment of Fruit and Vegetables.* COST'94 Workshop, 14–15 September 1993, Leuven, Belgium, pp. 247–253.

Otma, E.C. (1989) Controlled atmosphere storage and film wrapping of red bell peppers *Capsicum annuum* L. *Acta Horticulturae* 258, 515–522.

Oudit, D.D. (1976) Polythene bags keep cassava tubers fresh for several weeks at ambient temperature. *Journal of the Agricultural Society of Trinidad and Tobago* 76, 297–298.

Overholser, E.L. (1928) Some limitations of gas storage of fruits. *Ice and Refrigeration* 74, 551–552.

Pal, R.K. and Buescher, R.W. (1993) Respiration and ethylene evolution of certain fruits and vegetables in response to CO_2 in controlled atmosphere storage. *Journal of Food Science and Technology Mysore* 30, 29–32.

Pala, M., Damarli, E. and Alikasifoglu, K. (1994) A study of quality parameters in green pepper packaged in polymeric films. *Commissions C2,D1,D2/3 of the International Institute of Refrigeration International Symposium*. 8–10 June, Istanbul, Turkey, pp. 305–316

Palma, T., Stanley, D.W., Aguilera, J.M. and Zoffoli, J.P. (1993) Respiratory behavior of cherimoya *Annona cherimola* Mill. under controlled atmospheres. *HortScience* 28, 647–649.

Pantastico, E.B. (ed.) (1975) *Postharvest Physiology, Handling and Utilization of Tropical and Sub-tropical Fruits and Vegetables.* AVI Publishing, Westpoint.

Pantastico, E.B., Mendoza, D.B. Jr and Abilay, R.M. (1969) Some chemical and physiological changes during storage of lanzones *Lansium domesticum* Correa. *Philippine Agriculture* 52, 505.

Park, N.P., Choi, E.H. and Lee, O.H. (1970) Studies on pear storage. II. Effects of polyethylene film packaging and CO_2 shock on the storage of pears, cv. Changsyprang. *Korean Journal of Horticultural Science* 7, 21–25.

Parkin, K.L. and Schwobe, M.A. (1990) Effects of low temperature and modified atmosphere on sugar accumulation and chip colour in potatoes *Solanum tuberosum*. *Journal of Food Science* 55, 1341–1344.

Parsons, C.S. (1960) Effects of temperature, packaging and sprinkling on the quality of stored celery. *Proceeding of the American Society for Horticultural Science* 75, 403.

Parsons, C.S., Anderson, R.E. and Penny, R.W. (1970) Storage of mature green tomatoes in controlled atmospheres. *Journal of the American Society for Horticultural Science* 95, 791–793.

Parsons, C.S., Anderson, R.E. and Penny, R.W. (1974) Storage of mature-green tomatoes in controlled atmospheres. *Journal of the American Society for Horticultural Science* 95, 791–794.

Parsons, C.S., Gates, J.E. and Spalding D.H. (1964) Quality of some fruits and vegetables after holding in nitrogen atmospheres. *American Society for Horticultural Science* 84, 549–556.

Parsons, C.S. and Spalding, D.H. (1972) Influence of a controlled atmosphere, temperature, and ripeness on bacterial soft rot of tomatoes. *Journal of the American Society for Horticultural Science* 97, 297–299.

Passam, H.C. (1982) Experiments on the storage of eddoes and tannias (*Colocasia* and *Xanthosoma* spp.) under ambient conditions. *Tropical Science* 24, 39–46.

Paull, R.E. and Rohrbach, K.G. (1985) Symptom development of chilling injury in pineapple fruit. *Journal of the American Society for Horticultural Science* 110, 100–105.

Peacock, B.C. (1988) Simulated commercial export of mangoes using controlled atmosphere container technology. *ACIAR Proceedings* 23, 40–44.

Pelleboer, H. (1983) A new method of storing Brussels sprouts shows promise. *Bedrijfsontwikkeling* 14, 828–831.

Pelleboer, H. (1984) A future for CA storage of open grown vegetables. *Groenten en Fruit* 39, 62–63.

Pelleboer, H. and Schouten, S.P. (1984) New method for storing Chinese cabbage is a success. *Groenten en Fruit* 40, 16, 51.

Pendergrass, A. and Isenberg, F.M.R. (1974) The effect of relative humidity on the quality of stored cabbage. *HortScience* 9, 226–227.

Perez Zungia, F.J., Muñoz Delgado, L. and Moreno, J. (1983) Conservacion frigoric de melon cv. 'Tendral Negro' en atmosferas normal y controlada. *Primer Congreso Nacional* II, 985–994.

Pesis, E., Ampunpong, C., Shusiri, B., Hewett, E.W. and Pech, J.C. (1993a) High carbon dioxide treatment before storage as inducer or reducer of ethylene in apples. *Cellular and Molecular Aspects of the Plant Hormone Ethylene. Proceedings of the International Symposium on Cellular and Molecular Aspects of Biosynthesis and Action of the Plant Hormone Ethylene, Agen, France.* August 31 to September 4, 1992. *Current Plant Science and Biotechnology in Agriculture* 16, pp. 152–153.

Pesis, E., Levi, A., Sonego, L. and Ben Arie, R. (1986) The effect of different atmospheres in polyethylene bags on deastringency of persimmon fruits. *Alon Hanotea* 40, 1149–1156.

Pesis, E., Marinansky, R., Zauberman, G. and Fuchs, Y. (1993b) Reduction of chilling injury symptoms of stored avocado fruit by pre-storage treatment with high nitrogen atmosphere. *Acta Horticulturae* 343, 251–255.

Pesis, E. and Sass, P. (1994) Enhancement of fruit aroma and quality by acetaldehyde or anaerobic treatments before storage. *Acta Horticulturae* 368, 365–373.

Peters, P., Jeglorz, J. and Kastner, B. (1986) Investigations over several years on conventional and cold storage of Chinese cabbage. *Gartenbau* 33, 298–301.

Peters, P. and Seidel, P. (1987) Gentle harvesting of Brussels sprouts and recently developed cold storage methods for preservation of quality. Tagungsbericht, Akademie der Landwirtschaftswissenschaften der DDR. Moglichkeiten und Aufgaben zur umfassenden Intensivierung der Feldgemuseproduktion. *Proceedings of a Conference held in Grossbeeren, German Democratic Republic.* 15–18 June. 262, pp. 301–309.

Platenius, H., Jamieson, F.S. and Thompson, H.C. (1934) Studies on cold storage of vegetables. *Cornell University Agricultural Experimental Station Bulletin* 602.

Plumbley, R.A. and Rickard, J.E. (1991) Post-harvest deterioration of cassava. *Tropical Science* 31, 295–303.

Polderdijk, J.J., Boerrigter, H.A.M., Wilkinson, E.C., Meijer, J.G. and Janssens, M.F.M. (1993) The effects of controlled atmosphere storage at varying levels of relative humidity on weight loss, softening and decay of red bell peppers. *Scientia Horticulturae* 55, 315–321.

Poma Treccarri, C. and Anoni, A. (1969) Controlled atmosphere packaging of polyethylene and defoliation of the stalks in the cold storage of artichokes. *Riv. Octoflorofruttic. Hal* 53, 203.

Prange, R.K. and Lidster, P.D. (1991) Controlled atmosphere and lighting effects on storage of winter cabbage. *Canadian Journal of Plant Science* 71, 263–268.

Pujantoro, L., Tohru, S. and Kenmoku, A. (1993) The changes in quality of fresh shiitake *Lentinus edodes* in storage under controlled atmosphere conditions. *Proceeding of ICAMPE '93. 19–22 October. KOEX, Seoul, Korea, the Korean Society for Agricultural Machinery*, pp. 423–432.

Quazi, H.H. and Freebairn, H.T. (1970) The influence of ethylene, oxygen and carbon dioxide on the ripening of bananas. *Botanical Gazette* 131, 5.

Raghavan, G.S.V., Gariepy, Y., Theriault, R., Phan, C.T. and Lanson, A. (1984) System for controlled atmosphere long term cabbage storage. *International Journal of Refrigeration* 7, 66–71.

Raghavan, G.S.V., Tessier, S., Chayet, M., Norris, E.G. and Phan, C.T. (1982) Storage of vegetables in a membrane system. *Transactions of the American Society of Agricultural Engineers* 25, 433–436.

Ragnoi, S. (1989) Development of the market for Thai lychee in selected European countries. MSc thesis, Silsoe College, Cranfield Institute of Technology.

Rahman, A.S.A., Huber, D. and Brecht, J.K. (1993a) Physiological basis of low O_2 induced residual respiratory effect in bell pepper fruit. *Acta Horticulturae* 343, 112–116.

Rahman, A.S.A., Huber, D. and Brecht, J.K. (1993b) Respiratory activity and mitochondrial oxidative capacity of bell pepper fruit following storage under low O_2 atmosphere. *Journal of the American Society for Horticultural Science* 118, 470–475.

Rahman, A.S.A., Huber, D.J. and Brecht, J.K. (1995) Low-O_2-induced post-storage suppression of bell pepper fruit respiration and mitochondrial oxidative activity. *Journal of the American Society for Horticultural Science* 120, 1045–1049.

Reichel, M. (1974) The behaviour of Golden Delicious during storage as influenced by different harvest dates. *Gartenbau* 21, 268–270.

Renault, P., Houal, L., Jacquemin, G. and Chambroy, Y. (1994) Gas exchange in modified atmosphere packaging. 2. Experimental results with strawberries. *International Journal of Food Science and Technology* 29, 379–394.

Renel, L. and Thompson, A.K. (1994) Carambola in controlled atmosphere *Inter-American Institute for Co-operation on Agriculture, Tropical Fruits Newsletter* 11, 7.

Resnizky, D. and Sive, A. (1991) Storage of different varieties of apples and pears cv. Spadona in 'ultra-ultra' low O_2 conditions. *Alon Hanotea* 45, 861–871.

Reust, W., Schwarz, A. and Aerny, J. (1984) Essai de conservation des pommes de terre en atomsphere controlee. *Potato Research* 27, 75–87.

Reyes, A.A. (1988) Suppression of *Sclerotinia sclerotiorum* and watery soft rot of celery by controlled atmosphere storage. *Plant Disease* 72, 790–792.

Reyes, A.A. (1989) An overview of the effects of controlled atmosphere on celery diseases in storage. *International Controlled Atmosphere Conference Fifth*

Proceedings, 14–16 June 1989, Wenatchee, Washington, USA. Vol. 2. *Other Commodities and Storage Recommendations*, pp. 57–60.

Reyes, A.A. and Smith, R.B. (1987) Effect of O_2, CO_2, and carbon monoxide on celery in storage. *HortScience* 22, 270–271.

Richard, P. C. and Wickens, R. (1977) The effect of the time of harvesting of spring sown dry bulb onions on their yield keeping ability and skin quality. *Horticulture* 29, 45–51.

Richardson, D.G. and Meheriuk, M. (1982) Controlled atmospheres for storage and transport of perishable agricultural commodities. *Proceedings of the Third International Controlled Atmosphere Research Conference*. Timber Press, Beaverton.

Richardson, D.G. and Meheriuk, M. (1989) CA recommendations for pears including Asian pears. *International Controlled Atmosphere Conference Fifth Proceedings*. 14–16 June 1989, Wenatchee, Washington, USA. Vol. 2. *Other Commodities and Storage Recommendations*, pp. 285–302.

Risse, L.A. (1989) Individual film wrapping of Florida fresh fruit and vegetables. *Acta Horticulturae* 258, 263–270.

Robbins, J.A. and Fellman, J.K. (1993) Postharvest physiology, storage and handling of red raspberry. *Postharvest News and Information* 4, 53N–59N.

Roberts, R. (1990) An overview of packaging materials for MAP. *International Conference on Modified Atmosphere Packaging*. Part 1. Campden Food and Drinks Research Association, Chipping Campden.

Robinson, J.E., Brown, K.M. and Burton, W.G. (1975) Storage characteristics of some vegetables and soft fruits. *Annals of Applied Biology* 81, 339–408.

Robitaille, H.A. and Badenhop, A.F. (1981) Mushroom response to postharvest hyperbaric storage. *Journal of Food Science* 46, 249–253.

Roe, M.A., Faulks, R.M. and Belsten, J.L. (1990) Role of reducing sugars and amino acids in fry colour of chips from potato grown under different nitrogen regimes. *Journal of the Science of Food and Agriculture* 52, 207–214.

Roelofs, F. (1992) Supplying red currants until Christmas. *Fruitteelt Den Haag* 82, 11–13.

Roelofs, F. (1993a) CA storage of plums. Storage for longer than three weeks gives too much wastage. *Fruitteelt Den Haag* 83, 18–19.

Roelofs, F. (1993b) Choice of cultivar is partly determined by storage experiences. *Fruitteelt Den Haag* 83, 20–21.

Roelofs, F. (1993c) Research results of red currant storage trials 1992: CO_2 has the greatest influence on the storage result. *Fruitteelt Den Haag* 83, 22–23.

Roelofs, F. (1994) Experience with storage of red currants in 1993: unexpected quality problems come to the surface. *Fruitteelt Den Haag* 84, 16–17.

Roelofs, F. and Breugem, A. (1994) Storage of plums. Choose for flavour, choose for CA. *Fruitteelt Den Haag* 84, 12–13.

Roelofs, F. and Van-de Waart, A.J.P. (1993) Long-term storage of red currants under controlled atmosphere conditions. *Acta Horticulturae* 352, 217–222.

Romo Parada, L., Willemot, C., Castaigne, F., Gosselin, C. and Arul, J. (1989) Effect of controlled atmospheres low O_2, high CO_2 on storage of cauliflower *Brassica oleracea* L., *Botrytis group*. *Journal of Food Science* 54, 122–124.

Rosen, J.C. and Kader, A.A. (1989) Postharvest physiology and quality maintenance of sliced pear and strawberry fruits. *Journal of Food Science* 54, 656–659.

Roy, S., Anantheswaran, R.C. and Beelman, R.B. (1995) Fresh mushroom quality as affected by modified atmosphere packaging. *Journal of Food Science* 60, 334–340.

Rukavishnikov, A.M., Strel'tsov, B.N., Stakhovskii, A.M. and Vainshtein, I.I. (1984) Commercial fruit and vegetable storage under polymer covers with gas-selective membranes. *Khimiya v Sel'skom Khozyaistve* 22, 26–28.

Rutherford, P.P. and Whittle, R. (1982) The carbohydrate composition of onions during long term cold storage. *Journal of Horticultural Science* 57, 349–356.

Ryall, A.L. (1963) *Proceedings of the Seventeenth National Conference on Handling Perishables*. Purdue, USA, 11–14 March.

Ryall, A.L. and Lipton, W.J. (1972) *Handling, Transportation and Storage of Fruits and Vegetables*. AVI Publishing, Westpoint, Connecticut.

Saijo, R. (1990) Post harvest quality maintenance of vegetables. *Tropical Agriculture Research Series* 23, 257–269.

Saltveit, M.E. (1989) A summary of requirements and recommendations for the controlled and modified atmosphere storage of harvested vegetables. *International Controlled Atmosphere Conference Fifth Proceedings*. 14–16 June 1989. Wenatchee, Washington, USA. Vol. 2. *Other Commodities and Storage Recommendations*, pp. 329–352.

Salunkhe, D.K. and Wu, M.T. (1973) Effects of low oxygen atmosphere storage on ripening and associated biochemical changes of tomato fruits. *Journal of the American Society for Horticultural Science* 98, 12–14.

Salunkhe, D.K. and Wu, M.T. (1974) Subatmospheric storage of fruits and vegetables. *Lebensmittel Wissenschaft und Technologie* 7, 261–267.

Salunkhe, D.K. and Wu, M.T. (1975) Subatmospheric storage of fruits and vegetables. In: Haard, N.F. and Salunkhe, D.K. (eds) *Postharvest Biology and Handling of Fruits and Vegetables*, AVI Publishing, Westpoint, Connecticut, pp. 153–171.

Samisch, R.M. (1937) Observations on the effect of gas storage upon valencia oranges. *Proceedings of the American Society for Horticultural Science* 34, 103–106.

Sankat, C.K. and Basanta, A. (1997) Controlled atmosphere storage of the pomerac. *Seventh International Controlled Atmosphere Research Conference, 13–18 July 1997*. University of California, Davis, California 95616, USA (abstract) 120.

Sanket, C.K. and Maharaj, R. (1989) Controlled atmosphere storage of papayas. *International Controlled Atmosphere Conference Fifth Proceedings*. 14–16 June 1989, Wenatchee, Washington, USA. Vol. 2. *Other Commodities and Storage Recommendations*, pp. 161–170.

Sarananda, K.H. and Wilson Wijeratnam R.S. (1997) Changes in susceptibility to crown rot during maturation of Embul bananas and effect of low oxygen and high carbon dioxide on extent of crown rot. *Seventh International Controlled Atmosphere Research Conference, 13–18 July 1997*. University of California, Davis, California 95616 USA (abstract) 110.

Saray, T. (1988) Storage studies with Hungarian paprika *Capsicum annuum* L. var. *annuum* and cauliflower *Brassica cretica* convar. *Botrytis* used for preservation. *Acta Horticulturae* 220, 503–509.

Satyan, S., Scott, K.J. and Graham, D. (1992) Storage of banana bunches in sealed polyethylene tubes. *Journal of Horticultural Science* 67, 283–287.

Schaik, A. (1994) CA-storage of Elstar. Elstar can be stored longer with retention of quality. *Fruitteelt Den Haag* 84, 10–11.

Schaik, A. van, Bevers, N. and Van-Schaik, A. (1994) O_2 content for storage of Conference cannot be lowered further. Possibilities for storage of Jonathan in 0.9% O_2. *Fruitteelt Den Haag* 84, 18–19.

Schaik, A. van and Van-Schaik, A. (1994) Influence of a combined scrubber/separator on fruit quality: percentage of storage disorders ap67 to be reduced slightly. *Fruitteelt Den Haag* 84, 14–15.

Schales, F.D. (1985) Harvesting, packaging, storage and shipping of greenhouse vegetables. In: Savage, A.J. (ed.) *Hydroponics Worldwide: State of the Art in Soilless Crop Production.* Proceedings of a Conference, Hawaii, February 1985, pp. 70–76.

Schallenberger, R.S., Smith, O. and Treadaway, R.H. (1959) Role of sugars in the browning reaction in potato chips. *Journal of Agricultural and Food Chemistry* 7, 274.

Schlimme, D.V. and Rooney, M.L. (1994) Packaging of minimally processed fruits and vegetables. In: Wiley, R.C. (ed.) *Minimally Processed Refrigerated Fruits and Vegetables.* Chapman and Hall, New York, pp. 135–182.

Schmitz, S.M. (1991) Investigation on alternative methods of sprout suppression in temperate potato stores. MSc thesis Silsoe College, Cranfield Institute of Technology.

Schomer, H.A. and Sainsbury, G.F. (1957) *Controlled Atmosphere Storage of Starking Delicious Apples in the Pacific Northwest.* US Department of Agriculture, Agricultural Marketing Service 178, March.

Schouten, S.P. (1985) New light on the storage of Chinese cabbage. *Groenten en Fruit* 40, 60–61.

Schouten, S.P. (1992) Possibilities for controlled atmosphere storage of ware potatoes. *Aspects of Applied Biology* 33, 181–188.

Schouten, S.P. (1994) Increased CO_2 concentration in the store is disadvantageous for the quality of culinary potatoes. *Kartoffelbau* 45, 372–374.

Schouten, S.P. (1997) Improvement of quality of Elstar apples by dynamic control of ULO conditions. *Seventh International Controlled Atmosphere Research Conference, 13–18 July 1997.* University of California, Davis, California 95616, USA (abstract) 7.

Schulz, F.A. (1974) The occurrence of apple storage rots under controlled conditions. *Zeitschrift fur Pflanzenkrankheiten und Pflanzenschutz* 81, 550–558.

Scott, K.J., Blake, J.R., Strachan, G., Tugwell, B.L. and McGlasson, W.B. (1971) Transport of bananas at ambient temperatures using polyethylene bags. *Tropical Agriculture Trinidad* 48, 245–253.

Scott, K.J. and Wills, R.B.H. (1974) Reduction of brown heart in pears by absorption of ethylene from the storage atmosphere. *Australian Journal of Experimental Agriculture and Animal Husbandry* 14, 266–268.

SeaLand (1991) *Shipping Guide to Perishables.* SeaLand Services Inc., PO Box 800, Iselim, New Jersey 08830, USA.

Seymour, G.B., Thompson, A.K., Hughes, P.A. and Plumbley, R.A. (1981) The influence of hydrocooling and plastic box liners on the market quality of capsicums. *Acta Horticulturae* 116, 191–196.

Seymour, G.B., Thompson, A.K. and John, P. (1987) Inhibition of degreening in the peel of bananas ripened at tropical temperatures. I. The effect of high temperature changes in the pulp and peel during ripening. *Annals of Applied Biology* 110, 145–151.

Sfakiotakis, E., Niklis, N., Stavroulakis, G. and Vassiliadis, T. (1993) Efficacy of controlled atmosphere and ultra low O_2 – low ethylene storage on keeping quality and scald control of 'Starking Delicious' apples. *Acta Horticulturae* 326, 191–202.

Shan-Tao, Z. and Liang, Y. (1989) The application of carbon molecular sieve generator in CA storage of apple and tomato. *International Controlled Atmosphere Conference Fifth Proceedings*. 14–16 June 1989, Wenatchee, Washington, USA. Vol. 2. *Other Commodities and Storage Recommendations*, pp. 241–248.

Sharples, R.O. (1986) Obituary Cyril West. *Journal of Horticultural Science* 61, 555.

Sharples, R.O. (1989) Kidd, F. and West, C. In: Janick, J. (ed.) *Classical Papers in Horticultural Science*. Prentice Hall, New Jersey, pp. 213–219.

Sharples, R.O. and Johnson, D.S. (1987) Influence of agronomic and climatic factors on the response of apple fruit to controlled atmosphere storage. *HortScience* 22, 763

Sharples, R.O., Reid, M.S. and Turner, N.A. (1979) The effects of postharvest mineral element and lecithin treatments on the storage disorders of apple. *Journal of Horticultural Science* 54, 299–304.

Sharples, R.O. and Stow, J.R. (1986) Recommended conditions for the storage of apples and pears. *Report of the East Malling Research Station for 1985*, pp. 165–170.

Shipway, M.R. (1968) The refrigerated storage of vegetables and fruits. *Ministry of Agriculture Fisheries and Food UK*, 324.

Shorter, A.J., Scott, K.J. and Graham, D. (1987) Controlled atmosphere storage of bananas in bunches at ambient temperatures. *CSIRO Food Research Queensland* 47, 61–63.

Silva, J.L. and White, T.D. (1994) Bacteriological and color changes in modified atmosphere-packaged refrigerated channel catfish. *Journal of Food Protection* 57, 715–719.

Singh, A.K., Kashyap, M.M., Gupta, A.K. and Bhumbla, V.K. (1993) Vitamin-C during controlled atmosphere storage of tomatoes. *Journal of Research, Punjab Agricultural University* 30, 199–203.

Sitton, J.W., Fellman, J.K. and Patterson, M.E. (1997) Effects of low-oxygen and high-carbon dioxide atmospheres on postharvest quality, storage and decay of 'Walla Walla' sweet onions. *Seventh International Controlled Atmosphere Research Conference*. 13–18 July 1997. University of California, Davis, California 95616, USA (abstract) 60.

Sive, A. and Resnizky, D. (1979) Extension of the storage life of 'Red Rosa' plums by controlled atmosphere storage. *Bulletin de l'Institut International du Froid* 59, 1148.

Sive, A. and Resnizky, D. (1985) Experiments on the CA storage of a number of mango cultivars in 1984. *Alon Hanotea* 39, 845–855.

Skrzynski, J. (1990) Black currant fruit storability in controlled atmospheres. I. Vitamin C content and control of mould development. *Folia Horticulturae* 2, 115–124.

Smith, R.B. (1992) Controlled atmosphere storage of 'Redcoat' strawberry fruit. *Journal of the American Society for Horticultural Science* 117, 260–264.

Smith, R.B. and Reyes, A.A. (1988) Controlled atmosphere storage of Ontario grown celery. *Journal of the American Society for Horticultural Science* 113, 390–394.

Smith, R.B. and Skog, L.J. (1992) Postharvest CO_2 treatment enhances firmness of several cultivars of strawberry. *HortScience* 27, 420–421.

Smith, R.B., Skog, L.J., Maas, J.L. and Galletta, G.J. (1993) Enhancement and loss of firmness in strawberries stored in atmospheres enriched with CO_2. *Acta Horticulturae* 348, 328–333.

Smith, W.H. (1952) *The commercial storage of vegetables*. Department of Scientific and industrial Research London UK, Food Investigation Leaflet 15.

Smith, W.H. (1957) Storage of black currents. *Nature* 179, 876.

Smittle, D. A. (1988) Evaluation of storage methods for 'Granex' onions. *Journal of the American Society for Horticultural Science* 113, 877–880.

Smittle, D.A. (1989) Controlled atmosphere storage of Vidalia onions. *International Controlled Atmosphere Conference Fifth Proceedings*. 14–16 June 1989, Wenatchee, Washington, USA. Vol. 2. *Other Commodities and Storage Recommendations*, pp. 171–177.

Smock, R.M. (1938) The possibilities of gas storage in the United States. *Refrigeration Engineering* 36, 366–368.

Smock, R.M. (1956) Marketing controlled atmosphere apples. *Cornell University, Department of Agricultural Economics* 1028, 4–7.

Smock, R.M., Mendoza Jr, D.B. and Abilay, R.M. (1967) Handling bananas. *Philippines Farms and Gardens* 4, 12–17.

Smock, R.M. and Van Doren, A. (1938) Preliminary Studies on the gas storage of McIntosh and Northwestern Greening. *Ice and Refrigeration* 95, 127–128.

Smock, R.M. and Van Doren, A. (1939) Studies with modified atmosphere storage of apples. *Refrigerating Engineering* 38, 163–166.

Smoot, J.J. (1969) Decay of Florida citrus fruits stored in controlled atmospheres and in air. *Proceedings of the First International Citrus Symposium* 3, 1285–1293.

Snowdon, A.L. (1990) *A Colour Atlas of Postharvest Diseases and Disorders of Fruits and Vegetables*. Vol. 1. *General Introduction and Fruits*. Wolfe Scientific, London.

Snowdon, A.L. (1992) *A Colour Atlas of Postharvest Diseases and Disorders of Fruits and Vegetables*. Vol. 2. *Vegetables*. Wolfe Scientific, London.

Son, Y.K., Yoon, I.W., Han, P.J and Chung, D.S. (1983) Studies on storage of pears in sealed polyethylene bags. *Research Reports, Office of Rural Development, S. Korea, Soil Fertilizer, Crop Protection, Mycology and Farm Products Utilization* 25, pp. 182–187.

Spalding, D.H. and Reeder, W.F. (1972) Postharvest disorders of mangoes as affected by fungicides and heat treatments. *Plant Disease Reporter* 56, 751–753.

Spalding, D.H. and Reeder, W.F. (1974a) Quality of 'Tahiti' limes stored in a controlled atmosphere or under low pressure. Proceedings of the Tropical Region, *American Society for Horticultural Science* 18, 128–135.

Spalding, D.H. and Reeder, W.F. (1974b) Current status of controlled atmosphere storage of four tropical fruits. *Proceedings of the Florida State Horticultural Society* 87, 334–337.

Spalding, D.H. and Reeder, W.F. (1975) Low-oxygen, high carbon dioxide controlled atmosphere storage for the control of anthracnose and chilling injury of avocados. *Phytopathology* 65, 458–460.

Spencer, D. (1957) Proposed legislation to regulate modified atmosphere storage. *Proceedings of the One Hundred and Second Meeting of the New York State Horticultural Society*, pp. 218–219.

Staby, G. L. (1976) Hypobaric storage – an overview. *Combined Proceedings of the International Plant Propagation Society* 26, 211–215.

Stahl, A.L. and Cain, J.C. (1937) Cold storage studies of florida citrus fruit. III. The relation of storage atmosphere to the keeping quality of citrus fruit in cold storage. *Florida Agricultural Experiment Station*, Bulletin 316, October.

Stenvers, N. (1977) Hypobaric storage of horticultural products. *Bedrijfsontwikkeling* 8, 175–177.

Stewart, J.K. and Uota, M. (1971) CO$_2$ injury and market quality of lettuce held under controlled atmosphere. *Journal of the American Society for Horticultural Science* 96, 27–31.

Stoll, K. (1972) Largerung von Früchten und Gemusen in kontrollierter Atmosphäre. *Mitteilungen Eidgenossische Forschungsansalt für Obst, Wein und Gartenbau, Wädenswil, Flugschrift* 77.

Stoll, K. (1974) Storage of vegetables in modified atmospheres. *Acta Horticulturae* 38, 13–23.

Stoll, K. (1976) Storage of the pear cultivar Louise Bonne. *Schweizerische Zeitschrift für Obst und Weinbau* 112, 304–309.

Stow, J.R. (1989a) Effects of O$_2$ concentration on ethylene synthesis and action in stored apple fruits. *Acta Horticulturae* 258, 97–106.

Stow, J.R. (1989b) Low ethylene, low O$_2$ CA storage of apples. *International Controlled Atmosphere Conference Fifth Proceedings*. 14–16 June 1989, Wenatchee, Washington, USA. Vol. 1. *Pome Fruits*, pp. 325–332.

Stow, J.R. (1996a) The effects of storage atmosphere on the keeping quality of 'Idared' apples. *Journal of Horticultural Science* 70, 587–595.

Stow, J.R. (1996b) Gala breaks through the storage barrier. *Grower* 126, 26–27.

Streif, J. (1989) Storage behaviour of plum fruits. *Acta Horticulturae* 258, 177–184.

Streif, J., Retamales, J., Cooper, T. and Sass, P. (1994) Preventing cold storage disorders in nectarine. *Acta Horticulturae* 368, 160–165.

Strempfl, E., Mader, S. and Rumpolt, J. (1991) Trials of storage suitability of important apple cultivars in a controlled atmosphere. *Mitteilungen Klosterneuburg, Rebe und Wein, Obstbau und Fruchteverwertung* 41, 20–26.

Strop, I. (1992) Effects of plastic film wraps on the marketable life of asparagus and broccoli. MSc thesis, Silsoe College, Cranfield Institute of Technology.

Suhonen, I. (1969) On the storage life of white cabbage in refrigerated stores. *Acta Agriculturae Scandanavia* 19, 18.

Tamas, S. (1992) Cold storage of watermelons in a controlled atmosphere. *Elelmezesi Ipar* 46, 234–239, 242.

Tan, S.C. and Mohamed, A.A. (1990) The effect of CO$_2$ on phenolic compounds during the storage of 'Mas' banana in polybag, *Acta Horticulturae* 269, 389.

Tan, S.C., Mohamed, A.A. and Tan, S.C. (1990) The effect of CO$_2$ on phenolic compounds during the storage of 'Mas' banana in polybags. *Acta Horticulturae* 269, 389.

Tataru, D. and Dobreanu, M. (1978) Research on the storage of several vegetables in a controlled atmosphere. *Lucrari Stiintifice, Institutul de Cercetari pentru Valorificarea Legumelor si Fructelor* 9, 13–20.

Testoni, A. and Eccher Zerbini, P. (1989) Picking time and quality in apple storage. *Acta Horticulturae* 258, 445–454.

Testoni, A. and Eccher Zerbini, P. (1993) Controlled atmosphere storage trials with kiwifruits, prickly pears and plums. *Controlled atmosphere storage of fruit and vegetables, Proceedings of a Workshop,* 22–23 April 1993, Milan, Italy COST 94, pp. 131–136.

Testoni, A., Eccher Zerbini, P. and Zerbini, P.E. (1989) Picking time and quality in apple storage. *Acta Horticulturae* 258, 445–454.

Thatcher, R.W. (1915) Enzymes of apples and their relation to the ripening process. *Journal of Agricultural Research* 5, 103–105.

Thompson, A.K. (1971) The storage of mango fruit. *Tropical Agriculture Trinidad* 48, 63–70.

Thompson, A.K. (1972) Report on an assignment on secondment to the Jamaican Government as food storage advisor October 1970 to October 1972. *Tropical Products Institute Report,* R278.

Thompson, A.K. (1974) Post harvest studies on oranges, mangoes and cantaloupes. *Food and Agricultural Organization of the United Nations, Report Sud 70/543,* 14.

Thompson, A.K. (1975) West Indies: handling of some tropical crops. In: Pantastico, E.B. (ed.) *Postharvest Physiology, Handling and Utilization of Tropical and Sub-tropical Fruits and Vegetables.* AVI Publishing, Westpoint, USA, pp. 542–545.

Thompson, A.K. (1981) Reduction of losses during the marketing of arracacha. *Acta Horticulturae* 116, 55–60.

Thompson, A.K. (1996) *Postharvest Technology of Fruit and Vegetables.* Blackwell Science, Oxford.

Thompson, A.K. and Arango, L.M. (1977) Storage and marketing cassava in plastic films. *Proceedings of the Tropical Region of the American Horticultural Science* pp. 21, 30–33.

Thompson, A.K., Been, B.O. and Perkins, C. (1972) Handling, storage and marketing of plantains. *Proceedings of the Tropical Region of the American Society of Horticultural Science* 16, 205–212.

Thompson, A.K., Been, B.O. and Perkins, C. (1973) Reduction of wastage in stored yam. *Proceedings of the Third Symposium of the International Society for Tropical Root Crops, International Institute for Tropical Agriculture, Nigeria,* pp. 43–449

Thompson, A.K., Been, B.O. and Perkins, C. (1974a) Storage of fresh breadfruit. *Tropical Agriculture Trinidad* 51, 407–415.

Thompson, A.K., Been, B.O. and Perkins, C. (1974b) Prolongation of the storage life of breadfruits. *Proceedings of the Caribbean Food Crops Society* 12, 120–126.

Thompson, A.K., Falla, L. and Arango, L.M. (1978) Reduction of marketing losses of plantains and cassava. *Third International Seminar on Food Technology. Instituto de Investigaciones Tecnologicas, Bogota,* pp. 197–207.

Thompson, A.K. and Lee, G.R. (1971) Factors affecting the storage behaviour of papaya fruits. *Journal of Horticultural Science* 46, 511–516.

Thompson, A.K., Magzoub, Y. and Silvis, H. (1974c) Preliminary investigations into desiccation and degreening of limes for export. *Sudan Journal Food Science Technology* 6, 1–6.

Thompson, A.K., Mason, G.F. and Halkon, W.S. (1971) Storage of West Indian seedling avocado fruits. *Journal of Horticultural Science* 46, 83–88.

Thompson, A.K. and Stenning, B.C. (1994) The state of art of cold storage of fruit and vegetables. *Commissions C2, D1, D2/3 of the International Institute of Refrigeration International Symposium.* 8–10 June, Istanbul, Turkey, pp. 19–28.

Thornton, N.C. (1930) The use of carbon dioxide for prolonging the life of cut flowers. *American Journal of Botany* 17, 614–626.

Tiangco, E.L., Agillon, A.B. and Lizada, M.C.C. (1987) Modified atmosphere storage of 'Saba' bananas. *ASEAN Food Journal* 3, 112–116.

Tomkins, R.B. and Sutherland, J. (1989) Controlled atmospheres for seafreight of cauliflower. *Acta Horticulturae* 247, 385–389.

Tomkins, R.G. (1957) Peas kept for 20 days in gas store. *Grower* 48, 5, 226.

Tomkins, R.G. (1966) The storage of mushrooms. *Mushroom Growers Association Bulletin* 202, 534, 537, 538, 541.

Tongdee, S.C. (1988) Banana postharvest handling improvements. *Report of the Thailand Institute of Science and Technology Research, Bangkok.*

Tongdee, S.C., Suwanagul, A. and Neamprem, S. (1990) Durian fruit ripening and the effect of variety, maturity stage at harvest and atmospheric gases. *Acta Horticulturae* 269, 323–334.

Tonini, G., Brigati, S. and Caccioni, D. (1989) CA storage of kiwifruit: influence on rots and storability. *International Controlled Atmosphere Conference Fifth Proceedings.* 14–16 June 1989, Wenatchee, Washington, USA. Vol. 2. *Other Commodities and Storage Recommendations,* pp. 69–76.

Tonini, G., Caccioni, D. and Ceroni, G. (1993) CA storage of stone fruits: effects on diseases and disorders. *Controlled atmosphere storage of fruit and vegetables, Proceedings of a Workshop.* 22–23 April 1993, Milan, Italy COST 94, pp. 95–105.

Tonini, G. and Tura, E. (1997) New CA storage strategies for reducing rots (*Botrytis cinerea* and *Phialophora* spp.) and softening in kiwifruit. *Seventh International Controlled Atmosphere Research Conference, 13–18 July 1997.* University of California, Davis, California 95616, USA (abstract) 104.

Truter, A.B. and Combrink, J.C. (1992) Controlled atmosphere storage of peaches, nectarine and plums. *Journal of the Southern African Society for Horticultural Sciences* 2, 10–13.

Truter, A.B. and Combrink, J.C. (1997) Controlled atmosphere storage of South African plums. *Seventh International Controlled Atmosphere Research Conference, 13–18 July 1997.* University of California, Davis, California 95616, USA (abstract) 47.

Truter, A.B., Combrick, J.C., Fourie, P.C. and Victor, S.J. (1994) Controlled atmosphere storage prior to processing of some canning peach and apricot cultivars in South Africa. *Commissions C2,D1,D2/3 of the International Institute of Refrigeration International Symposium,* 8–10 June, Istanbul, Turkey, pp. 243–254.

Truter, A.B. and Eksteen, G.J. (1986) Controlled amosphere storage of avocados and bananas in South Africa. *IFF/IIR-FRIGAIR '86, Commissions C2, E1, D1, Pretoria, RSA.*

Tsay, L.M. and Wu, M.C. (1989) Studies on the postharvest physiology of sugar apple. *Acta Horticulturae* 258, 287–294.

Tucker, W.G. (1977) The sprouting of bulb onions in storage. *Acta Horticulturae* 485–492.

Tucker, W.G. and Drew, R.L.K. (1982) Post harvest studies on autumm-drilled bulb onions. *Journal of Horticultural Science* 57, 339–348.

Tugwell, B. and Chvyl, L. (1995) Storage recommendations for new varieties. *Pome Fruit Australia*, May, 4–5.

Turbin, V.A. and Voloshin, I.A. (1984) Storage of table grape varieties in a controlled gaseous environment. *Vinodelie i Vinogradarstvo SSSR* 8, 31–32.

Urban, E. (1995) Postharvest storage of apples. *Erwerbsobstbau* 37, 145–151.

Van der Merve, J.A., Combrick, J.C., Truter, A.B. and Calitz, F.J. (1997) Effect of initial low oxygen stress treatment and controlled atmosphere storage at increased carbon dioxide levels on the post-storage quality of South African-grown 'Granny Smith' and 'Topred'apples. *Seventh International Controlled Atmosphere Research Conference, 13–18 July 1997*. University of California, Davis, California 95616, USA (abstract) 8.

Van der Merwe, J.A. (1996) Controlled and modified atmosphere storage. In: Combrink, J.G. (ed.) *Integrated Management of Post-harvest Quality*. South Africa Infruitec ARC/LNR, pp. 104–112.

Van Doren, A. (1940) Physiological studies with McIntosh Apples in modified atmosphere cold storage. *Proceedings of the American Society for Horticultural Science* 37, 453–458.

Van Doren, A. (1952) The storage of golden delicious and Red Delicious apples in modified atmospheres. *Proceedings of the Forty-eighth Annual Meeting of the Washington State Horticultural Association*, pp. 91–95.

Van Doren, A., Hoffman, M.B. and Smock, R.M. (1941) CO_2 treatment of strawberries and cherries in transit and storage. *Proceedings of the American Society for Horticultural Science* 38, 231–238.

Van Leeuwen, G. and Van de Waart, A. (1991) Delaying red currants is worthwhile. *Fruitteelt Den Haag* 81, 14–15.

Veierskov, B. and Hansen, M. (1992) Effects of O_2 and CO_2 partial pressure on senescence of oat leaves and broccoli miniflorets. *New Zealand Journal of Crop and Horticultural Science* 20, 153–158.

Vidigal, J.C., Sigrist, J.M.M., Figueiredo, I.B. and Medina, J.C. (1979) Cold storage and controlled atmosphere storage of tomatoes. *Boletim do Instituto de Tecnologia de Alimentos, Brasil* 16, 421–442.

Vigneault, C. and Raghavan, G.S.V. (1991) High pressure water scrubber for rapid O_2 pull-down in controlled atmosphere storage. *Canadian Agricultural Engineering* 33, 287–294.

Vijaysegaran, S. (1993) Preharvest fruit fly control: strategies for the tropics. *International Conference on Postharvest Handling of Tropical Fruit, Chiang Mai Thailand*, July 1993.

Vilasachandran, T., Sargent, S.A. and Maul, F. (1997) Controlled atmosphere storage shows potential for maintaining postharvest quality of fresh litchi. *Seventh International Controlled Atmosphere Research Conference. 13–18 July 1997*. University of California, Davis, California 95616, USA (abstract) 54.

Viraktamath, C.S. *et al.* (1963) Pre-packaging studies on fresh produce III. Brinjal eggplant *Solanum melongena. Food Science Mysore* 12, 326–331. (Horticultural Abstracts 1964.)

Visai, C., Vanoli, M., Zini, M. and Bundini, R. (1994) Cold storage of Passa Crassana pears in normal and controlled atmosphere. *Commissions C2, D1, D2/3 of the*

International Institute of Refrigeration International Symposium. 8–10 June, Istanbul, Turkey, pp. 255–262.

Voisine, R., Hombourger, C., Willemot, C., Castaigne, D. and Makhlouf, J. (1993) Effect of high CO_2 storage and gamma irradiation on membrane deterioration in cauliflower florets. *Postharvest Biology and Technology* 2, 279–289.

Wade, N.L. (1974) Effects of O_2 concentration and ethephon upon the respiration and ripening of banana fruits. *Journal of Experimental Botany* 1974, 25, 955–964.

Wade, N.L. (1979) Physiology of cold storage disorders of fruit and vegetables. In: Lyons, J.M., Graham, D. and Raison, J.K. (eds) *Low Temperature Stress in Crop Plants.* Academic Press, New York.

Wade, N.L. (1981) Effects of storage atmosphere, temperature and calcium on low temperature injury of peach fruit. *Scientia Horticulturae* 15, 145–154.

Waelti, H., Zhang, Q., Cavalieri, R.P. and Patterson, M.E. (1992) Small scale CA storage for fruits and vegetables. *American Society of Agricultural Engineers Meeting Presentation*, Paper No. 926568.

Walsh, J.R., Lougheed, E.C., Valk, M. and Knibbe, E.A. (1985) A disorder of stored celery. *Canadian Journal of Plant Science* 65, 465–469.

Wang, C.Y. (1983) Postharvest responses of Chinese cabbage to high CO_2 treatment or low O_2 storage. *Journal of the American Society for Horticultural Science* 108, 125–129.

Wang, C.Y. (1985) Effect of low O_2 atmospheres on postharvest quality of Chinese cabbage, cucumbers and eggplants. *Proceedings of the Fourth National Controlled Atmosphere Research Conference, North Carolina State University, Horticultural Report*, 126, p. 142.

Wang, C.Y. (1990) Physiological and biochemical effects of controlled atmosphere on fruit and vegetables. In: Calderon, M. and Barkai-Golan, R. (eds) *Food Preservation by Modified Atmospheres* CRC Press, Boca Raton, pp. 197–223.

Wang, C.Y. and Ji, Z.L. (1988) Abscissic acid and 1-aminocyclopropane 1-carboxylic acid content of Chinese cabbage during low O_2 storage. *Journal of the American Society for Horticultural Science* 113, 881–883.

Wang, C.Y. and Ji, Z.L. (1989) Effect of low O_2 storage on chilling injury and polyamines in zucchini squash. *Scientia Horticulturae* 39, 1–7.

Wang, C.Y. and Kramer, G.F. (1989) Effect of low O_2 storage on polyamine levels and senescence in Chinese cabbage, zucchini squash and McIntosh apples. *Proceedings of the Fifth International Controlled Atmosphere Research Conference.* Wenatchee, Washington, USA. 14–16 June, 2, pp. 19–27.

Wang, G.X., Han, Y.S. and Yu, L. (1994) Studies on ethylene metabolism of kiwifruit after harvest. *Acta Agriculturae Universitatis Pekinensis* 20, 408–412.

Ward, C.M. and Tucker, W.G. (1976) Respiration of maleic hydrazide treated and untreated onion bulbs during storage. *Annals of Applied Biology* 82, 135–141.

Wardlaw, C.W. (1938) Tropical fruits and vegetables: an account of their storage and transport. *Low Temperature Research Station, Trinidad Memoir* 7 (reprinted from *Tropical Agriculture Trinidad* 14).

Weber, J. (1988) The efficiency of the defence reaction against soft rot after wound healing of potato tubers. 1. Determination of inoculum densities that cause infection and the effect of environment. *Potato Research* 31, 3–10.

Wei, Y. and Thompson, A.K. (1993) Modified atmosphere packaging of diploid bananas *Musa* AA. *Post-harvest Treatment of Fruit and Vegetables.* COST'94 Workshop. 14–15 September 1993, Leuven, Belgium, pp. 235–246.

Weichmann, J. (1973) The influence of different CO_2 partial pressures on respiration in carrots. *Gartenbauwissenschaft* 38, 243–252.

Weichmann, J. (1981) CA storage of horseradish, *Armoracia rusticana* Ph. Gartn. B. Mey et Scherb. *Acta Horticulturae* 116, 171–181.

Westercamp, P. (1995) Storage of President plums – influence of the harvest date on storage of the fruits. *Infos Paris* 113, 34–37.

Wigginton, M.J. (1974) Effects of temperature, O_2 tension and relative humidity on the wound-healing process in the potato tuber. *Potato Research* 17, 200–214.

Wilcke, C. and Buwalda, J.G. (1992) Model for predicting storability and internal quality of apples. In: Atkins, T.A. (ed.) *Third International Symposium on Computer Modelling in Fruit Research and Orchard Management*, Palmerston North, New Zealand, 11–14 February 1992 (*Acta-Horticulturae* 313, 115–124).

Wild, B.L., McGlasson, W.B. and Lee, T.H. (1977) Long term storage of lemon fruit. *Food Technology in Australia* 29, 351–357.

Wild de, H. and Roelofs, F. (1992) Plums can be stored for 3 weeks. *Fruitteelt Den Haag* 82, 20–21.

Wilkinson, B.G. (1972) Fruit storage. *East Malling Reseach Station Annual Report for 1971*, pp. 69–88.

Wilkinson, B.G. and Sharples, R.O. (1973) Recommended storage conditions for the storage of apples and pears. *East Malling Reseach Station Annual Report*, p. 212.

Willaert, G.A., Dirinck, P.J., Pooter, H.L. and Schamp, N.N. (1983) Objective measurement of aroma quality of Golden Delicious apples as a function of controlled-atmosphere storage time. *Journal of Agricultural and Food Chemistry* 31, 809–813.

Williams, M.W. and Patterson, M.E. (1964) Non-volatile organic acids and core breakdown of 'Bartlett' pears. *Journal of Agriculture and Food Chemistry* 12, 89.

Wills, R.B.H. (1990) Postharvest technology of banana and papaya in Association of Southeast Asian Nations: an overview. *Association of Southeast Asian Nations Food Journal* 5, 47–50.

Wills, R.B.H., Klieber, A., David, R., and Siridhata, M. (1990) Effect of brief premarketing holding of bananas in nitrogen on time to ripen. *Australian Journal of Experimental Agriculture* 30, 579–581.

Wills, R.B.H., Pitakserikul, S. and Scott, K.J. (1982) Effects of pre-storage in low O_2 or high CO_2 concentrations on delaying the ripening of bananas, *Australian Journal of Agricultural Research* 33, 1029–1036.

Wilson, L.G. (1976) Handling of postharvest tropical fruit. *Horticultural Science* 11, 120–121.

Woltering, E.J., Schaik, A.C.R.-van. and Jongen, W.M.F. (1994) Physiology and biochemistry of controlled atmosphere storage: the role of ethylene. *COST 94. The Post-harvest Treatment of Fruit and Vegetables: Controlled Atmosphere Storage of Fruit and Vegetables. Proceedings of a Workshop, Milan, Italy.* 22–23 April 1993, pp. 35–42.

Woodruff, R.E. (1969) Modified atmosphere storage of bananas. *Proceedings of the National Controlled Atmosphere Research Conference, Michigan State University, Horticultural Report* 9, 80–94.

Woodward, J.R. and Topping, A.J. (1972) The influence of controlled atmospheres on the respiration rates and storage behaviour of strawberry fruits. *Journal of Horticultural Science* 47, 547–553.

Workman, M.N. and Twomey, J. (1969) The influence of storage atmosphere and temperature on the physiology and performance of Russet Burbank seed potatoes. *Journal of the American Society for Horticultural Science* 94, 260.

Worrel, D.B. and Carrington, C.M.S. (1994) Post-harvest storage of breadfruit. *Inter-American Institute for Co-operation on Agriculture, Tropical Fruits Newsletter*, 11, 5.

Wu, M.T., Jadhav, S.J. and Salunkhe, D.K. (1972) Effects of sub atmospheric pressure storage on ripening of tomato fruits. *Journal of Food Science* 37, 952–956.

Xue, Y.B., Yu, L. and Chou, S.T. (1991) The effect of using a carbon molecular sieve nitrogen generator to control superficial scald in apples. *Acta Horticulturae Sinica* 18, 217–220.

Yahia, E.M. (1989) CA storage effect on the volatile flavor components of apples. *Proceedings of the Fifth International Controlled Atmosphere Research Conference*. Wenatchee, Washington, USA, 14–16 June 1989 1, 341–352.

Yahia, E.M. (1991) Production of some odor-active volatiles by 'McIntosh' apples following low-ethylene controlled-atmosphere storage. *HortScience* 26, 1183–1185.

Yahia, E.M. (1995) The current status and the potential use of modified and controlled atmospheres in Mexico. In: Kushwaha, L., Serwatowski, R. and Brook, R. (eds) *Proceedings of a Conference held in Guanajuato Mexico*. 20–24 February 1995. *Harvest and Postharvest Technologies for Fresh Fruits and Vegetables*, pp. 523–529.

Yahia, E.M. and Kushwaha, L. (1995) Insecticidal atmospheres for tropical fruits. Harvest and postharvest technologies for fresh fruits and vegetables. *Proceedings of the International Conference, Guanajuato, Mexico*. 20–24 February, pp. 282–286.

Yahia, E.M., Medina, F. and Rivera, M. (1989) The tolerance of mango and papaya to atmospheres containing very high levels of CO_2 and/or very low levels of O_2 as a possible insect control treatment. *International Controlled Atmosphere Conference Fifth Proceedings*. 14–16 June 1989, Wenatchee, Washington, USA. *Other Commodities and Storage Recommendations* 2, 77–89.

Yahia, E.M., Rivera, M. and Hernandez, O. (1992) Responses of papaya to short-term insecticidal O_2 atmosphere. *Journal of the American Society for Horticultural Science* 117, 96–99.

Yahia, E.M. and Tiznado Hernandez, M. (1993) Tolerance and responses of harvested mango to insecticidal low O_2 atmospheres. *HortScience* 28, 1031–1033.

Yang, Y.J. and Henze, J. (1987) Influence of CA storage on external and internal quality characteristics of broccoli *Brassica oleracea* var. *italica*. I. Changes in external and sensory quality characteristics. *Gartenbauwissenschaft* 52, 223–226.

Yang, Y.J. and Henze, J. (1988) Influence of controlled atmosphere storage on external and internal quality features of broccoli *Brassica oleracea* var. *italica*. II. Changes in chlorophyll and carotenoid contents. *Gartenbauwissenschaft* 53, 41–43.

Young, N., deBuckle, T.S., Castel Blanco, H., Rocha, D. and Velez, G. (1971) Conservacion of yuca fresca. *Instituto Investigacion Tecnologia Bogata Colombia Report.*

Zagory, D. (1990) Application of computers in the design of modified atmosphere packaging to fresh produce. In: *International Conference on Modified Atmosphere Packaging*, Part 1. Campden Food and Drinks Research Association, Chipping Campden.

Zagory, D., Ke, D. and Kader, A.A. (1989) Long term storage of 'Early Gold' and 'Shinko' Asian pears in low oxygen atmospheres. *Proceedings of the Fifth International Controlled Atmosphere Research Conference.* Wenatchee, Washington, USA, 14–16 June, 1989 1, pp. 353–357.

Zagory, D. and Reid, M.S. (1989) Controlled atmosphere storage of ornamentals. *International Controlled Atmosphere Conference Fifth Proceedings.* 14–16 June 1989, Wenatchee, Washington, USA. Vol. 2. *Other Commodities and Storage Recommendations*, pp. 353–358.

Zanon, K. and Schragl, J. (1988) Storage experiments with white cabbage. *Gemuse* 24, 14–17.

Zhang, D. and Quantick, P.C. (1997) Preliminary studies on effects of modified atmosphere packaging on postharvest storage of longan fruit. *Seventh International Controlled Atmosphere Research Conference, 13–18 July 1997.* University of California, Davis, California 95616, USA (abstract) 55.

Zhao, H. and Murata, T. (1988) A study on the storage of muskmelon 'Earl's Favourite'. *Bulletin of the Faculty of Agriculture, Shizuoka University* 38, 713 (Abstract).

Zheng, Y.H. and Xi, Y.F. (1994) Preliminary study on colour fixation and controlled atmosphere storage of fresh mushrooms. *Journal of Zhejiang Agricultural University* 20, 165–168.

Zhou, L.L., Yu, L. and Zhou, S.T. (1992a) The effect of garlic sprouts storage at different O_2 and CO_2 levels. *Acta Horticulturae Sinica* 19, 57–60.

Zhou, L.L., Yu, L., Zhao, Y.M., Zhang, X. and Chen, Z.P. (1992b) The application of carbon molecular sieve generators in the storage of garlic sprouts. *Acta Agriculturae Universitatis Pekinensis* 18, 47–51.

Zoffoli, J.P., Rodriguez, J., Aldunce, P. and Crisosto, C. (1997) Development of high concentration carbon dioxide modified atmosphere packaging systems to maintain peach quality. *Seventh International Controlled Atmosphere Research Conference.* 13–18 July 1997. University of California, Davis, California 95616, USA (abstract) 45.

Zong, R.J. (1989) Physiological aspects of film wrapping of fruits and vegetables. *Proceedings of the Fifth International Controlled Atmosphere Research Conference.* Wenatchee, Washington, USA, 14–16 June 1989, 2, pp. 29–39.

Zong, R.J., Morris, L. and Cantwell, M. (1995) Postharvest physiology and quality of bitter melon (*Momordica charantia* L.). *Postharvest Biology and Technology* 6, 65–72.

Index